BIODEGRADATION AND BIOREMEDIATION

BIODEGRADATION
AND BIOREMEDIATION

MARTIN ALEXANDER
Department of Crop, Soil, and Atmospheric Sciences
College of Agriculture and Life Sciences
Cornell University
Ithaca, New York

ACADEMIC PRESS
A Division of Harcourt Brace & Company
San Diego   New York   Boston   London   Sydney   Tokyo   Toronto

Front cover photo: © Comstock, Inc., Cameron Davidson.
Photo shows polluted wastewater from Florida citrus.

This book is printed on acid-free paper. ∞

Academic Press
525 B Street, Suite 1900, San Diego, California 92101-4495

United Kingdom Edition published by
Academic Press Limited
24–28 Oval Road, London NW1 7DX

Library of Congress Cataloging-in-Publication Data

Alexander, Martin,   [Date]
    Biodegradation and bioremediation / Martin Alexander.
        p.   cm.
    Includes index.
    ISBN 0-12-049860-X
    1. Bioremediation.   I. Title.
TD192.5.A43   1994
629.5'2--dc20   93-37670
                                                              CIP

PRINTED IN THE UNITED STATES OF AMERICA
97 BC 9 8 7 6 5 4 3

CONTENTS

PREFACE

Biodegradation of individual compounds has been the subject of active concern for more than 40 years. The initial interest was in the fate and persistence of pesticides in soils; however, the field has expanded enormously in recent years to encompass a wide variety of chemicals and a broad array of issues. Moreover, technologies have been developed that markedly enhance biodegradation or that result in microbial destruction of organic pollutants that otherwise would persist at the sites of contamination. These bioremediation technologies have led to the cleanup of many polluted groundwaters and soils, and they have fostered the development of a new bioremediation industry.

This book is designed to present the basic principles of biodegradation and to show how those principles relate to bioremediation. It considers some of the microbiological, chemical, environmental, engineering, and technological aspects of biodegradation and bioremediation, but it does not cover all facets. The field is too large and diverse, and its knowledge base is expanding too rapidly to be covered in a single text. Nevertheless, there are key general principles that underlie the science and the technology, and these can be, and hopefully here are, presented within a single volume.

An adequate introduction to biodegradation and bioremediation requires knowledge not just of one or two disciplines; information from many disciplines is needed. The processes are microbiological, the behavior of the compounds follows chemical principles, changes in hazard and exposure represent topics of concern in environmental toxicology, the areas containing the pollutants represent environments with unique properties, and the technologies are based on approaches common in environmental engineering. Thus, the book is addressed to—and should be of value to—microbiologists, chemists, toxicologists, environmental scientists, and environmental engineers. Individual readers may be unhappy with the lack of more extensive coverage of one or another topic, particularly those that are included within their own disciplines, but it is hoped that the references appended to each chapter will provide an adequate guide to further information.

Acronyms, abbreviations, and sometimes trade names are used for many of the compounds or groups of compounds. Because these terms are widely accepted and are part of the vocabulary of professionals in biodegradation and bioremediation, they are used here. In instances in which two terms apply to the same chemical, the more widely accepted term or the term approved by the relevant technical society (as with pesticides) is adopted. Each term, acronym, or abbreviation is defined or its formula is given in the appendix.

The book could not have been written without the understanding, patience, and encouragement of my wife, Renee. To her, my expressions of thanks will never suffice.

The book is dedicated to Laura, Jeremy, Anna, and all other children that they may live in a cleaner and healthier environment.

Martin Alexander

1 INTRODUCTION

The application of highly sensitive analytical techniques to environmental analysis has provided society with disturbing information. The air we breathe, the water we drink and bathe in, the soil in which our crops are grown, and the environments in which populations of animals and plants grow are contaminated with a variety of synthetic chemicals. In agricultural areas and adjacent ground and surface waters, some of these chemicals are pesticides or products generated from pesticides. Many are industrial chemicals that have been deliberately or inadvertently discharged into waters or onto soils following their intended use. Others are by-products of manufacturing operations that do not utilize waste-treatment facilities or by-products that were inadequately treated. Some are probably formed in nature from synthetic compounds, and a few are generated by reaction of natural organic materials with Cl_2 used for the treatment of water for human consumption.

As a rule, these organic compounds are not found individually but rather in simple or complex mixtures. The mixtures may be associated with the release, storage, or transport of many chemicals in surface or groundwaters, waste-treatment systems, soils, or sediments. The number of chemicals found to date is enormous, and the types of mixtures are similarly countless. Moreover, the concentrations of individual compounds vary appreciably, and they may be higher than 1.0 g per liter of water or per kilogram of soil at sites subject to spills from tank cars or trucks, to discharge of industrial waste, or to leakages from storage or disposal facilities for industrial chemicals. In contrast, the concentration may be lower than 1.0 μg per liter of water or per kilogram of soil at some distance from the point of release, spill, or storage. Even at these low concentrations, some chemicals are toxic, or risk analyses suggest that exposure of large populations to the low levels will result in deleterious effects to a few individuals; in addition, some chemicals at low concentrations are subject to biomagnification and may reach levels that have deleterious effects on humans, animals, or plants.

1

It is not surprising that synthetic chemicals are present in the human environment, for example, in areas used for food and feed production, and environments supporting natural populations of animals and plants. Modern society relies on a striking array of organic chemicals, and the quantities used are staggering. Values for the annual production of organic compounds in the United States alone show the vast tonnages that are part of human activities (Table 1.1). Although many of these chemicals are consumed or destroyed, a high percentage are released into the air, water, and soil. The quantity released varies with the compound and its particular use, but regulatory agencies in industrialized countries have found that significant percentages of the total quantity consumed by industry, agriculture, and domestic pursuits do, indeed, find their way into air, water, and soil.

Predicting the hazards of an organic compound to humans, animals, and plants requires information not only on its toxicity to living organisms but also on the degree of exposure of the organisms to the compound. The mere discharge of a chemical does not, in itself, constitute a hazard: the individual human, animal, or plant must also be exposed to it. In evaluating exposure, the transport of the chemical and its fate must be considered. A molecule that is not subject to environmental transport is not a health or environmental problem except to species at the specific point of release, so that information on dissemination of the chemical from the point of its release to the point where it could have an effect is of great relevancy. However, the chemical may be modified structurally or totally destroyed during its transport, and the fate of the compound during transport, that is, its modification or destruction, is crucial to defining the exposure. A compound that is modified to yield products that

TABLE 1.1
Production of Organic Chemicals in the United States in 1992
(in kg $\times$ 10^9)[a]

Ethylene	18.33	p-Xylene	2.57
Propylene	10.25	Terephthalic acid	2.56
Ethylene dichloride	7.23	Ethylene oxide	2.52
Vinyl chloride	6.00	Ethylene glycol	2.32
Benzene	5.45	Cumene	2.07
Ethylbenzene	4.99	Phenol	1.68
Methyl t-butyl ether	4.93	Butadiene	1.44
Styrene	4.06	Acrylonitrile	1.28
Methanol	3.96	Propylene oxide	1.22
Formaldehyde	3.17	Vinyl acetate	1.21
Xylene	2.89	Acetone	1.08
Toluene	2.74	Cyclohexane	1.00

[a]From Reisch (1993).

are less or more toxic, or that is totally degraded or is biomagnified—these being factors associated with the fate of the molecule—represents greater or lesser hazard to the species that are potentially subject to injury.

At the specific site of discharge or during its transport, the organic molecule may be acted on by abiotic mechanisms. Photochemical transformations occur in the atmosphere and at or very near the surfaces of water, soil, and vegetation, and these processes may totally destroy or appreciably modify a number of different types of organic chemicals. Nonenzymatic, nonphotochemical reactions are also prominent in soil, sediment, and surface and groundwater, and these may bring about significant changes; however, such processes rarely if ever totally convert organic compounds to inorganic products in nature, and many of these nonenzymatic reactions only bring about a slight modification of the molecule so that the product is frequently similar in structure, and often in toxicity, to the precursor compound.

On the other hand, biological processes may modify organic molecules at the site of their discharge or during their transport. Such biological transformations, which involve enzymes as catalysts, frequently bring about extensive modification in the structure and toxicological properties of pollutants or potential pollutants. These biotic processes may result in the complete conversion of the organic molecule to inorganic products, cause major changes that result in new organic products, or occasionally lead to only minor modifications. Plants and, to a lesser extent, animals in natural or man-modified environments may cause a number of changes in a wide array of chemicals, and these are of enormous importance in protecting or in sometimes increasing the toxicity of the chemical to the plant or animal that is exposed. Nevertheless, the available body of information suggests that the major agents causing the biological transformations in soil, sediment, wastewater, surface and groundwater, and many other sites are the microorganisms that inhabit these environments. These microfloras are thus frequently the major agents affecting the fate of chemicals at the sites of their release or in the environments through which they pass.

Biodegradation can be defined as the biologically catalyzed reduction in complexity of chemicals. In the case of organic compounds, biodegradation frequently, although not necessarily, leads to the conversion of much of the C, N, P, S, and other elements in the original compound to inorganic products. Such a conversion of an organic substrate to inorganic products is known as *mineralization*. *Ultimate* biodegradation is a term sometimes used as a synonym for mineralization. Thus, in the mineralization of organic C, N, P, S, or other elements, CO_2 or inorganic forms of N, P, S, or other elements are released by the organism and enter the surrounding environment. Plant and animal respiration are mineralization processes that destroy numerous organic molecules of living organisms, but the mineralization of synthetic chemicals by biological processes appears to result largely or, in some environments, entirely from microbial activity. Indeed, frequently microorganisms are the sole means, biological or nonbiological, of converting synthetic chemicals to inorganic products. Few nonbiological reactions in nature bring about comparable changes. It is because of their ability to mineralize anthropogenic compounds that microorganisms play a large role in soils, waters, and sediments.

Many synthetic molecules discharged into these environments are directly toxic or become hazardous following biomagnification. Because mineralization results in the total destruction of the parent compound and its conversion to inorganic products, such processes are beneficial. In contrast, nonbiological and many biological processes, although degrading organic compounds, convert them to other organic products. Some of these products are toxic, but others evoke no untoward response. Nevertheless, the accumulation in nature of an organic product is still cause for some concern inasmuch as a material not presently known to be harmful may, with new techniques or measurements of new toxicological manifestations, reveal itself to be undesirable. The literature of toxicology contains examples in which the increasing base of knowledge or new procedures or approaches have revealed that chemicals previously deemed to be safe were in fact harmful. Thus, mineralization is especially important in ridding natural environments of actual or possible hazards to humans, animals, and plants.

Microorganisms carry out biodegradation in many different types of environments. Of particular relevance for pollutants or potential pollutants are sewage-treatment systems, soils, underground sites for the disposal of chemical wastes, groundwater, surface waters, oceans, sediments, and estuaries. Microbial processes in the various kinds of aerobic and anaerobic systems for treating industrial, agricultural, and municipal wastes are extremely important because these treatment systems represent the first point of the discharge of many chemicals into environments of importance to humans or other living organisms. Microbial processes have long been known to be important in sewage and wastewater for the destruction of a large number of synthetic compounds. Soils also receive countless synthetic molecules from farming operations, land spreading of industrial wastes, accidental spills, or sludge disposal, and the degradation of natural materials in soils was recognized even in prehistoric times. In this century, the disposal of industrial wastes on or below the surface of the land became widespread before the evidence of groundwater pollution became prominent, but the sites adjacent to these points of chemical disposal contain microbial communities that, should they not be directly affected by the toxicity of the wastes, destroy many of the organic compounds. Groundwater adjacent to these waste-disposal sites, lakes and rivers that receive inadvertent or deliberate discharges of chemicals, and the oceans and estuaries similarly contain highly diverse and often very active communities of bacteria, fungi, and protozoa that, directly or indirectly, destroy many natural products as well as various synthetics. In addition, a variety of pollutants are retained by the sediments below fresh or marine waters, and these sediments also contain large and metabolically active communities of heterotrophic microorganisms.

Natural communities of microorganisms in these various habitats have an amazing physiological versatility. They are able to metabolize and often mineralize an enormous number of organic molecules. Probably every natural product, regardless of its complexity, is degraded by one or another species in some particular environment; if not, such compounds would, this long after the appearance of life on earth, have accumulated in enormous amounts. The lack of significant accumulation of natural products in oxygen-containing ecosystems, in itself, is an indication that the indigenous microfloras act on an astounding array of natural products. A particular species metabolizes only a small number from this array, but another species in the same habitat is able to make up for the deficiencies of its neighbor. Although certain bacteria and fungi act on a broad range of organic compounds, no organism known to date is sufficiently omnivorous to destroy a very large percentage of the natural chemicals that are formed by plants, animals, and other microorganisms.

Similarly, communities of bacteria and fungi metabolize a multitude of synthetic chemicals. The number of such molecules that can be degraded has yet to be counted, but literally thousands are known to be destroyed as a result of microbial activity in one or another environment. It is not clear how many of the millions of known organic molecules synthesized in the laboratory or made industrially can be modified in these ways, but of the list of chemicals presently regarded as pollutants and that are derived from the activities of human society, many clearly can be modified and often are mineralized by actions of these natural communities. Because too few of the known organic compounds have been tested, however, it is not yet certain to what degree the impressive microbial versatility applies to all organic compounds, but at least this versatility has been amply demonstrated with regard to many of the environmental pollutants of current concern.

Several conditions must be satisfied for biodegradation to take place in an environment. These include the following: (a) An organism must exist that has the necessary enzymes to bring about the biodegradation. The mere existence of an organism with the appropriate catabolic potential is necessary but not sufficient for biodegradation to occur. (b) That organism must be present in the environment containing the chemical. Although some microorganisms are present in essentially every environment near the earth's surface, particular environments may not contain an organism with the necessary enzymes. (c) The chemical must be accessible to the organism having the requisite enzymes. Many chemicals persist even in environments containing the biodegrading species simply because the organism does not have access to the compound that it would otherwise metabolize. Inaccessibility may result from the substrate being in a different microenvironment from the organism, in a solvent not miscible with water, or sorbed to solid surfaces. (d) If the initial enzyme bringing about the degradation is extracellular, the bonds acted upon by that enzyme must be exposed for the catalyst to function. This is not always the case because of sorption of many organic molecules. (e) Should the enzymes catalyzing the initial degradation be intracellular, that molecule must penetrate the surface of the cell to the internal sites where the enzyme acts. Alternatively, the products of an extracellular reaction must penetrate the cell for the transformation to proceed further. (f) Because the population or biomass of bacteria or fungi acting on many synthetic compounds is initially small, conditions in the environment must be conducive to allow for proliferation of the potentially active microorganisms (Alexander, 1973).

Because microorganisms are frequently the major and occasionally the sole means for degradation of particular compounds, the absence of a microorganism from a particular environment, or its inability to function, frequently means that the compound disappears very slowly. If microorganisms are the sole agents of destruction, the chemical will not be destroyed at all. If any of the conditions mentioned are not met, the chemical similarly will be long-lived. Hence, the frequent finding that organic pollutants are persisting is evidence that microorganisms are not functioning, they are acting very slowly, or no microorganism exists with the capacity to modify the molecule. It is not certain at the present time how many compounds persist in one or another environment because of the absence of microorganisms in that site, the occurrence of conditions not conducive for microbial biodegradation, or the complete absence in nature of species having the capacity to bring about the transformation. Monitoring programs have revealed that many chlorinated hydrocarbons used in industry and agriculture, compounds containing substituents other than halogens, and other categories of materials endure for long periods, but this very persistence shows either that microorganisms are not omnipotent or that particular environmental conditions prevent appreciable biological activity. Microbial successes are clearly evident because the organic molecule is destroyed; in contrast, their failings are also evident because the chemical endures.

During the mineralization of an organic molecule, not all of the C or other elements is necessarily converted to inorganic forms. Some may be incorporated into the cells of the active population, and some may appear as products that are typical of the conversions of naturally occurring organic materials.

REFERENCES

Alexander, M., *Biotechnol. Bioeng.* **15,** 611–647 (1973).
Reisch, M. S., *Chem. Eng. News* **71**(15), 10–13 (1993).

2 GROWTH-LINKED BIODEGRADATION

Microorganisms use naturally occurring and many synthetic chemicals for their growth. They use these molecules as a source of C, energy, N, P, S, or another element needed by the cells. Most attention has been focused on the acquisition of C and energy to sustain the growth of bacteria and fungi. For the synthetic substrates that are extensively degraded, the molecule is simply another organic substrate from which the population can obtain the needed elements or the energy required for biosynthetic reactions.

A common research procedure that relies on the ability of microorganisms to use organic compounds as sources of C and energy for growth is known as the enrichment-culture technique. The method is based on the selective advantage gained by an organism that is able to use a particular test compound as a C and energy source in a medium containing inorganic nutrients but no other sources of C and energy. Under these conditions, a species that is able to grow by utilizing that chemical will multiply. Few other bacteria and fungi will proliferate in this medium. However, species that use products excreted by the populations acting on the added organic nutrient will also flourish, and thus the final isolation of a microorganism in pure culture requires plating on an agar medium so that individual colonies can be selected. That agar medium is also made selective by having a single source of C and energy. Repeated transfer of the enrichment through solutions that contain the test compound and inorganic nutrients further increases the degree of selectivity before plating because organic materials and unwanted species from the original environmental sample are diluted by the serial transfers.

The enrichment-culture technique has been the basis for the isolation of pure cultures of bacteria and fungi that are able to use a large number of organic molecules as C and energy sources. However, attempts to obtain microorganisms that are able to grow on a variety of other organic compounds have met with failure. Undoubtedly, many of the failures can be attributed to misuse of the technique or errors in the approach of the

8

investigator; for example, sometimes the concentration of the organic nutrient may be too low to give detectable turbidity in the enrichment solution or too high so that the microorganisms fail to develop because of the toxicity. In other instances, the failure results from the absence from the selective medium of the growth factors essential for the organisms degrading the compound. Nevertheless, when the failure to isolate a microorganism by enrichment culture agrees with the prolonged persistence of the chemical in nature, it is likely that the compound is not used by microorganisms as a source of C and energy.

Members of a large number of genera of bacteria and fungi have been isolated that grow on one or more synthetic compounds. Much of the early literature deals with sugars, amino acids, other organic acids, and other cellular or tissue constituents of living organisms, but a variety of pesticides have also been shown to support the growth of one or another bacterium or fungus. Under these conditions, bacteria increase in numbers and fungi increase in biomass in culture media. At the same time, the chemical disappears, typically at a rate that parallels the increase in cell number or biomass. As the concentration of the C source declines, the rate of cell or biomass increase diminishes until, when all the substrate is consumed, the population rise ends.

As a rule, mineralization of organic compounds is characteristic of growth-linked biodegradation, in which the organism converts the substrate to CO_2, cell components, and products typical of the usual catabolic pathways. It is likely, however, that mineralization in nature occasionally may not be linked to growth but instead results from nonproliferating populations. Conversely, some species growing at the expense of a C compound may still not mineralize and produce CO_2 from the substrate; however, if O_2 is present, the organic products excreted by one species probably will be converted to CO_2 by another species, so that even if the initial population does not produce CO_2, the second species will. The net effect is still one of mineralization.

A compound, such as many environmental pollutants, that represents a novel C and energy source for a particular population still is transformed by the metabolic pathways that are characteristic of heterotrophic microorganisms. For the organism to grow on the compound, it must thus be converted to the intermediates that characterize these major metabolic sequences. If the compound cannot be modified enzymatically to yield such intermediates, it will not serve as a C and energy source because the energy-yielding and biosynthetic processes cannot function. The initial phases of the biodegradation thus involve modification of the novel substrate to yield a product that is itself an intermediate or, following further metabolism, is converted to an intermediate in these ubiquitous metabolic sequences. This need to convert the synthetic molecule to intermediates is characteristic of both aerobes and anaerobes as they derive C and energy from the substrate.

It should be stressed, however, that an organic compound need not be a substrate for growth in order for it to be metabolized by microorganisms. Two categories of transformations exist. In the first, the biodegradation provides C and energy to support growth, and the process therefore is growth-linked. In the second, the biodegradation is not linked to multiplication; the reasons will be considered in the following.

Several studies have demonstrated that the number of microbial cells or the biomass of the species acting on the chemical of interest increases as degradation proceeds. During a typical growth-linked mineralization brought about by bacteria, the cells use some of the energy and C of their organic substrate to make new cells, and this increasingly large population causes an increasingly rapid mineralization. In these instances, the mineralization reflects the population changes. During the decomposition of 2-, 3-, or 4-chlorobenzoate or 3,4-dichlorobenzoate in sewage, for example, bacteria acting on these compounds multiply, and the increase in cell numbers parallels the destruction of the chemicals that serve as their source of C (DiGeronimo et al., 1979). Similarly, bacteria capable of metabolizing 4-nitrophenol proliferate in sewage samples as the chemical disappears from the water phase (Fig. 2.1). Bacteria using 2,4-D similarly increase in numbers as the microbial community of soil destroys this herbicide (Kunc and Rybarova, 1983). Many observations have been made that pure cultures grow as they destroy synthetic chemicals, for example, during the decomposition of the herbicide IPC by Arthrobacter sp. (Clark and Wright, 1970). It also has been reported frequently that synthetic molecules, such as the herbicide endothal (Sikka and Saxena, 1973), are converted to typical constituents of microbial cells as the chemicals are used as C and energy sources.

ASSIMILATION OF CARBON

Many measurements have been made of the percentage of the C in the organic substrate that is converted into the cells that are carrying out the biodegradation. The values reflect the biological efficiency of converting the substrate into biomass, with the higher values characterizing the more efficient organisms. Such measurements are simple and straightforward in liquid media with water-soluble substrates since the biomass is particulate and thus can be readily distinguished from C in solution. The measurements in soils, wastewater, sewage, or sediments, in contrast, are compli-

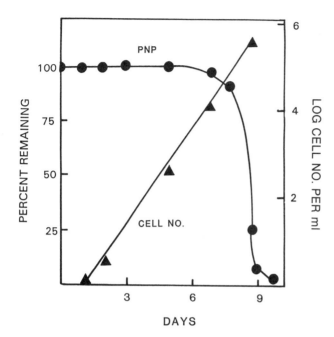

FIG. 2.1 Growth of bacteria degrading 4-nitrophenol (PNP) in sewage amended with 2 mg of 4-nitrophenol per liter. (From Wiggins *et al.*, 1987. Reprinted with permission from the American Society for Microbiology.)

cated because other particulate matter is present in addition to the cells and because complex water-insoluble products are often formed that must be distinguished from the cell material. In samples of such environments, therefore, C assimilation is estimated as

$$C_{assimilated} = C_{substrate} - C_{mineralized}.$$

The C that is assimilated is further mineralized as the cells metabolizing the original substrate are themselves decomposed or consumed by protozoa or other predators.

The values from the measurements in pure cultures of microorganisms are often expressed as *growth yield*, which is the weight of biomass formed divided by the weight of substrate used. The values for pure cultures may also be given as a *molar growth yield*, which is the weight of biomass formed divided by moles of substrate metabolized.

The values for such estimates of the efficiency of biomass production vary appreciably, for both aerobes and anaerobes. Some species are efficient in capturing the energy in the organic substrate and converting the

C to cells, but others are notably inefficient. Typical values are presented in Table 2.1. Some of the values listed are not actual estimates of C assimilated but rather are quantities of substrate-C not mineralized; such figures represent the total C in biomass and in products. It is immediately evident that the percentages of substrate-C converted to cells, the amount not mineralized, or both vary enormously. In some instances, little biomass is formed. In others, the yield of cells plus products is very high.

Under certain conditions in fresh and wastewaters, and possibly in other environments as well, essentially all the C is mineralized, and little or none accumulates in the biomass. This is surprising and as yet unexplained because mineralization generates energy, and the metabolic pathways leading to the formation of CO_2 are assumed to involve biochemical sequences that result in C assimilation. In one study, for example, 93 to 98% of benzoate, benzylamine, aniline, phenol, and 2,4-D added to samples of lake water or sewage at levels below 300 μg per liter was converted to CO_2, and direct measurements revealed no C assimilation during the mineralization of 24 ng to 250 μg of benzylamine per liter (Subba-Rao et al., 1982). Similarly, only 1.2% of the C of 2,4-D added to stream water was converted to particulate form, the particle fraction in waters containing the

TABLE 2.1
Percentages of Substrate-C Converted to Cells or Mineralized

Organism or environmental sample	Substrate	% of substrate-C converted to cells	Reference
Bacillus acido- caldarius	Glucose	15–47	Farrand et al. (1983)
Candida utilis	Glucose	39[a]	Johnson (1967)
Candida utilis	Acetate	56[a]	Johnson (1967)
Arthrobacter sp.	Glucose	21–28[a]	Cacciari et al. (1983)
River water	Biphenyl	<20–40[b]	Bailey et al. (1983)
Pond water	Phenol	20–25	Chesney et al. (1985)
Lake water	Aniline	40–60[b]	Hoover et al. (1986)
Lake water	4-Nitrophenol	<10	Hoover et al. (1986)
Soil	Maleic hydrazide	44[b]	Helweg (1975)
Soil	Phenthoate	39[b]	Iwata et al. (1977)
Soil	Several	20–40	Kassim et al. (1981)
Soil	Glucose	17–53[b]	Martin and Haider (1979)
Soil	Several	>50[b]	Scow et al. (1986)
Soil	Acetate	>70[b]	Stevenson and Ivarson (1964)
Soil	2,4-D	19–92[b]	Stott et al. (1983)
Sewage inoculum	Phthalate esters	2–44	Sugatt et al. (1984)

[a] Assumes cells contain 50% C.
[b] The value represents substrate-C converted to cells and organic products (i.e., it is the percentage of C not mineralized).

microbial cells (Boethling and Alexander, 1979). This lack of significant C assimilation may be a result of the inability of the organisms to obtain C and energy for biosynthetic purposes at these low concentrations, the immediate use of the C for respiration in order for the cells to maintain their viability (i.e., for maintenance energy), or the rapid decomposition and mineralization of the cells and their constituents.

In contrast, a high percentage of the C in other compounds or in similar compounds in different environments is incorporated and accumulates in the biomass, even at low substrate concentrations. With some bacteria, moreover, the efficiency of incorporation of substrate-C into cells is essentially the same from 43 ng to 100 mg of glucose-C per liter. This constancy is especially surprising at substrate concentrations so low that presumably all the C is being diverted to respiration for the organisms to maintain their viability (maintenance metabolism), although it is possible that bacteria use other organic molecules in their environment for maintenance and not the compound whose biodegradation is being determined (Seto and Alexander, 1985).

The percentage of the substrate that is either mineralized or incorporated depends on the species carrying out the transformation, the identity of the substrate, its concentration, temperature, and probably other environmental factors. In one investigation, only 15% of NTA was found to be mineralized in lake water at 100 ng/liter, but the value was more than 90% at 10 μg or 1.0 mg/liter; on the other hand, the values were 59, 78, and 12% for IPC at 400 ng, 10 μg, and 1.0 mg/liter, respectively, so it is not possible to generalize that the percentage of mineralization increases or decreases with increasing concentration (Hoover et al., 1986). An effect of concentration on the percentage mineralized is also evident in soil (Sielicki et al., 1978). On the other hand, the percentage of substrate mineralized by some bacteria may not change over enormous ranges of substrate concentration (Seto and Alexander, 1985). Temperature also affects the percentage of substrate-C that is incorporated into biomass or mineralized by sediment microfloras and cultures of individual bacterial species (Tison and Pope, 1980).

In natural communities, the cells that grow on the chemical of interest themselves are decomposed by other species or grazed upon and the C respired as CO_2 by protozoa or other predators. Hence, the percentage of substrate-C incorporated into the biomass of natural communities declines and the percentage mineralized increases with time, at least in the presence of O_2, and the values initially reflect the populations active on the organic compound but, with time, reflect the activities of the community of microorganisms. Thus, patterns of mineralization have a characteristic initial phase that, to a significant degree, represents the species acting on the parent molecule. Thereafter, a slower phase of mineralization is evident as the original cells, as well as their excretions, are destroyed and converted to CO_2 and other products.

In soil and undoubtedly other environments, a small or a large part of the substrate-C is also converted to high-molecular-weight complexes that are resistant to rapid biodegradation. Such humic substances may contain much of the C originally added to that environment, and this organic matter is only very slowly converted to CO_2 (Stott et al., 1983).

ASSIMILATION OF OTHER ELEMENTS

Synthetic molecules may be used as sources of required elements other than C. Microorganisms need N, P, S, and a variety of other elements, and these nutrient requirements may be satisfied as the responsible species degrade the compound of interest. It is common for the element that is in organic complex to be converted to the inorganic form before it becomes incorporated into cell components. For example, *Klebsiella pneumoniae* uses bromoxynil as a N source, but it does so only after converting the nitrile to NH_3, which is then assimilated (McBride et al., 1986). Similarly, a strain of *Pseudomonas* sp. that uses 2,6-dinitrophenol as a N source for growth first cleaves the nitro groups to free nitrite that, presumably after reduction of the nitrite to NH_3, sustains multiplication of the bacteria (Bruhn et al., 1987). Bacteria are also able to use a large number of organophosphorus insecticides (Rosenberg and Alexander, 1979), alkyl phosphates and phosphonates (Cook et al., 1978), and the herbicide glyphosate (Balthazor and Hallas, 1986) as P sources. Sulfur may also be extracted from organic molecules and then support multiplication, as shown by the use of O,O-diethylphosphorothioate and O,O-diethylphosphorodithioate as S sources by *Pseudomonas acidovorans* (Cook et al., 1980). Although organic substrates may contain more than one of the elements needed for growth, the organism frequently is able to use the chemical as a source of only one of its constituent elements.

For heterotrophic microorganisms in most natural ecosystems, the limiting element is generally C, and usually sufficient N, P, S, and other nutrient elements are present to satisfy the microbial demand. Because C is limiting and because it is the element for which there is intense competition, a species with the unique ability to grow on synthetic molecules has a selective advantage. No such selective advantage exists for an organism using an organic compound as the source of an element that already is available in abundant supply. Hence, it is unlikely that microorganisms obtaining other nutrient elements from synthetic molecules are selectively enhanced in such environments. Nevertheless, as the organisms use the molecules as C or energy sources, the biodegradative process usually will still lead to the mineralization of the other elements in the chemical.

REFERENCES

Bailey, R. E., Gonsior, S. J., and Rhinehart, W. L., *Environ. Sci. Technol.* **17**, 617–621 (1983).
Balthazor, T. M., and Hallas, L. E., *Appl. Environ. Microbiol.* **51**, 432–434 (1986).
Boethling, R. S., and Alexander, M., *Appl. Environ. Microbiol.* **37**, 1211–1216 (1979).
Bruhn, C., Lenke, H., and Knackmuss, H.-J., *Appl. Environ. Microbiol.* **53**, 208–210 (1987).
Cacciari, I., Lippi, D., Ippoliti, S., and Pietrosanti, W., *Can. J. Microbiol.* **29**, 1136–1140 (1983).
Chesney, R. H., Sollitti, P., and Rubin, H. E., *Appl. Environ. Microbiol.* **49**, 15–18 (1985).
Clark, C. G., and Wright, S. J. L., *Soil Biol. Biochem.* **2**, 19–26 (1970).
Cook, A. M., Daughton, C. G., and Alexander, M., *Appl. Environ. Microbiol.* **36**, 668–672 (1978).
Cook, A. M., Daughton, C. G., and Alexander, M., *Appl. Environ. Microbiol.* **39**, 463–465 (1980).
DiGeronimo, M. J., Nikaido, N., and Alexander, M., *Appl. Environ. Microbiol.* **37**, 619–625 (1979).
Farrand, S. G., Jones, C. W., Linton, J. D., and Stephenson, R. J., *Arch. Microbiol.* **135**, 276–283 (1983).
Helweg, A., *Weed Res.* **15**, 53–58 (1975).
Hoover, D. G., Borgonovi, G. E., Jones, S. H., and Alexander, M., *Appl. Environ. Microbiol.* **51**, 226–232 (1986).
Iwata, Y., Ittig, M., and Gunther, F. A., *Arch. Environ. Contam. Toxicol.* **6**, 1–12 (1977).
Johnson, M. J., *Science* **155**, 1515–1519 (1967).
Kassim, G., Martin, J. P., and Haider, K., *Soil Sci. Soc. Am. J.* **45**, 1106–1112 (1981).
Kunc, F., and Rybarova, J., *Soil Biol. Biochem.* **15**, 141–144 (1983).
Martin, J. P., and Haider, K., *Soil Sci. Soc. Am. J.* **43**, 917–920 (1979).
McBride, K. E., Kenny, J. W., and Stalker, D. M., *Appl. Environ. Microbiol.* **52**, 325–330 (1986).
Rosenberg, A., and Alexander, M., *Appl. Environ. Microbiol.* **37**, 886–891 (1979).
Scow, K. M., Simkins, S., and Alexander, M., *Appl. Environ. Microbiol.* **51**, 1028–1035 (1986).
Seto, M., and Alexander, M., *Appl. Environ. Microbiol.* **50**, 1132–1136 (1985).
Sielicki, M., Focht, D. D., and Martin, J. P., *Appl. Environ. Microbiol.* **35**, 124–128 (1978).
Sikka, H. C., and Saxena, J., *J. Agric. Food Chem.* **21**, 402–406 (1973).
Stevenson, I. L., and Ivarson, K. C., *Can. J. Microbiol.* **10**, 139–142 (1964).
Stott, D. E., Martin, J. P., Focht, D. D., and Haider, K., *Soil Sci. Soc. Am. J.* **47**, 66–70 (1983).
Subba-Rao, R. V., Rubin, H. E., and Alexander, M., *Appl. Environ. Microbiol.* **43**, 1139–1150 (1982).
Sugatt, R. H., O'Grady, D. P., Banerjee, S., Howard, P. H., and Gledhill, W. E., *Appl. Environ. Microbiol.* **47**, 601–606 (1984).
Tison, D. L., and Pope, D. H., *Appl. Environ. Microbiol.* **39**, 584–587 (1980).
Wiggins, B. A., Jones, S. H., and Alexander, M., *Appl. Environ. Microbiol.* **53**, 791–796 (1987).

3 ACCLIMATION

Prior to the degradation of many organic compounds, a period is noted in which no destruction of the chemical is evident. This time interval is designated an *acclimation period* or, sometimes, an adaptation or lag period. It may be defined as the length of time between the addition or entry of the chemical into an environment and evidence of its detectable loss. During this interval, no change in concentration is noted, but then the disappearance becomes evident and the rate of destruction often becomes rapid (Fig. 3.1).

This acclimation phase may be of considerable public health or ecological significance because the chemical is not destroyed. Hence, the period of exposure of humans, animals, and plants is prolonged, and the possibility of an undesirable effect is increased. Furthermore, if the chemical is present in flowing waters above or below ground, it may be widely disseminated laterally or vertically because of the absence of detectable biodegradation. In the case of toxicants, such increased dispersal may result in the exposure of susceptible species at distant sites before the harmful substance is destroyed.

Acclimation periods have been reported for many compounds that are introduced into soil, fresh water, sediment, and sewage. Among the chemicals for which such a phase has been described, either aerobically or anaerobically, are the following.

(a) Herbicides: 2,4-D, MCPA, Mecoprop, 4-(2,4-DB), TCA, amitrole, dalapon, monuron, chlorpropham, endothal, pyrazon, and DNOC.
(b) Insecticides: methyl parathion and azinphosmethyl.
(c) Quaternary ammonium compounds: dodecyltrimethylammonium chloride.
(d) Polycylic aromatic hydrocarbons: naphthalene and anthracene.
(e) Others: phenol, 4-chlorophenol, 4-nitrophenol, chlorobenzene, 1,2- and 1,4-dichlorobenzene, 3,5-dichlorobenzoic acid, PCP, diphenyl-methane, and NTA.

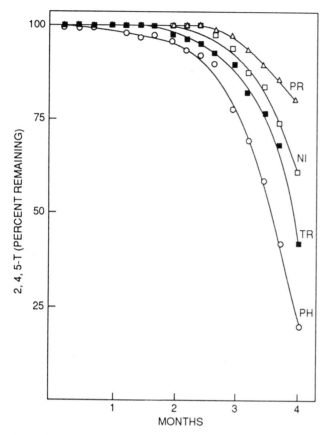

FIG. 3.1 Disappearance of 2,4,5-T in soil from the Philippines (PH), Trinidad (TR),
Nigeria (NI) and Puerto Rico (PR). (From Rosenberg and Alexander, 1980. Reprinted
with permission from the American Chemical Society.)

The length of the acclimation period varies enormously. It may be less
than 1 h or many months. The duration varies among chemicals and
environments, and it also depends on the concentration of the compound
and a number of environmental conditions. Some typical values are given
in Table 3.1. The values shown are not fixed, and the acclimation period
for any one of the chemicals may be longer or shorter than the times shown,
depending on the concentration, the environment, the temperature, the
aeration status, and other, often undefined factors. The time period may
be especially long in anaerobic environments for some compounds, such
as chlorinated molecules (Linkfield *et al.*, 1989). Especially disturbing is
the inability to predict accurately the duration of the acclimation phase
for most chemicals in nearly all environments.

TABLE 3.1
Lengths of Acclimation Phases for Several Organic Compounds

Chemical	Environment	Length of acclimation phase	Reference
Several aromatics	Soil	10–30 h	Kunc and Macura (1966)
Dodecyltrimethyl- ammonium chloride	Fresh water	24 h	Ventullo and Larson (1986)
4-Nitrophenol	Water–sediment	40–80 h	Spain and Van Veld (1983)
Amitrole	Soil	7 days	Riepma (1962)
Chlorinated benzenes	Biofilm	10 days–5 months	Bouwer and McCarty (1984)
DNOC	Soil	16 days	Hurle and Rademacher (1970)
PCP	Stream water	21–35 days	Pignatello et al. (1986)
Mecoprop	Enrichments	30–37 days	Lappin-Scott et al. (1986)
NTA	Estuary	50 days	Pfaender et al. (1985)
Halobenzoates	Sediment (anaerobic)	3 weeks–6 months	Linkfield et al. (1989)
2,4,5-T	Soil	4–10 weeks	Rosenberg and Alexander (1980)
Several	Groundwater	>16 weeks	Wilson et al. (1986)

The acclimation phase is considered to end at the onset of the period of detectable biodegradation. After the acclimation, the rate of metabolism of the chemical may be slow or rapid, but if a second addition of the chemical is made during this time of active metabolism, the loss of the second increment characteristically occurs with little or no acclimation (Fig. 3.2). The disappearance or marked reduction in the acclimation period has been noted in soils amended with 2,4-D (Audus, 1949), DNOC (Hurle and Pfefferkorn, 1972), amitrole (Riepma, 1962), methomyl (Fung and Uren, 1977), and CIPC (Kaufman and Kearney, 1965), soil suspensions amended with 4-(2,4-DB) (Whiteside and Alexander, 1960), river water supplemented with 4-nitrophenol (Spain et al., 1980), and marine

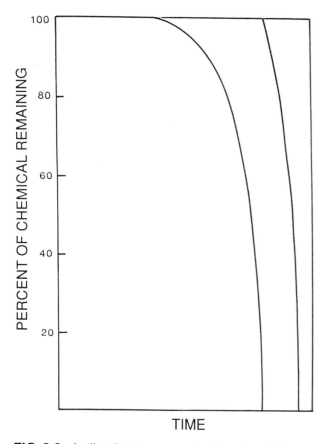

TIME

FIG. 3.2 Acclimation phase preceding the microbial destruction of an organic chemical after its first addition, and the absence of a delay period and rapid metabolism of the chemical following its second addition.

waters containing 4-chlorophenol (Kuiper and Hanstveit, 1984). It is generally assumed that biodegradation is detected immediately following the second introduction of the chemical because the organisms responsible for the transformation became numerous as they grew on the organic chemical following its first introduction.

The rate of biodegradation of the second addition may be the same as the final rate evident during the active phase of breakdown of the first addition (Kaufman and Kearney, 1965). However, it is far more common to have a greater rate of biodegradation, which is usually measured as the loss of parent compound or the formation of $^{14}CO_2$ from labeled compound, following the second than after the first application. The rate is further enhanced with still more additions. This enhancement of rate upon repeated additions of chemical has been reported frequently for pesticides added to soil. For example, the rate of parathion loss and its conversion to CO_2 rises as soil receives additional monthly treatments with the insecticide (de Andrea et al., 1982). The degradation of iprodione and vinclozolin similarly becomes more rapid as a result of prior additions of these fungicides to soil (Walker et al., 1986). In soil to which EPTC or butylate is applied, the rate of mineralization increases as a result of prior treatments with these herbicides (Obrigawitch et al., 1982, 1983). Greater rates of disappearance of the nematicide enthoprop and diphenamid are evident following the second than after the first introduction into soil (Kaufman et al., 1985). Not all of these instances show an acclimation phase, but such acclimations are common prior to the period of rapid pesticide breakdown in soil. Similar changes occur in water. Thus, dodecyltrimethylammonium chloride is rapidly mineralized in fresh water after an acclimation period, and the rate is faster following the second than after the first addition of the quaternary ammonium compound (Ventullo and Larson, 1986).

The greater rate on subsequent additions probably results from increases in numbers of degrading organisms following repeated treatment with the chemical. Consider a simple illustration. Assume that a chemical is added repeatedly at a concentration of 1.0 μg per unit volume, that its mineralization is a result of bacterial action, and that each bacterial cell that is formed destroys 1 pg (10^{-12} g) of the organic molecule. A considerable amount of time would elapse until 1.0 $\times$ 10^6 cells appear to destroy the first increment. However, far less time would be required for those 1.0 $\times$ 10^6 cells to destroy the next 1.0 μg, but the population size would then grow to 2.0 $\times$ 10^6. Still less time would elapse for those 2.0 $\times$ 10^6 bacteria to destroy the following 1.0 μg, but they would then grow to give 3.0 $\times$ 10^6 cells, etc. The example is an oversimplification in many ways, one of which is that not all cells produced would survive because they might die or be destroyed by protozoa or other predators before the next addition is made, but it does illustrate the greater rate to be expected of the larger populations that may be produced with successive introductions of a C source they use for growth.

Once the indigenous community of microorganisms has become acclimated to the degradation of a chemical and the activity becomes marked, the community may retain its active state for some time, that is, the potential for activity may continue to remain higher than in comparable soils, waters, or sewage-treatment systems that have not acquired this capacity. For example, 2-4-D disappears more quickly from soils treated a year previously with the herbicide than from untreated soils (Newman and Thomas, 1949), and the rate of degradation of isofenphos is also more rapid in soil treated with this insecticide a year earlier than in untreated soil (Chapman *et al.*, 1986b). An effect of pretreatment with TCA is evident even after about 3 years (McGrath, 1976), Conversely, the effect of prior treatment to acclimate the microbial community may be short-lived, as indicated by the loss of the higher activity on 4-nitrophenol after 7 weeks in a fresh water–sediment mixture (Spain and Van Veld, 1983). Too little information is presently available to permit generalizations among chemicals on the duration of the beneficial influence of prior additions of the compound. It is not presently clear why a microbial community that has acclimated to a particular substrate loses that activity; it could result from a decline in numbers or biomass of the responsible microorganisms or a loss of the metabolic activity in the absence of the specific chemical.

FACTORS AFFECTING ACCLIMATION

Acclimation of a microbial community to one substrate frequently results in the simultaneous acclimation to some, but not all, structurally related molecules. Because individual species often act on several structurally similar substrates, the species favored by the first addition may then quickly destroy the analogues. For example, when the microbial community of soil becomes acclimated to destroy 2,4-dichloro- or 4-chloro-2-methylphenoxyacetic acid, it simultaneously acquires the capacity for more rapid destruction of the other herbicide (Soulas *et al.*, 1983); if the soil becomes enriched with organisms that bring about the rapid degradation of EPTC, a more rapid degradation of the structurally similar herbicide butylate will occur (Obrigawitch *et al.*, 1983). Similarly, stimulation of the organisms that metabolize phenol in water, following acclimation, will result in enhanced metabolism of 4-chlorophenol, 3-aminophenol, and *m*-cresol (Shimp and Pfaender, 1987). Analogous simultaneous acclimations for transformation of several polycyclic aromatic hydrocarbons occur in slurries of marine sediments following acclimation to other polycyclics or benzene (Bauer and Capone, 1988). The length of the acclimation is affected by several environmental factors. Temperature has a major impact on the duration of the period before the active phase, as indicated by the longer interval before the onset of rapid oil biodegradation at lower than at higher temperatures (Atlas and Bartha, 1972). The pH and aeration status of an environment also appear to affect the duration of the acclimation for some compounds. The concentration of N, P, or both may be important in some environments, as in some natural waters, in which their concentrations are so low that they may limit microbial growth (Lewis et al., 1986; Wiggins et al., 1987). Conversely, the acclimation prior to the biodegradation of P- or N-containing organic compounds could be extended because of high levels of P or N if the responsible organisms use the inorganic phosphate or N from the environment in preference to that which would be released as a result of cleavage of the organic molecule (Daughton et al., 1979).

The concentration of the compound that is being metabolized greatly affects the length of time before one can detect a decline in its concentration. The rate of biodegradation of chemicals at trace levels increases with concentration, but because chemical loss is usually determined and not CO_2 or product formation, the low precision of analyses leads to data indicating a longer acclimation at higher concentration. Thus, based on measurements of loss of the test chemical, the acclimation period lengthens as the concentration of mecoprop in soil increases from 1 to 40 mg/kg (Amrein et al., 1981), the level of picloram rises from 0.25 to 1.0 mg/kg (Grover, 1967), and the concentration of 4-nitrophenol in sewage effluent increases from 1 to 25 mg/liter (Nyholm et al., 1984).

The duration of the acclimation period is not fixed even at a single concentration but varies from site to site, and some microbial communities acclimate to a particular chemical whereas others do not. Such variation in the occurrence of acclimation has been noted for 4-nitrophenol added to fresh and marine waters (Spain and Van Veld, 1983) and IPC and 2,4-D added to lake waters (Hoover et al., 1986).

As stated earlier, many chemicals are rapidly degraded only after an acclimation period, and the second addition is destroyed more readily than the first. On the other hand, the destruction of some compounds by microorganisms does not show such patterns. For example, NTA mineralization in estuarine waters is occasionally slow and does not become more rapid with time (Pfaender et al., 1985). Similarly, although many thiocarbamate herbicides are destroyed more readily following their second than after their first addition to soil, not all thiocarbamates behave in this fashion (Gray and Joo, 1985). Other herbicides also are not destroyed at rates that become faster following two or more applications to soil, for example, monolinuron and simazine (Paeschke et al., 1978).

There appear to be concentrations of some chemicals below which no acclimation occurs. A typical case is 4-nitrophenol, which is destroyed in samples containing sediments and natural waters at concentrations above but not below 10 μg/liter (Spain and Van Veld, 1983). Similarly, second applications of 2,4-D to soil are mineralized more rapidly than the first if the two additions are at 3.3 or 33 but not at 0.33 mg/kg (Fournier et al., 1981), and analogous enhanced decomposition of carbofuran occurs in soil treated with 1.0 or 10 mg/kg but not at 0.01 or 0.1 mg/kg (Chapman et al., 1986a). On the other hand, microorganisms in fresh or marine waters may acclimate to destroy compounds at levels below which they can use single compounds as sole carbon sources for growth (i.e., below the threshold), for example, at 1.0 μg of IPC per liter (Hoover et al., 1986), below 2 μg/liter for dodecyltrimethylammonium chloride in stream water (Shimp et al., 1989), and 0.7 μg/liter for toluene in seawater (Button and Robertson, 1985).

Acclimation for the metabolism of a particular compound may be needed in some environments but not in others. Thus, the mineralization of 1 to 50 μg of 4-nitrophenol, 4 μg of 2,4-dichlorophenol, and 20 μg of NTA per kilogram in soil may proceed with little or no acclimation period (Scow et al., 1986), but acclimations are characteristic of the decomposition of such chemicals at higher concentrations or in other soils.

ACCELERATED PESTICIDE BIODEGRADATION

Farmers commonly grow the same crops in particular areas either continuously or at regular intervals in a crop-rotation sequence. This often results in the reappearance of the same pests each time the individual crop is grown. These pests may be insects, weeds, or plant pathogens. To reduce the severity or prevent the occurrence of large pest populations, the farmer, each year or several times in a single growing season, typically applies pesticides—insecticides for insects, herbicides for weeds, and fungicides for many plant pathogens. Many of the pesticides are applied to the soil, and they may be added before planting and subsequent emergence of the crop plants (preemergent pesticides) or after the plants emerge from the soil (postemergent pesticides). If the pest-control chemical is added before planting, it must persist sufficiently long to be present at concentrations high enough to control the insects or plant pathogens that harm the plants that later appear above ground or to prevent growth of the weeds that appear some time after the pesticide is applied. Obviously, no soil-applied chemical would be used unless it persisted for the requisite time.

With some useful pesticides, however, a change in their persistence and their consequent suppression of pests occurs with the passage of time. In the years immediately after the introduction and widespread use of these compounds, the control of harmful insects, weeds, and plant pathogens is adequate. Therefore, farmers continue to apply the chemicals as part of their usual field operations. With repeated use of the pesticides, however, the pests unexpectedly are no longer controlled and cause a marked reduction in crop yield. The pesticides appear to be losing effectiveness with time. Sometimes the inability of the original product to continue to control pests is a result of species acquiring resistance to the pesticides. In other cases, the loss of effectiveness is directly attributable to the more rapid degradation of the chemical as a result of its use season after season. This problem, often termed accelerated pesticide degradation or enhanced microbial degradation, is a consequence of a change in the length of the acclimation phase, the rate of biodegradation, or both as a direct consequence of the repeated use. The problem was initially recognized because of the loss of control by carbofuran of the rootworm that affects corn, whose eggs hatch 3 to 5 weeks after planting (Felsot et al., 1981), and the declining effectiveness of EPTC applied as a preemergence herbicide to control weeds growing in corn fields (Kaufman et al., 1985). Enhanced microbial degradation of carbofuran may also result in the inability to control other insects, including those affecting a variety of crops (Wilde and Mize, 1984). The accelerated degradation of EPTC has a spotty distribution, being a problem in some fields but not in nearby locations (Roeth et al., 1989).

A major practical impact of the inability to control certain insects with carbofuran and many weeds with EPTC has been to stimulate considerable research on their enhanced degradation. Loss of insect control by carbofuran is noted if the compound has been used for the previous 2 to 4 years (Felsot et al., 1981). In laboratory studies, even a single application of EPTC to soil leads to its more rapid mineralization than in soil not previously treated with the herbicide. In this instance, it appears that the differences are in the rate of mineralization and not the lengths of the acclimation phase associated with the two treatments (Obrigawitch et al., 1982).

Apart from the practical implications, which are particularly pronounced only with a few pesticides, the phenomenon of accelerated pesticide degradation is characteristic of many pest-control agents. It was early recognized for such herbicides as 2,4-D, MCPA (Torstensson *et al.*, 1975), and DNOC (Hurle and Pfefferkorn, 1972), but the list of affected compounds includes the herbicides TCA (McGrath, 1976), vernolate, and butylate (Gray and Joo, 1985), the insecticides chlorfenvinphos (Hommes and Pestemer, 1985) and fensulfothion (Read, 1983), and a variety of other important pesticides (Gray and Joo, 1985; Avidov *et al.*, 1988; Slade *et al.*, 1992). Moreover, a soil treated with one pesticide may show enhanced biodegradation and thus shorter periods of effectiveness of structurally related chemicals, for example, among the thiocarbamate herbicides EPTC, butylate, and vernolate (Wilson, 1984).

The enhanced degradation is the result of microbial action on many if not all of these compounds, as evidenced by findings that little or no mineralization or loss of pesticidal activity or of the chemical itself occurred in sterile soil, at least with those substances tested. Not all pesticides are subject to accelerated transformation following their regular use for control of insects, weeds, or plant pathogens, for example, atrazine, simazine, isoproturon, and chlorpyrifos (Racke *et al.*, 1990; Walker and Welch, 1991).

The enhanced or accelerated pesticide biodegradation is not surprising. Considerable research on the acclimation phase and on enhanced rates of transformation following repeated treatments had been undertaken. What is surprising is the many years that elapsed before it was reported to be a major practical problem. The reason for the enhancement may be an increase, following the first addition, in population or biomass of the microorganisms able to degrade the compound and use it as a source of C and energy; if the population or biomass is still large when the soil again receives the pesticide, the disappearance will then occur without an acclimation period. If the population size has diminished by the time of the second addition or if the compound is metabolized but does not serve as a C source, no acceleration should be evident (B. K. Robertson and M. Alexander, unpublished data). On the other hand, evidence exists that the accelerated degradation of EPTC and carbofuran in soil is not attributable to increased numbers of microorganisms but rather to a greater activity per cell (Moorman, 1988; Scow *et al.*, 1990).

To enhance the persistence of the chemicals and thus to allow for the control of pests that would no longer be suppressed, another chemical may be added together with the pest-control agent. The second compound, sometimes called an extender because it extends the life of the pesticide, acts by inhibiting the biodegrading populations. Extenders to increase the duration of effectiveness of EPTC and butylate are fonofos and dietholate (Rudyanski *et al.*, 1987).

EXPLANATIONS FOR THE ACCLIMATION PHASE

Many explanations have been proposed for the acclimation of microbial communities to the biodegradation of organic compounds in natural waters, soils, or wastewaters. Many of these were proposed based on early studies of pure cultures of bacteria growing in media containing single organic substrates, often at cell densities far higher than is common for individual species of bacteria in nature. Some were based on investigations of the biochemistry or genetics of individual species acting in pure culture on very high concentrations of sugars, amino acids, or other natural products that can be metabolized by a diverse array of microbial species. Few of the explanations, however, were derived from studies of natural microbial communities acting on synthetic compounds at environmentally relevant concentrations, and hence the original emphasis placed on certain of these hypotheses must be considered with skepticism. On the other hand, more recent studies have been designed to evaluate these hypotheses as they relate to natural communities as contrasted to pure cultures, to cell densities more characteristic of natural ecosystems than those bacterial densities commonly used in tests of pure cultures, to synthetic compounds acted on by only a few rather than a diversity of microbial genera or species, and to concentrations that are characteristic of environmental pollutants rather than of organic nutrients included in culture media.

These inquiries have often led not to the rejection of all the earlier hypotheses but rather to the establishment of somewhat different mechanisms as the most common causes of the acclimation. These explanations are related to: (a) proliferation of small populations; (b) presence of toxins; (c) predation by protozoa; (d) appearance of new genotypes; and (e) diauxie.

Proliferation of Small Populations

Soils, natural waters, sewage, and wastewaters typically contain small populations of microorganisms acting on many of the synthetic organic compounds that are capable of supporting growth. The population is so small that one, two, three, or several more doublings of the cell number would not bring about an appreciable loss of the chemical. For example, if the initial density of bacteria is 10^2 cells per unit volume and each bacterium, as it divides, destroys 1.0 pg of organic substrate, a decline in concentration of a compound initially present at 0.1 μg per unit volume would not be detected initially even as the bacteria grow (Fig. 3.3). There would be an apparent acclimation phase simply because the precision of

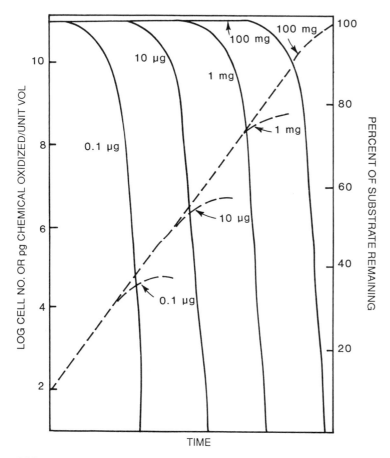

FIG. 3.3 Decline in concentration of organic substrate (solid lines), increase in amount of chemical oxidized (or product formed) on a logarithmic scale (dashed lines), and increase in logarithm of cell density (also dashed lines) per unit volume. The estimates shown are based on the assumptions that the bacteria are growing exponentially by using the substrate as the sole C source, that each cell consumes 1 pg of substrate and forms ca. 1 pg of CO_2 as it multiplies, and that the rate of growth does not change markedly as the substrate concentration declines.

analyses would not detect the loss of the 100 pg destroyed by 10^2 cells (0.1 μg or 10^5 pg less the 10^2 pg metabolized) or 10^3 cells (10^5 pg less the 10^3 pg metabolized). Only when the bacteria have undergone many cell divisions would a decline in concentration of parent chemical be detected. From Fig. 3.3, it is evident that even longer apparent acclimation periods would be evident as a population of such initial size is exposed to concentrations of 10 μg, 1 mg, and 100 mg of organic chemical per unit volume; the bacteria growing exponentially must reach ever greater abundance, and hence more time elapses, before a decline in concentration is evident.

In the circumstances in which this mechanism applies, the acclimation would not appear to be as long, or would not even exist, if analyses for product formation were performed (rather than loss of substrate). Thus, if one could detect 100 pg of product (e.g., $^{14}CO_2$ from a ^{14}C-labeled substrate), then the activity of 100 cells (each of which might form ca. 1 pg of CO_2) might be detected, although the loss of 100 or 1000 pg of substrate might not be measurable when the initial substrate concentration is 10^5 pg per unit volume. From Fig. 3.3, it would seem that the amount of substrate-C oxidized (equivalent to product-C generated) would be detectable by *sensitive* methods long before the loss of parent compound by *precise* techniques would be noted.

It is not presently clear how often acclimation, which in this case is more apparent than real, results simply from the time required for a small population to become sufficiently large to give a detectable loss of the organic substrate. Certainly many investigations have suggested its importance without providing supporting data. Yet, some research has in fact verified that the numbers of bacteria growing at the expense of specific organic chemicals rise as those compounds are degraded, for example, dodecyltrimethylammonium chloride and phenol in aquatic environments (Ventullo and Larson, 1986; Shimp and Pfaender, 1987), 2,4-D in lake water (Chen and Alexander, 1989), 4-nitrophenol in natural water plus sediment (Spain et al., 1980), and isofenphos in soil (Racke and Coats, 1987). In lake water amended with 2.0 mg of 4-nitrophenol per liter, for example, the number of cells able to metabolize the compound increases shortly after the addition, but loss of the chemical can only be detected at about 8 days, at which time the density of cells acting on 4-nitrophenol has reached ca. 10^5 per milliliter (Fig. 2.1). In many of these studies, the cell counts are less than those that might be expected based on the initial chemical concentration, but many of the cells that are formed undoubtedly are dying continuously or are consumed by predators such as protozoa. If a metabolized compound that does not support microbial growth has an acclimation phase, this explanation obviously does not apply.

In instances in which acclimation is solely a reflection of the time for the population size to become large enough to effect a detectable change, any factor that enhances (or diminishes) the growth rate would shorten (or lengthen) the acclimation. Thus, in environments in which the concentrations of N, P, or possibly other inorganic nutrients are low, the acclimation phase may be longer than in similar environments having higher levels, and the addition of N to N-poor environments and P to P-deficient environments may shorten the acclimation phase. Such effects have been noted during the degradation of p-cresol by a mixture of aquatic microorganisms (Lewis *et al.*, 1986) and during the mineralization of 4-nitrophenol in sewage and lake water (Wiggins *et al.*, 1987; Jones and Alexander, 1988).

From an examination of Fig. 3.3, it is evident why the acclimation may appear to be longer at higher substrate concentrations: more time is required for the cell density to become large enough to give a detectable loss at higher than at lower substrate concentrations. Other reasons may explain the longer acclimation at high than at low concentrations (e.g., toxicity at the higher levels), but difference in time to give the necessary cell size is surely one likely cause. From Fig. 3.3 it is also apparent why second additions of a chemical may often be destroyed with little detectable delay: the population became sufficiently large as a result of the first addition. On the other hand, if appreciable time elapsed between the day all the chemical was destroyed and the day when an additional increment is introduced, an acclimation may be evident; many of the cells may have died or been consumed by predators or possibly parasites once the unique C supply for that population was exhausted.

The phenomenon of accelerated pesticide degradation may have a similar explanation, as indicated earlier. The population size rises following the first treatment of soil with the pesticide. If the cell numbers or biomass remains large from season to season or between pesticide applications in a single season, the second increment disappears more readily than the first. If the population declines to about its original size before the next addition, a long acclimation will again be evident. Because many pesticides are added annually to soil, accelerated pesticide degradation may become evident with time if the repeated annual use of the chemical results, at the start of each growing season, in a population size that is larger than that at the start of the previous growing season (even if many cells die after all the chemical is metabolized); this ever larger number of cells would result in an ever shorter acclimation phase.

Presence of Toxins

Two circumstances are common in which the presence of inhibitors may affect the length of the period prior to the onset of rapid biodegradation. First, the chemical of interest may be present at such high levels that few of the biodegrading microorganisms present may be able to grow or metabolize, and biodegradation will not be detectable until the rare species, either present or transported to the site, is able to multiply to reach a biomass sufficient to cause appreciable chemical loss. Such high concentrations are known in waste discharges from chemical manufacturing or in accidental spills. Second, many sites containing toxic chemicals have a mixture of compounds, and one or more of the mixture may be inhibitory to the organisms destroying the test substance. When that

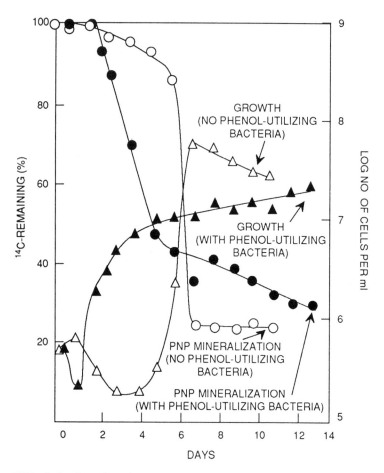

FIG. 3.4 Growth and 4-nitrophenol mineralization by a strain of *Pseudomonas* in phenol-containing media in the presence or absence of phenol-degrading bacteria. (From Murakami and Alexander, 1989. Copyright by John Wiley & Sons and reprinted with permission by the copyright holder.)

toxicant disappears by biodegradation, nonenzymatic destruction, sorption, or volatilization, the period in which no destruction of the test substance is evident is replaced by a period in which the destruction becomes marked (Fig. 3.4). The inhibition in many hazardous waste sites may be complete so that even readily utilizable substrates are not metabolized, but degradation will occur as the pollutant plume in the groundwater moves away from the source and the inhibitors become diluted.

The toxicant may act in several ways. (a) In the first and second circumstances just indicated, it may merely act to slow the growth rate of the degrading species and hence lengthen the period during which no loss of the chemical being measured is evident. Such toxicants may be organic molecules, or they may be inorganic. The role of the latter is shown by the finding that the acclimation phase prior to rapid NTA biodegradation in activated sludge is lengthened if the sludge is rich in heavy metals (Stephenson et al., 1984). (b) The toxicant may be eliminated so that the degraders are able to proliferate, and the acclimation period then represents the sum of the time for lowering the level of the antimicrobial agent to a noninhibitory concentration plus the subsequent time for multiplication of the degrading species to a density adequate for significant chemical loss. For example, in wastewater containing both 4-nitrophenol and 2,4-dinitrophenol, the acclimation phase for the mineralization of the first compound resulted from the toxicity of the second to the 4-nitrophenol-degrading species, but when the dinitro compound was biodegraded, the 4-nitrophenol utilizers proliferated and the acclimation phase soon ended (Wiggins and Alexander, 1988b). Evidence that part of the acclimation period for oil biodegradation reflects the time for volatilization of its antimicrobial constituents has been obtained by Atlas and Bartha (1972). Conversely, the acclimation period for 4-nitrophenol mineralization in the presence of high phenol concentrations is a result of the antimicrobial effect of the second chemical, but this period is markedly shortened as phenol is mineralized (Murakami and Alexander, 1989). (c) The toxicant may suppress the faster-growing species that usually predominate in a mixture of species capable of metabolizing the contaminants, but a resistant and slower-growing species will then have a selective advantage that it did not previously have—and the longer acclimation is simply a reflection of the longer time needed for the appearance of the large biomass of the slow-growing organisms. (d) The toxins may not be present initially, but they may be generated during biodegradation. This possibility is suggested by the observation that phenol mineralization may be suppressed by products generated microbiologically from 4-nitrophenol that is present together with phenol (Murakami and Alexander, 1989).

Predation by Protozoa

A number of natural ecosystems and aerated waste-treatment systems are characterized by large and active populations of protozoa. These microscopic animals feed and multiply because they prey on the bacteria in these environments. This feeding reduces the abundance of bacteria if their densities become especially high and probably keeps the bacterial density lower than might otherwise be expected based on the supply of readily utilizable organic nutrients.

Although little attention has been given to the role that protozoa play in governing acclimation, it appears that in sewage, and probably other environments in which bacteria are actively destroying synthetic molecules, these unicellular animals are quite important. The evidence comes from studies in which comparisons were made of the mineralization of 4-nitrophenol in samples of sewage containing protozoa and those in which protozoa were suppressed by additions of inhibitors selective for eucaryotes (Fig. 3.5). Because protozoa are the chief eucaryotes present in many wastewaters, the use of such selective inhibitors largely eliminates protozoan grazing on bacteria. When the predatory activities are thus

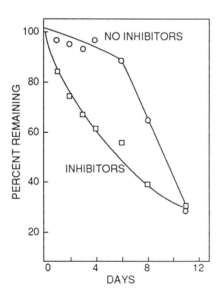

FIG. 3.5 Effect of protozoa on the biodegradation of 2 mg of 4-nitrophenol per liter of sewage. Protozoa were suppressed in half of the flasks by addition of the eucaryotic inhibitors cycloheximide and nystatin. (Reprinted with permission from Wiggins and Alexander, 1988a.)

suppressed, the normally long acclimation interval is markedly shortened; that is, the protozoan consumption of bacteria is directly related to the longer acclimation (Wiggins and Alexander, 1988a). Presumably, the protozoa act to keep the density of bacteria responsible for the degradation so low that no appreciable chemical loss is detectable. With time, the density of total bacteria falls to a low level, because of both this grazing as well as the reduced supply of readily utilizable organic matter, and then the protozoa become less active. As a result, bacteria growing on synthetic compounds proliferate by using the synthetic molecules whose presence gives these species a selective advantage; at this time, the decline in concentration of the compound becomes evident, and the acclimation phase ends (Fig. 3.6). Acclimation is less likely to be attributable to protozoa in environments in which predation is not marked, such as many natural waters (Wiggins et al., 1987).

Appearance of New Genotypes

Bacteria and fungi may undergo genetic change as a result of a mutant appearing in the population or the transfer of genetic information from one species to another. Such events occur at low frequency, so only a few cells in a population represent a new genotype with a particular set of new phenotypic traits. However, if the new genotype possesses physiological characteristics that give it a selective advantage, it will multiply. The possession of enzymes that degrade a novel substrate, like a synthetic molecule, and that give energy and C to the cells synthesizing these enzymes is clearly a selective advantage if other members of the microbial community are unable to grow at the expense of that molecule. Thus, acclimation could reflect the sum of the times for the mutation or gene transfer to occur and for the resulting organism to multiply to reach the requisite high population density.

The ease of showing the occurrence of mutations for acquisition of certain traits in pure cultures of bacteria and of gene transfer between large populations of two bacterial species has prompted frequent suggestions that acclimation is often a result of the appearance of new genotypes. However, little information exists to support these contentions. This is not to say that genetic changes followed by population growth are not involved in the acclimation prior to biodegradation of some chemicals, but only that the view is not well supported by experimental observations.

One line of evidence that mutations are not the cause of many acclimations is based on the random occurrence of mutations. An event that occurs randomly should occur sporadically and only in occasional replicate samples of the same environment collected at different times, and

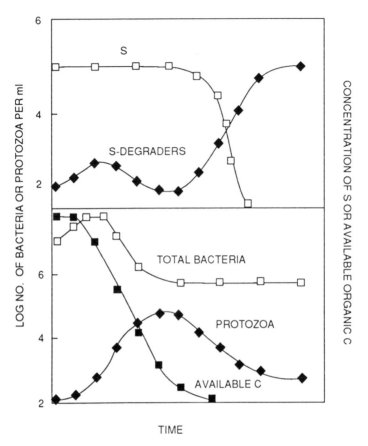

TIME

FIG. 3.6 Changes in populations of total bacteria, a bacterium degrading a test compound (S), and protozoa and the disappearance of total available C and the test compound in wastewater.

the acclimation phase that results from the time for a mutation to take place should vary in length among replicate samples from the environment; hence, the observations that the acclimation periods for the mineralization of 4-nitrophenol in replicate samples of sewage, 2,4-D in lake water, and halobenzoates in anaerobic sediments are essentially the same for each compound suggest that a mutation (followed by multiplication) does not account for the acclimation phase with these compounds (Wiggins *et al.*, 1987; Chen and Alexander, 1989; Linkfield *et al.*, 1989). The reproducible duration of the acclimation period prior to the degradation of other chemicals or in samples from several environments also has been used to argue against the significance of mutation (Spain *et al.*, 1980; Fournier *et al.*,

1981). Conversely, the marked difference in the times for onset of biodegradation of 3- and 5-nitrosalicylates and 4-nitroaniline in sewage and lake water suggests that mutations had occurred and that the acclimation reflected the time for the mutant to arise and for the population of the new genotype to multiply (Wiggins and Alexander, 1988b). Mutants of *Pseudomonas putida* may also have appeared in a microbial mixture growing in a medium containing the herbicide dalapon (Senior *et al.*, 1976).

The transfer of genes involved in biodegradation has been shown to take place in media containing two different bacteria. For example, the capacity to degrade chlorocatechols can be transferred from a strain of *Pseudomonas* to a strain of *Alcaligenes,* and the resulting new genotype is able to metabolize 2-, 3-, and 4-chlorophenols, a property possessed by neither of the parent bacteria (Schwien and Schmidt, 1982). Similarly, the transfer of genes coding for steps in the metabolism of 3-chlorobenzoate may take place between dissimilar strains of *Pseudomonas* (Rubio *et al.*, 1986). However, gene transfer leading to the evolution of new genotypes involved in biodegradation in natural environments or samples of such environments brought to the laboratory has yet to be confirmed.

Diauxie

Pure cultures of bacteria growing in media containing relatively high concentrations of two C sources often do not show the single exponential phase that is characteristic of the same organisms multiplying in media with a single C source. Instead, they have two exponential growth phases, which are separated by an interval with little or no growth (Fig. 3.7). During

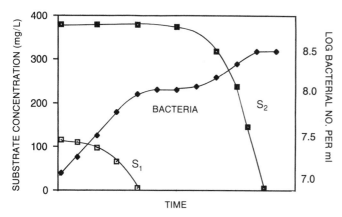

FIG. 3.7   Biphasic growth of a bacterium in pure culture with S_1 and S_2 as carbon sources.

the first exponential period, only one of the substrates is metabolized by the organism to support growth, and the second exponential period corresponds to growth on and degradation of the second organic compound. This biphasic growth and utilization of two substrates in sequence is known as *diauxie*. The C source that permits faster growth usually is metabolized first. Diauxie is characterized by repression of synthesis of the enzymes concerned with initial steps in the metabolism of the second C source as the bacterium uses the first (Harder *et al.*, 1984).

This utilization of one organic compound in preference to a second has been advanced as an explanation for acclimation; that is, the chemical of interest is the second substrate, and its loss does not begin until the supply of the first is depleted. For example, it has been suggested that bacteria in seawater only attack 4-chlorophenol after certain naturally occurring organic constituents of the water are consumed by the bacteria (Kuiper and Hanstveit, 1984). Enrichment cultures that are probably dominated by a single bacterial type, and thus behave much like a pure culture, also show sequential destruction of organic substrates and diauxic growth (Stumm-Zollinger, 1966). Diauxic utilization of substrates, and thus an apparent acclimation before use of the second substrates, may also occur with P compounds, as suggested by a report that *Pseudomonas testosteroni* uses inorganic phosphate before methylphosphonate as a P source for growth, resulting in an initial period when the phosphonate is not being degraded. Only when the inorganic phosphate is consumed does the disappearance of the second P source begin (Daughton *et al.*, 1979).

Direct tests for the occurrence of diauxie in biodegradation or its role in acclimation in natural ecosystems or in waste-treatment systems are scarce. Diauxie does not appear to account for the acclimation prior to mineralization of low concentrations of 4-nitrophenol in sewage or lake water (Wiggins *et al.*, 1987), but it may be important in other environments, especially those in which the one or both of the compounds needed for diauxic biodegradation are present at high concentrations. However, in a heterogeneous microbial community, it is likely that two readily available compounds will be used by different species, rather than one of the molecules persisting while the bacteria active on the first are growing and using their preferred nutrient.

ENZYME INDUCTION AND LAG PHASE

Microorganisms produce many enzymes regardless of whether the substrates for those enzymes are present. These are known as *constitutive enzymes*. In contrast, *inducible enzymes* are formed in appreciable amounts only when the substrate, or sometimes a structurally related chemical or metabolite, is present. The inducer is the specific molecule that, when provided to the cells, is involved in the process of induction. The inducible enzyme may be detectable in the absence of the inducer, but the level is not high. The process of induction has been extensively studied and is known to be a complex process, which typically involves an increase in the rate of formation of the degradative enzymes. The enzymatic activity of a population may also be controlled by *catabolite repression,* in which products generated during the catabolism of one substrate repress the synthesis of enzymes concerned with the degradation of a second substrate that itself would be converted to the same products.

Many of the enzymes involved in one or more of the early steps in the breakdown of synthetic compounds are inducible, for example, many of the dehalogenases that remove chlorine from halogenated molecules and release ionic chloride. However, because enzyme induction is usually largely complete in minutes or hours (Richmond, 1968) and acclimation phases often are weeks in duration (Table 3.1), only the very early portion of the usually far longer acclimation period would involve the time for induction of catabolic enzymes. An acclimation period associated with the time for induction of the enzymes metabolizing 2,4-D by a bacterium is depicted in Fig. 3.8. Except as it may be implicated in diauxie (Stumm-Zollinger, 1966), moreover, catabolite repression also does not seem to be a factor governing much of the time entailed for acclimation.

If the first steps in the metabolism of a compound require the biosynthesis of inducible enzymes and the conditions preclude such synthesis, the compound will not be degraded. In this light, it is noteworthy that a threshold appears to exist for the induction of certain enzymes, at least by some bacteria. For example, induction of the enzymes concerned with early steps in the catabolism of 3- and 4-chlorobenzoates by *Acinetobacter calcoaceticus* occurs at concentrations above but not below 1 μM (Reber, 1982), and the amidase that is necessary for a gram negative bacterium to cleave certain phenylurea herbicides is induced at 50 but not 10 μM linuron (Lechner and Straube, 1984).

The lag phase in the bacterial growth cycle is evident when a bacterium is transferred into fresh medium, even in a medium identical to that in which it previously grew. During this period, the organism does not multiply. Subsequently, the organism initiates rapid growth and, in rich media, enters into the exponential phase of growth. The lag phase is characteristic, or at least well studied, of inocula containing low cell densities, and the small population size coupled with the absence of multiplication denote little or no substrate loss. Because the lag phase in the bacteria studied to date lasts several hours at most and acclimation periods are often far

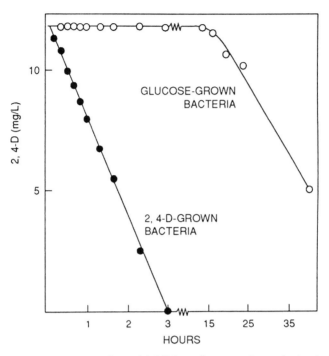

FIG. 3.8 Metabolism of 2,4-D by cell suspensions of a bacterium grown on 2,4-D or glucose. (Reprinted with permission from Chen and Alexander, 1989.)

longer, the lag phase per se does not explain the longer acclimations. However, as with enzyme induction, the very initial part of the acclimation may be associated with the true bacterial lag.

REFERENCES

Amrein, J., Hurle, K., and Kirchhoff, J., Z. Pflanzenkr. Pflanzenschutz, Sonderh. 9, 329–341 (1981).
Atlas, R. M., and Bartha, R., Can. J. Microbiol. 18, 1851–1855 (1972).
Audus, L. J., Plant Soil 2, 31–36 (1949).
Avidov, E., Aharonson, N., and Katan, J., Weed Sci. 36, 519–523 (1988).
Bauer, E. J., and Capone, D. G., Appl. Environ. Microbiol. 54, 1649–1655 (1988).
Bouwer, E. J., and McCarty, P. L., Ground Water 22, 433–440 (1984).
Button, D. K., and Robertson, B. R., Mar. Ecol.: Prog. Ser. 26, 187–193 (1985).
Chapman, R. A., Harris, C. R., and Harris, C., J. Environ. Sci. Health, Part B B21, 125–141 (1986a).
Chapman, R. A., Harris, C. R., Moy, P., and Henning, K., J. Environ. Sci. Health, Part B B21, 269–276 (1986b).

Chen, S., and Alexander, M., *J. Environ. Qual.* **18**, 153–156 (1989).

Daughton, C. G., Cook, A. M., and Alexander, M., *Appl. Environ. Microbiol.* **37**, 605–609 (1979).

de Andrea, M. M., Lord, K. A., Bromilow, R. H., and Ruegg, E. F., *Environ. Pollut., Ser. A* **27**, 167–177 (1982).

Felsot, A., Maddox, J. V., and Bruce, W., *Bull. Environ. Contam. Toxicol.* **26**, 781–788 (1981).

Fournier, J. C., Codaccioni, P., and Soulas, G., *Chemosphere* **10**, 977–984 (1981).

Fung, K. K. H., and Uren, N. C., *J. Agric. Food Chem.* **25**, 966–969 (1977).

Gray, R. A., and Joo, G. K., *Weed Sci.* **33**, 698–702 (1985).

Grover, R., *Weed Res.* **7**, 61–67 (1967).

Harder, W., Dijkhuisen, L., and Veldkamp, H., *in* "The Microbe" (D. P. Kelly and N. G. Carr, eds.), Part II, pp. 51–95. Cambridge Univ. Press, Cambridge, UK, 1984.

Hommes, M., and Pestemer, W., *Meded. Fac. Landbouwwet., Rijksuniv. Gent* **50**(2B), 643–650 (1985).

Hoover, D. G., Borgonovi, G. E., Jones, S. H., and Alexander, M., *Appl. Environ. Microbiol.* **51**, 226–232 (1986).

Hurle, K., and Pfefferkorn, V., *Proc. Br. Weed Control Conf., 11th, 1972*, Vol. 2, pp. 806–810 (1972).

Hurle, K., and Rademacher, B., *Weed Res.* **10**, 159–164 (1970).

Jones, S., and Alexander, M., *Appl. Environ. Microbiol.* **54**, 3177–3179 (1988).

Kaufman, D. D., and Kearney, P. C., *Appl. Microbiol.* **13**, 443–446 (1965).

Kaufman, D. D., Katan, Y., Edwards, D. F., and Jordan, E. G., *in* "Agricultural Chemicals of the Future" (J. L. Hilton, ed.), pp. 437–451. Rowman & Allenheld, Totowa, NJ, 1985.

Kuiper, J., and Hanstveit, A. O., *Ecotoxicol. Environ. Saf.* **8**, 15–33 (1984).

Kunc, F., and Macura, J., *Folia Microbiol. (Prague)* **11**, 248–256 (1966).

Lappin-Scott, H. M., Greaves, M. P., and Slater, J. G., *in* "Microbial Communities in Soil" (V. Jensen, A. Kjøller, and L. H. Sørensen, eds.), pp. 211–217. Elsevier Applied Science, London, 1986.

Lechner, U., and Straube, G., *Z. Allg. Mikrobiol.* **24**, 581–584 (1984).

Lewis, D. L., Kollig, H. P., and Hodson, R. E., *Appl. Environ. Microbiol.* **51**, 598–603 (1986).

Linkfield, T. G., Suflita, J. M., and Tiedje, J. M., *Appl. Environ. Microbiol.* **55**, 2773–2778 (1989).

McGrath, D., *Weed Res.* **16**, 131–137 (1976).

Moorman, T. B., *Weed Sci.* **36**, 96–101 (1988).

Murakami, Y., and Alexander, M., *Biotechnol. Bioeng.* **33**, 832–838 (1989).

Newman, A. S., and Thomas, J. R., *Soil Sci. Soc. Am. Proc.* **14**, 160–164 (1949).

Nyholm, N., Lindgaard-Jørgensen, P., and Hansen, N., *Ecotoxicol. Environ. Saf.* **8**, 451–470 (1984).

Obrigawitch, T., Wilson, R. G., Martin, A. R., and Roeth, F. W., *Weed Sci.* **30**, 175–181 (1982).

Obrigawitch, T., Martin, A. R., and Roeth, F. W., *Weed Sci.* **31**, 187–192 (1983).

Paeschke, R. R., Ebing, W., and Heitefuss, R., *Z. Pflanzenkr. Pflanzenschutz* **85**, 280–297 (1978).

Pfaender, F. K., Shimp, R. J., and Larson, R. J., *Environ. Toxicol. Chem.* **4**, 587–593 (1985).

Pignatello, J. J., Johnson, L. K., Martinson, M. M., Carlson, R. E., and Crawford, R. L., *Can. J. Microbiol.* **32**, 38–46 (1986).

Racke, K. D., and Coats, J. R., *J. Agric. Food Chem.* **35**, 94–99 (1987).

Racke, K. D., Laskowski, D. A., and Schultz, M. R., *J. Agric. Food Chem.* **38**, 1430–1436 (1990).

Read, D. C., *Agric. Ecosyst. Environ.* **10**, 37–46 (1983).

Reber, H. H. *Eur. J. Appl. Microbiol. Biotechnol.* **15**, 138–140 (1982).

Richmond, M. H., *Essays Biochem.* **4**, 105–154 (1968).

Riepma, P., *Weed Res.* **2**, 41–50 (1962).

Roeth, F. W., Wilson, R. G., Martin, A. R., and Shea, P. J., *Weed Technol.* **3**, 24–29 (1989).

Rosenberg, A., and Alexander, M., *J. Agric. Food Chem.* **28**, 705–709 (1980).

Rubio, M. A., Engesser, K.-H., and Knackmuss, H.-J., *Arch. Microbiol.* **145**, 116–122 (1986).

Rudyanski, W. J., Fawcett, R. S., and McAllister, R. S., *Weed Sci.* **35**, 68–74 (1987).

Schwien, U., and Schmidt, E., *Appl. Environ. Microbiol.* **44**, 33–39 (1982).

Scow, K. M., Simkins, S., and Alexander, M., *Appl. Environ. Microbiol.* **51**, 1028–1035 (1986).

Scow, K. M., Merica, R. R., and Alexander, M., *J. Agric. Food Chem.* **38**, 908–912 (1990).

Senior, E., Bull, A. T., and Slater, J. H., *Nature (London)* **263**, 476–479 (1976).

Shimp, R. J., and Pfaender, F. K., *Appl. Environ. Microbiol.* **53**, 1496–1499 (1987).

Shimp, R. J., Schwab, B. S., and Larson, R. J., *Environ. Toxicol. Chem.* **8**, 723–730 (1989).

Slade, E. A., Fullerton, R. A., Stewart, A., and Young, H., *Pestic. Sci.* **35**, 95–100 (1992).

Soulas, G., Codaccioni, P., and Fournier, J. C., *Chemosphere* **12**, 1101–1106 (1983).

Spain, J. C., and Van Veld, P. A., *Appl. Environ. Microbiol.* **45**, 428–435 (1983).

Spain, J. C., Pritchard, P. H., and Bourquin, A. W., *Appl. Environ. Microbiol.* **40**, 726–734 (1980).

Stephenson, T., Lester, J. N., and Perry, R., *Chemosphere* **13**, 1033–1040 (1984).

Stumm-Zollinger, E., *Appl. Microbiol.* **14**, 654–664 (1966).

Torstensson, N. T. L., Stark, J., and Goransson, B., *Weed Res.* **15**, 159–164 (1975).

Ventullo, R. M., and Larson, R. J., *Appl. Environ. Microbiol.* **51**, 356–361 (1986).

Walker, A., and Welch, S. J., *Weed Res.* **31**, 49–57 (1991).

Walker, A., Brown, P. A., and Entwistle, A. R., *Pestic. Sci.* **17**, 183–193 (1986).

Whiteside, J. S., and Alexander, M., *Weeds* **8**, 204–213 (1960).

Wiggins, B. A., and Alexander, M., *Can. J. Microbiol.* **34**, 661–666 (1988a).

Wiggins, B. A., and Alexander, M., *Appl. Environ. Microbiol.* **54**, 2803–2807 (1988b).

Wiggins, B. A., Jones, S. H., and Alexander, M., *Appl. Environ. Microbiol.* **53**, 791–796 (1987).

Wilde, G., and Mize, T., *Environ. Entomol.* **13**, 1079–1082 (1984).

Wilson, B. H., Smith, G. B., and Rees, J. F., *Environ. Sci. Technol.* **20**, 997–1002 (1986).

Wilson, R. G., *Weed Sci.* **32**, 264–268 (1984).

4 DETOXICATION

The most important role of microorganisms in the transformation of pollutants is their ability to bring about detoxication. Detoxication (sometimes designated detoxification) refers to the change in a molecule that renders it less harmful to one or more susceptible species. The susceptible species may be humans, animals, plants, other microorganisms, or the detoxifying population itself. Particular attention, needless to say, is given to detoxications that make organic compounds less injurious to humans, but an abundant body of information also exists on detoxication reactions that alter the toxicity to animals and plants. In studies of environmental pollution, detoxications that reduce the harm to microorganisms have received little inquiry.

Detoxications result in inactivation, with the toxicologically active substance being converted to an inactive product. Because toxicological activity is associated with many chemical entities, substituents, and modes of action, detoxications similarly include a large array of different types of reactions.

A simple way of demonstrating detoxication is to measure the effect of environmental samples on the behavior, growth, or viability of susceptible species. This clearly would not be acceptable when humans are the suscepts, but it is a common procedure when destruction of compounds affecting plants or lower animals is of concern. For example, seeds introduced into soil containing a herbicide or other phytotoxin (Hill *et al.*, 1955) or insects added to a soil containing an insecticidal agent (Thompson, 1973) will not grow, and they often die. However, as detoxication occurs, seeds or insects introduced at progressively later periods into the amended soil will grow but poorly, or they will develop normally. If the tests are done with identical environmental samples that differ only in that they are sterile, the more rapid dissipation of the injurious effect in the nonsterile samples will show that the detoxication results from biological activity (Fig. 4.1). Bioassays are especially useful inasmuch as they reflect the loss of biological activity of a chemical, but they are frequently replaced

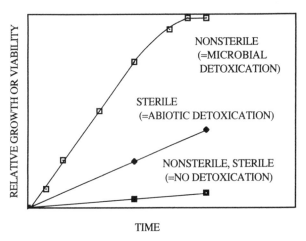

FIG. 4.1 Bioassays for destruction of a toxicant in an environmental sample by microorganisms or by an abiotic mechanism.

by direct chemical analysis showing the loss of the parent compound or the formation of products.

Detoxication is advantageous to the microorganisms carrying out the transformation if the concentration of the chemical is in the range that suppresses these species. If the reaction is the first step in a process by which organisms use the molecule as a C source, the reaction is also beneficial, not because it inactivates the molecule but by virtue of its helping the cell to acquire C. For the many microbial detoxication reactions involving substances that are toxic to humans, plants, and animals and provide no nutritional benefit to the microflora, however, the transformations are important in public health, agriculture, or natural biological communities but not for the microorganisms responsible for the conversion.

The enzymatic step or sequence that results in the conversion of the active molecule into the innocuous product usually occurs within the cell. The product may then undergo one of three fates: (a) it may be excreted; (b) after one or more additional enzymatic steps, it may be changed to a compound that enters the normal metabolic pathways within the cell and ultimately the C is excreted as an organic waste; or (c) it may be modified to a new molecule that becomes subject to these normal reaction sequences, and finally the C is released as CO_2 (Fig. 4.2). The product in the first case is structurally similar to the toxin, but it is harmless at the prevailing concentration. The last fate is mineralization, the mineralization of inhibitors being detoxications, but the actual detoxication step occurs

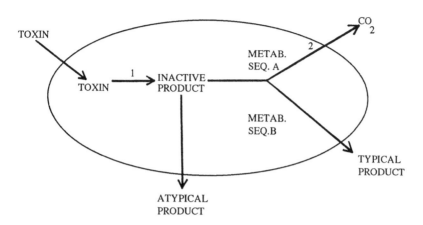

FIG. 4.2 Fate of chemicals that are detoxified.

at some early step in the catabolic sequence that finally yields CO_2. Cometabolic processes (which will be discussed later) often are detoxications, but the products of the transformations are structurally similar to the original substrate.

Several processes may result in detoxication. These processes represent only the first step in Fig. 4.2. These include: (a) hydrolysis, (b) hydroxylation, (c) dehalogenation, (d) demethylation or other dealkylations, (e) methylation, (f) nitro reduction, (g) deamination, (h) ether cleavage, (i) conversion of a nitrile to an amide, and (j) conjugation.

a. Hydrolysis. Cleavage of a bond by the addition of water is a common means by which microorganisms inactivate toxicants. Such reactions may involve a simple hydrolysis of an ester bond, as with the insecticide malathion (Walker and Oesch, 1983) by a carboxyesterase:

$$\overset{O}{\overset{\|}{R C O R'}} + H_2O \rightarrow \overset{O}{\overset{\|}{R C O H}} + HOR'$$

This hydrolysis of malathion is effected microbiologically (Paris *et al.*, 1975). Such reactions may entail hydrolytic cleavage of anilides by an amidase (Munnecke, 1981):

$$RNH\overset{\overset{\displaystyle O}{\|}}{C}CH_2R' + H_2O \rightarrow RNH_2 + HO\overset{\overset{\displaystyle O}{\|}}{C}CH_2R'$$

Other hydrolyses resulting in inactivation have been described.

b. Hydroxylation. The addition of OH to an aromatic or aliphatic molecule often makes it less harmful. Thus, the simple replacement of H by OH inactivates the fungicide MBC (Davidse, 1976) or the herbicide 2,4-D (Jensen, 1982):

$$RH \rightarrow ROH$$

The hydroxylation of the ring of 2,4-D similarly converts the parent herbicide to a nontoxic product (Owen, 1989). Microorganisms may bring about such a detoxication when they hydroxylate the ring in the 4-position, a process that leads to a migration of the chlorine to give 2,5-dichloro-4-hydroxyphenoxyacetic acid (Faulkner and Woodcock, 1964).

c. Dehalogenation. Many pesticides and hazardous industrial wastes contain Cl or other halogens, and removal of the halogen often converts the toxicant to an innocuous product. The enzymes are designated dehalogenases. These dehalogenations may involve replacement of the halogen by H (reductive dehalogenation),

$$RCl \rightarrow RH$$

or by OH (hydrolytic dehalogenation),

$$RCl \rightarrow ROH$$

or it may result in removal of the halogen and an adjacent H (dehydrodehalogenation),

$$RCH_2CHClR' \rightarrow RCH = CHR'$$

The halogen is released in these reactions as inorganic chloride, fluoride, bromide, or iodide. Three dehalogenations of pesticides are illustrated in Fig. 4.3. The dehydrodechlorinase that acts on DDT converts this insecticide to a nontoxic product (Walker and Oesch, 1983). Similarly, the microbial conversion of lindane to 2,3,4,5,6-pentachloro-1-cyclohexene detoxifies the insecticide (Francis *et al.*, 1975; Yule *et al.*, 1967), as does the conversion of the herbicide dalapon to pyruvic acid (Berry *et al.*, 1979). The cleavage of some of these C–halogen bonds is most surprising because of the strength of the bonds. The strength of a chemical bond is the amount of energy required to break that bond:

$$C-F + energy \rightarrow C\cdot + \cdot F$$

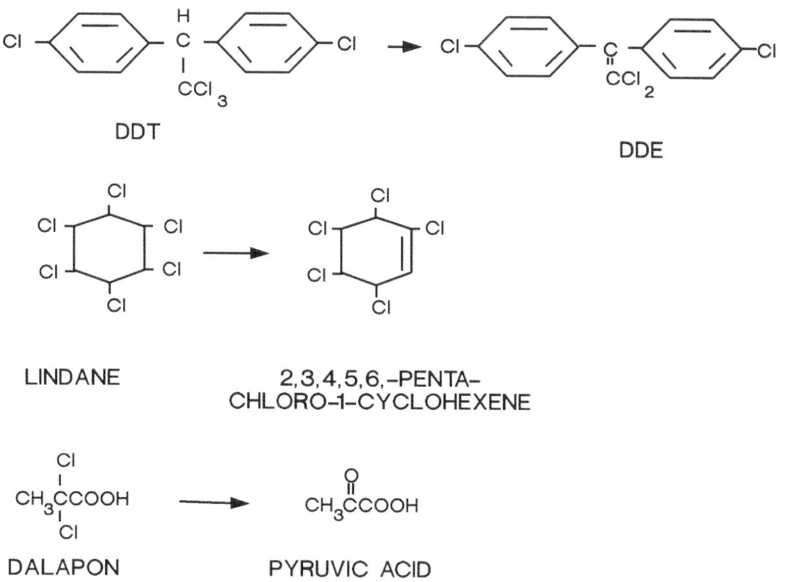

DDT

DDE

LINDANE

2,3,4,5,6,-PENTA-
CHLORO-1-CYCLOHEXENE

DALAPON

PYRUVIC ACID

FIG. 4.3 Dehalogenations that represent detoxications.

To break the C–F bond, for example, requires considerable energy because the bond energy for the bond between C and F is 116 kcal/mole (Speier, 1964).

d. Demethylation or other dealkylations. Many pesticides contain methyl or other alkyl substituents. These may be linked to N or O (*N*- or *O*-alkyl substitution). An *N*- or *O*-dealkylation catalyzed by microorganisms frequently results in loss of pesticidal activity. A number of herbicides that are structurally related to phenylurea become less active when microorganisms *N*-demethylate the molecules, as in the case of the conversion of diuron to the monomethyl derivative (Fig. 4.4). The subsequent removal of the second *N*-methyl group renders the molecule wholly nontoxic (Ellis and Camper, 1982; Jensen, 1982; Hathaway, 1986). Similar reactions leading to detoxications occur when an *s*-triazine herbicide like atrazine is dealkylated as it loses its *N*-ethyl or *N*-isopropyl groups (Jensen, 1982), a reaction that occurs in soil (Khan and Marriage, 1977), presumably as a result of microbial action. The microbial *O*-demethylation of chloroneb creates a nontoxic product, 2,5-dichloro-4-methoxyphenol (Fig. 4.4) (Hock and Sisler, 1969).

e. Methylation. The reverse reaction—the addition of a methyl group—may inactivate toxic phenols. Thus, penta- and tetrachlorophenols

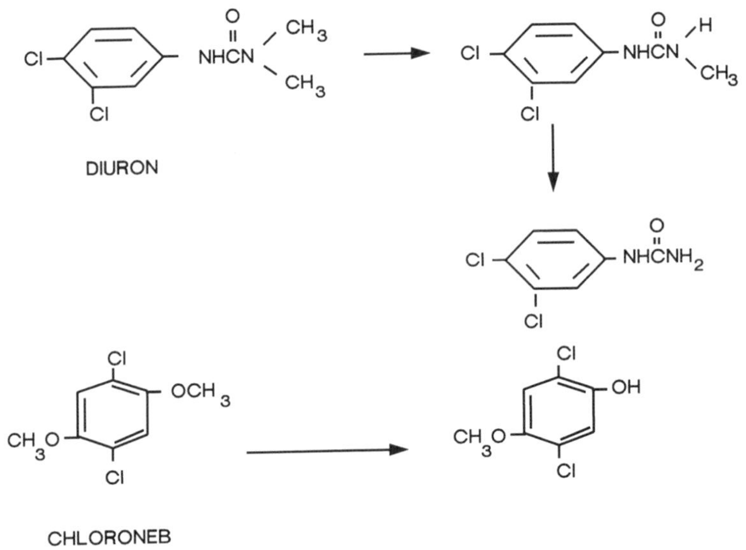

DIURON

CHLORONEB

FIG. 4.4 Dealkylations that detoxify pesticides.

are fungicides, the former being especially widely used, and these can be detoxified microbiologically by addition of a methyl group in a reaction that represents an O-methylation (Cserjesi, 1972; Cserjesi and Johnson, 1972):

$$ROH \rightarrow ROCH_3$$

Such reductions may result in loss or diminution of the harmful effects as microorganisms convert the broad-spectrum poison 2,4-dinitrophenol to 2-amino-4- and 4-amino-2-nitrophenol (Madhosingh, 1961), the fungicide pentachloronitrobenzene to pentachloroaniline (Nakanishi and Oku, 1969), and the insecticide parathion to aminoparathion (Mick and Dahm, 1970).

f. Nitro reduction. Nitro compounds are harmful to many types of organisms, both higher and lower. These may be rendered less toxic by reduction of the nitro to an amino group.

$$RNO_2 \rightarrow RNH_2$$

g. Deamination. The herbicide known as metamitron can be transformed microbiologically to yield a deaminated product (Engelhardt and Wallnöfer, 1978) that is nontoxic to plants (Hathaway, 1986) (Fig. 4.5).

h. Ether cleavage. Phenoxy herbicides contain ether linkages (C–O–C), and the cleavage of these linkages destroys the phytotoxicity of the molecule. This is illustrated by the cleavage of the ether bond in 2,4-D (Loos *et al.*, 1967) (Fig. 4.5). This microbial conversion is somewhat surprising because of the bond energy between C and O, which is 85.5 kcal/ mole (Speier, 1964), and thus the need of the microorganism to provide the energy to cleave the bond.

i. Conversion of nitrile to amide. A potent inhibitor of the growth of certain plants is 2-6-dichlorobenzonitrile, which is sold under the name of dichlobenil. However, when it is converted to 2,6-dichlorobenzamide, the molecule is rendered inactive (Ashton and Crafts, 1981):

$$R-C\equiv N \rightarrow R-\overset{\overset{\displaystyle O}{\|}}{C}-NH_2$$

This detoxication reaction is brought about by microorganisms in soil (Verloop, 1972).

j. Conjugation. A conjugation involves a reaction between a common intermediate in some natural metabolic pathway with a synthetic molecule. Products of the combination of a normal metabolite with a toxicant fre-

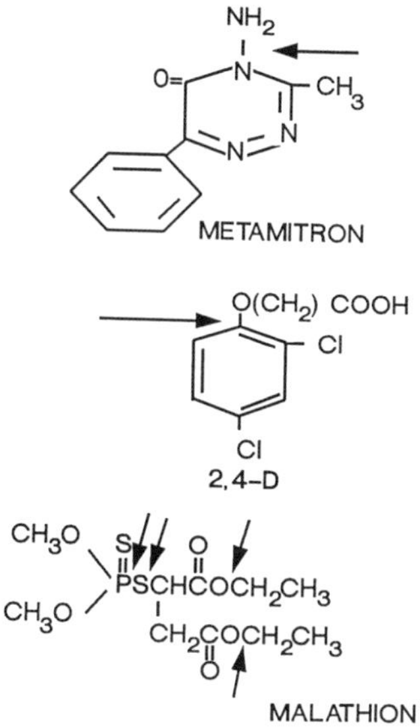

FIG. 4.5 Initial reactions that result in detoxications. Arrows indicate cleavage sites.

quently are harmless. Considerable effort has been directed to animal and plant conjugations that involve sugars, glutathione, and amino acids, but the possible role of microorganisms in conjugations leading to detoxication has received little attention. However, *Cunninghamella elegans* has been shown to conjugate pyrene with glucose to yield nontoxic glucoside conjugates (Cerniglia *et al.*, 1986), and several microbial species detoxify the fungicide sodium dimethyldithiocarbamate by converting it to the less toxic conjugate γ-(dimethylthiocarbamoylthio)-γ-aminobutyric acid (Kaars Sijpesteijn *et al.*, 1962).

A particular microorganism or a microbial community may detoxify a single toxicant in several ways. This is illustrated for the insecticide malathion in Fig. 4.5 (Walker and Stojanovic, 1974). Such multiple pathways are initiated by entirely different enzymes. Other pesticides are also acted on by several dissimilar enzymes, which may thus yield several inactive products.

The ten reaction types given here are not always detoxications, however. A compound acted on by one or another mechanism may yield a product no less toxic than its precursor. Indeed, several such reactions may yield products far more toxic than the original substrates. Furthermore, a reaction or a sequence that yields a product not injurious to one species may not result in detoxication for a second species. Thus, one cannot consider detoxication in a general sense: the sensitive species that is protected must be defined.

REFERENCES

Ashton, F. M., and Crafts, A. S., "Mode of Action of Herbicides." Wiley, New York, 1981.
Berry, E. K. M., Allison, N., Skinner, A. J., and Cooper, R. A., *J. Gen. Microbiol.* **110**, 39–45 (1979).
Cerniglia, C. E., Kelly, D. W., Freeman, J. P., and Miller, D. W., *Chem.-Biol. Interact.* **57**, 203–216 (1986).
Cserjesi, A. J., *Int. Biodeterior. Bull.* **8**, 135–138 (1972).
Cserjesi, A. J., and Johnson E. L., *Can. J. Microbiol.* **18**, 45–49 (1972).
Davidse, L. C., *Pestic. Biochem. Physiol.* **6**, 538–546 (1976).
Ellis, P. A., and Camper, N. D., *J. Environ. Sci. Health, Part B* **B17**, 277–289 (1982).
Engelhardt, G., and Wallnöfer, P. R., *Chemosphere* **7**, 463–466 (1978).
Faulkner, J. K., and Woodcock, D., *Nature (London)* **203**, 865 (1964).
Francis, A. J., Spanggord, R. J., and Ouchi, G. I., *Appl. Microbiol.*, **29**, 567–568 (1975).
Hathaway, D. E., *Biol. Rev. Cambridge Philos. Soc.* **61**, 435–486 (1986).
Hill, G. D., McGahen, J. W., Baker, H. M., Finnerty, D. W., and Bingeman, C. W., *Agron. J.* **47**, 93–104 (1955).
Hock, W. K., and Sisler, H. D., *J. Agric. Food Chem.* **17**, 123–128 (1969).

Jensen, K. I. N., *in* "Herbicide Resistance in Plants" (H. M. LeBaron and J. Gressel, eds.), pp. 133–162. Wiley (Interscience), New York, 1982.

Kaars Sijpesteijn, A., Kaslander, J., and van der Kerk, G. J. M., *Biochim. Biophys. Acta* **62**, 587–589 (1962).

Khan, S. U., and Marriage, P. B., *J. Agric. Food Chem.* **25**, 1408–1413 (1977).

Loos, M. A., Bollag, J. M., and Alexander, M., *J. Agric. Food Chem.* **15**, 858–860 (1967).

Madhosingh, C., *Can. J. Microbiol.* **7**, 553–567 (1961).

Mick, D. L., and Dahm, P. A., *J. Econ. Entomol.* **63**, 1155–1159 (1970).

Munnecke, D. M., *in* "Microbial Degradation of Xenobiotics and Recalcitrant Compounds" (T. Leisinger, A. M. Cook, R. Hütter, and J. Nüesch, eds.), pp. 251–269. Academic Press, London, 1981.

Nakanishi, T., and Oku, H., *Phytopathology* **59**, 1761–1762 (1969).

Owen, W. J., *in* "Herbicides and Plant Metabolism" (A. D. Dodge, ed.), pp. 171–198. Cambridge Univ. Press, Cambridge, UK, 1989.

Paris, D. F., Lewis, D. L., and Wolfe, N. L., *Environ. Sci. Technol.* **9**, 135–138 (1975).

Speier, J. L., *Chemistry* **37**(7), 6–11 (1964).

Thompson, A. R., *J. Econ. Entomol.* **66**, 855–857 (1973).

Verloop, A., *Residue Rev.* **43**, 55–103 (1972).

Walker, C. H., and Oesch, F., *in* "Biological Basis of Detoxication" (J. Caldwell and W. B. Jakoby, eds.), pp. 349–368. Academic Press, New York, 1983.

Walker, W. W., and Stojanovic, B. J., *J. Environ. Qual.* **3**, 4–10 (1974).

Yule, W. N., Chiba, M., and Morley, H. V., *J. Agric. Food Chem.* **15**, 1000–1004 (1967).

5 ACTIVATION

One of the more surprising, and possibly the most undesirable, aspects of microbial transformations in nature is the formation of toxicants. A large number of chemicals that are themselves innocuous can be, and often are, converted to products that may be harmful to humans, animals, plants, or microorganisms. By such means, the resident microflora creates pollutants where none was present previously. Hence, it is not sufficient to know that the parent compound has disappeared—the products may be the problem, not the parent. The process of forming toxic products from innocuous precursors is known as *activation*.

Activation is a major reason to study the pathways and products of breakdown of organic molecules both in natural ecosystems and in waste-disposal systems that lead to environmental discharges. Because the molecules thus synthesized may pose a problem where the precursors were benign, those metabolites must be identified. Activation occurs in soil, water, wastewater, and other environments in which microorganisms are active, and the products created thereby may have a short residence time or persist for long periods (Fig. 5.1). The conversion may represent a single reaction or a simple sequence in a cometabolic process. Alternatively, the harmful product may be an intermediate in mineralization, yet it may persist long enough to create a pollution problem. The consequences of activation include the biosynthesis of carcinogens, mutagens, teratogens, neurotoxins, phytotoxins, and insecticidal and fungicidal agents. Moreover, the mobility of the product of activation is sometimes far different from that of its precursor, so that the product may be transported to distant sites to a far greater or to a significantly smaller extent than the molecule from which it was formed.

Many different pathways, mechanisms, and enzymes are associated with activation. This should not be surprising in view of the differences in structure of chemicals that are toxic to one species or another, or that affect one or another physiological process or target site in the susceptible species. Therefore, a consideration of activation requires a review of a

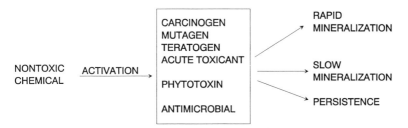

FIG. 5.1   General processes associated with activation.

variety of dissimilar mechanisms. The only common attribute among these mechanisms is the greater hazard of the product compared to the starting material.

Particular attention in research on activation has been given to pesticides. This emphasis arose naturally when it became evident that certain pesticides themselves were not especially harmful to the insects, weeds, or sometimes plant pathogens they were designed to control, but rather they were modified within the pest to the molecule that caused injury or death. Subsequently, it was learned that similar conversions were carried out by microorganisms in soils, natural waters, and other environments.

Even before the products of activation were identified, bioassays revealed that a number of pesticides were activated in soil. This became clear when the low level of toxicity in soil receiving the chemical increased with time, with the mortality of sensitive species rising as the molecule underwent some then unspecified transformation. This is depicted in Fig. 5.2 for soils amended with the insecticides zinophos, trichloronat, and

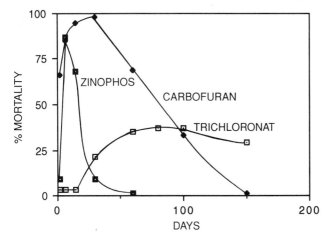

FIG. 5.2   Activation of zinophos, trichloronat, and carbofuran added to soil at 3 mg/ kg. (Reprinted with permission from Read, 1971a.)

carbofuran. Similar increases with time in toxicity to insects have been noted with phorate, diazinon, dasanit, dyfonate, and chlorfenvinphos (Read, 1969, 1971a,b). An analogous instance has been observed for phytotoxin formation; thus, red pine growing in soil treated with ipazine showed no symptoms of toxicity for 80 days, but symptoms of injury developed rapidly thereafter (Kozlowski, 1965). Such bioassay data indicate that several chemical classes may be activated. Some of the mechanisms of activation of these and of a variety of other substrates for microorganisms are considered in the following. Except as noted, not all the reactions have been confirmed to result from microbial metabolism, although most are probably a consequence of their activities.

MECHANISMS OF ACTIVATION

Dehalogenation

A major activation occurs during the microbial metabolism of trichloroethylene (TCE). This compound was once widely used and now represents a major contaminant of many aquifers. Because TCE is metabolized by many bacteria, its elimination by bioremediation is being actively pursued. Unfortunately, a major product that is frequently encountered is vinyl chloride, which is a potent carcinogen:

$$Cl_2C{=}CHCl \rightarrow ClHC{=}CH_2$$

The same carcinogen can also be formed during the anaerobic metabolism of 1,1- and trans-1,2-dichloroethylene (Wilson et al., 1986). Vinyl chloride has been found frequently in TCE-contaminated groundwaters and during tests of technologies to bring about TCE bioremediation by anaerobic bacteria.

TCE can also be converted in cultures of methanotrophs to 2,2,2-trichloroacetaldehyde, which is often called chloral hydrate (Newman and Wackett, 1991):

$$Cl_2C{=}CHCl \rightarrow Cl_3C{-}CHO$$

Chloral hydrate is a mutagen and is also acutely toxic, and if it is consumed together with an alcoholic beverage, it will lead to unconsciousness. The microbial conversion is not really a dehalogenation but rather causes a migration of the chlorine to the adjacent C. Differing from vinyl chloride, chloral hydrate is not known to be a problem in the field.

N-Nitrosation of Secondary Amines (Nitrosamine Formation)

Many activations involve chemicals that are used as pesticides, and coinciding with their widespread use, the activation reaction was described. In the instance of N-nitrosation, in contrast, the precursors for the toxicants had been widely used for long periods before the activation reaction was discovered. The precursors are secondary amines and nitrate. The former are common synthetic chemicals whose annual production is many millions of kilograms. The latter is an anion that is found in nearly all soils and in most natural waters. Furthermore, secondary amines are natural products that are present, sometimes in reasonable amounts, in plants, fish, decaying products, and other materials.

A secondary amine can be written as RNHR'. The N-nitrosation of such a secondary amine occurs in the presence of nitrite, which is formed microbiologically from nitrate. The product is an N-nitroso compound or, as it is commonly called, a nitrosamine:

$$\begin{array}{c} R \\ \diagdown \\ N{-}N{=}O \\ \diagup \\ R' \end{array}$$

In some instances, R and R' are identical. The reason for concern with nitrosamines is their potency: they are active at very low concentrations as carcinogens, teratogens, and mutagens.

Many secondary and tertiary amines are used in industry. The tertiary amine has the structure

$$\begin{array}{c} R \\ \diagdown \\ N{-}R'' \\ \diagup \\ R' \end{array}$$

Secondary or tertiary amines are found in common household products, and a number of pesticides are also secondary or tertiary amines. As a result of the wide use of synthetic amines and their occurrence in living organisms, secondary amines are present in river waters, wastewaters even following treatment (Sander et al., 1974; Neurath et al., 1977), as well as in soil (Vlasenko et al., 1981). Many secondary amines may exist in some rivers (Neurath et al., 1977). Secondary and tertiary amines also appear during the decay of plant residues (Fujii et al., 1972) as well as in the decomposition of creatinine, choline, and phosphatidylcholine in sewage. Certain pesticides are converted to secondary amines in soil (Tate and Alexander, 1974; Mallik *et al.*, 1981). In turn, the tertiary amines are converted in soil, sewage, and microbial cultures to secondary amines (Ayanaba and Alexander, 1973, 1974; Tate and Alexander, 1976). This conversion can be depicted as

$$\begin{array}{c} R \\ \diagdown \\ N—R'' \rightarrow \\ \diagup \\ R' \end{array} \quad \begin{array}{c} R \\ \diagdown \\ NH \\ \diagup \\ R' \end{array}$$

A well-studied example is the microbial transformation of trimethylamine to dimethylamine:

$$(CH_3)_3N \rightarrow (CH_3)_2NH$$

Such conversions of tertiary to secondary amines in nature are microbial.

Simple secondary and tertiary amines, and their complex nitrogenous precursors, rarely are hazardous. However, the stage is set for activation either once the secondary amine is added or as it is formed. The first reactant for activation is now present; the second reactant needed for activation is nitrite. Although nitrite is rarely found in significant amounts in nature, indeed its presence often goes undetected, it is generated by microorganisms when nitrate is present:

$$NO_3^- \rightarrow NO_2^-$$

The yield is often low, but even a small yield apparently is adequate for activation. The activation is the *N*-nitrosation of the secondary amine to give the highly toxic *N*-nitroso compound:

$$\begin{array}{c} R \\ \diagdown \\ NH + NO_2^- \rightarrow \\ \diagup \\ R' \end{array} \quad \begin{array}{c} R \\ \diagdown \\ N—N{=}O + OH^- \\ \diagup \\ R' \end{array}$$

Such nitrosations may occur in sewage, lake water (Ayanaba and Alexander, 1974; Yordy and Alexander, 1981), soil (Ayanaba *et al.*, 1973), and wastewater (Greene *et al.*, 1981), and the precursors may be such secondary amines as dimethylamine and diethanolamine.

Microorganisms can convert a number of secondary amines to the corresponding *N*-nitroso compounds (Ayanaba and Alexander, 1973; Hawksworth and Hill, 1971; Kunisaki and Hayashi, 1979). Moreover, microbial enzymes can also *N*-nitrosate several amines in the presence of nitrite (Ayanaba and Alexander, 1973). Nevertheless, the actual nitrosation step may be, to a lesser or greater extent, nonenzymatic and may result from a spontaneous reaction of the amine and nitrite with some metabolic product (Collins-Thompson *et al.,* 1972) or cell constituents (Mills and Alexander, 1976). The extent of conversion of the amine to the nitrosamine is nearly always small at the pH values common in nature, although the yield can be high in artificially acidified solutions. Hence, the microbial role in activation is the enzymatic formation of the secondary amine and nitrite and, enzymatically or otherwise, the actual *N*-nitrosation.

Such activations of secondary amines do occur. For example, a pesticide-manufacturing company in Elmira, Ontario, disposed of dimethylamine in an adjacent uncontrolled waste-disposal site. Several years later, the entire water supply of Elmira, which came from groundwaters, was found to contain *N*-nitrosodimethylamine, $(CH_3)_2N—N=O$, at levels far in excess of the regulatory level, which, because of the extreme potency of this carcinogen, was several parts per trillion. Although it is not certain whether the formation of *N*-nitrosodimethylamine was microbial or abiotic, the circumstances were suitable for microbial nitrosation and the nitrite precursor was likely of microbial origin.

The nitrosation reaction serves as a good illustration of the fallacy of assuming that dangers reside in synthetic chemicals and not in "natural" compounds. As pointed out earlier, plants and fish contain secondary or tertiary amines. Fish is a notable source of trimethylamine and trimethylamine *N*-oxide. Microbial transformations in the alimentary tract will convert the tertiary amines to secondary amines and the nitrate in drinking water and many vegetables to nitrite. Nitrosation is then a likely consequence, although the reaction may either be microbial or result from an abiotic conversion at the low pH in the stomach. The issue is not whether the chemical is made naturally or by industry, but rather the identity of the molecule itself.

Other examples of such activations are the detection of *N*-nitrosodiethylamine and *N*-nitrosodimethylamine in municipal sludge (Brewer *et al.,* 1980). These two carcinogens and *N*-nitrosomorpholine have been found in sewage-treatment operations, and *N*-nitrosodiethanolamine has been detected in the outlet of a cutting-fluid recovery plant (Richardson *et al.,* 1980). The latter nitrosamine probably is a result of the microbial metabolism of diethanolamine (Yordy and Alexander, 1981), which is a common constituent of cutting fluids and many other products.

Epoxidation

Microorganisms are able to form epoxides from several compounds having double bonds:

$$\text{—HC} = \text{CH—} \rightarrow \text{—HC} \overset{\displaystyle \overset{O}{\diagup \diagdown}}{\frown} \text{CH—}$$

In the case of several chemicals marketed as insecticides, this oxidation converts the precursor to a product that is more toxic to animals. This type of transformation is illustrated by the conversion of heptachlor to heptachlor epoxide in soil (Duffy and Wong, 1967) and in culture (Elsner *et al.*, 1972) and the analogous oxidation of aldrin to its epoxide, which is called dieldrin, in soil as a result of microbial action (Lichtenstein and Schulz, 1960) and in culture (Korte *et al.*, 1962).

The activation of aldrin and the consequent formation of dieldrin is at the heart of an environmental cleanup of enormous magnitude. The reason is because dieldrin is not only more toxic than aldrin but it is far more persistent. The disappearance of aldrin and the microbial synthesis and persistence of dieldrin are illustrated in Fig. 5.3. Dieldrin is known to be present in some soils more than 15 years after aldrin or dieldrin itself was introduced into soil to control soil-borne insects harmful to crops. The especially dramatic example involves a site near Denver, Colorado, known as the Rocky Mountain Arsenal. Here, in addition to chemicals being manufactured by the U.S. Army, a private company made aldrin and dieldrin for pest control. Unfortunately, that company disposed of residual insecticide at the site, and the chemical concentration became high at the disposal site but also became disseminated across the large land area of the arsenal grounds. Dieldrin remains there more than 20 years after its

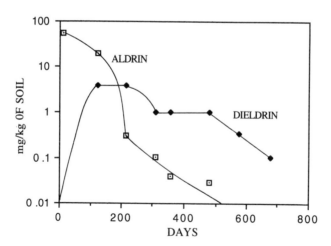

FIG. 5.3 Epoxidation of aldrin in soil leading to dieldrin. (Reprinted with permission from Singh *et al.*, 1991.)

disposal. Now, the cost of cleaning up the soil contaminated with dieldrin, as well as other chemicals in the groundwater, is estimated to exceed one billion U.S. dollars.

Conversion of Phosphorothionate to Phosphate

A widely used group of insecticides are the phosphorothionates. These have the common structure

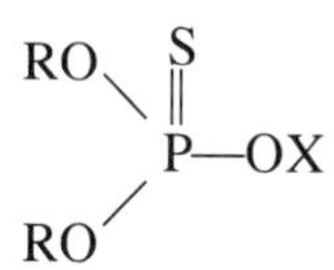

$$
\begin{array}{c}
RO \\
 \diagdown \\
 P\!-\!OX \\
 \diagup \\
RO
\end{array}
\quad
\overset{\displaystyle S}{\underset{}{\|}}
$$

in which R is a short alkyl chain, typically CH_3 or CH_3CH_2, and X may be one of a variety of different structures. These molecules have little toxicity, but when they are converted to the corresponding phosphates,

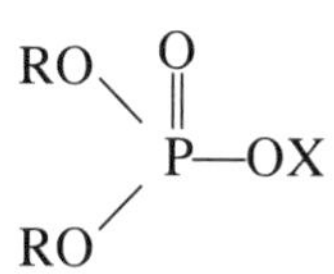

$$
\begin{array}{c}
RO \\
 \diagdown \\
 P\!-\!OX \\
 \diagup \\
RO
\end{array}
\quad
\overset{\displaystyle O}{\underset{}{\|}}
$$

they become potent insecticides. Many are also highly toxic to humans and other mammals. Although such activations occur within the animal, they can also be carried out by microorganisms in natural environments and in agricultural soils. These activations can take place, for example, in soil treated with chlorfenvinphos (Read, 1971b). This type of reaction is an oxidative desulfuration that leads to the creation of a potent inhibitor of cholinesterase activity. Anticholinesterase potency may increase by a factor of about 10,000-fold (O'Brien, 1960). A typical example is the conversion of parathion to its oxygen analogue (known as paraoxon) in soil (Sethunathan and Yoshida, 1973a) and microbial cultures (Munnecke and Hsieh, 1976).

A similar type of activation occurs with a group of compounds known as phosphorodithoates:

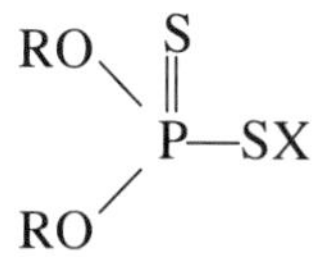

$$
\begin{array}{c}
RO \\
 \diagdown \\
 P\!-\!SX \\
 \diagup \\
RO
\end{array}
\quad
\overset{\displaystyle S}{\underset{}{\|}}
$$

They also become active when P=S is converted to P=O. A well-described case is the conversion of dimethoate to its oxygen analogue in soil (Duff and Menzer, 1973). This type of activation is shown in Fig. 5.4.

Metabolism of Phenoxyalkanoic Acids

A prominent herbicide is 2,4-D, which is itself a potent phytotoxin. However, a number of structurally related but inactive compounds may be converted in plants to 2,4-D, and thus they act as herbicides following the activation process. Similar conversions occur in soil. Collectively, these phenoxyalkanoic acids are ω-(2,4-dichlorophenoxy)alkanoic acids, the ω-linkage referring to the attachment of the last (ω) C of the fatty (alkanoic) acid to the 2,4-dichlorophenoxy moiety through the O. The transformation may thus be viewed as shown in Fig. 5.5 for 6-(2,4-dichlorophenoxy)hexanoic acid as the parent chemical (Gutenmann et al., 1964). The sequence is called β-oxidation because the steps in which each of the two carbons are removed initially involve the oxidation of the β-carbon of the aliphatic acid moiety:

$$RCH_2CH_2COOH \rightarrow RCH{=}CHCOOH \rightarrow R\overset{OH}{\underset{|}{C}}HCH_2COOH \rightarrow R\overset{O}{\underset{\parallel}{C}}CH_2COOH$$

$$\rightarrow R\overset{O}{\underset{\parallel}{C}}OH + CH_3COOH$$

Because similar reactions occur in bacterial cultures (Taylor and Wain, 1962) and the chemicals probably do not undergo such reactions in nature by abiotic mechanisms, this type of activation in soil appears to be microbial.

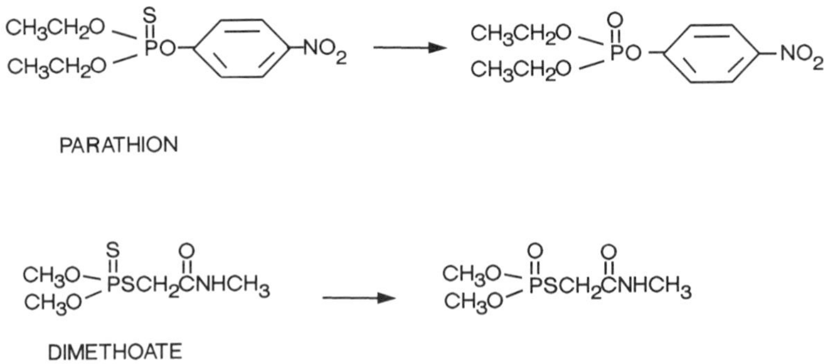

PARATHION

DIMETHOATE

FIG. 5.4 Conversion of parathion and dimethoate to their oxygen analogues.

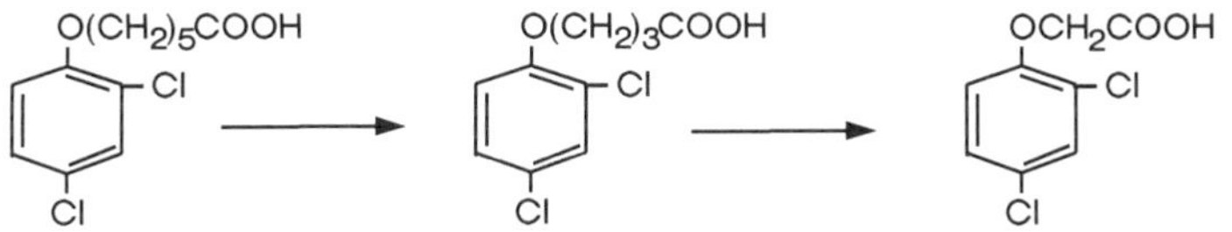

FIG. 5.5 Transformation of 6-(2,4-dichlorophenoxy)hexanoic acid to 4-(2,4-DB) and finally to the actual phytotoxin, 2,4-D.

A related activation is evident in the removal of the side chain of nonylphenol polyethoxylate surfactants by microorganisms in sewage sludge. The product is 4-nonylphenol (Fig. 5.6). The latter is important because it is highly toxic to fish and other aquatic organisms (Giger *et al.*, 1984). In this way, a component of detergents, which are widely used in the home and in commerce, is ultimately released to surface waters, where it may cause harm.

Oxidation of Thioethers

A number of chemicals containing thioether linkages (—C—S—C—) are sold as insecticides yet have only modest toxicity to insects, but they are activated and become more potent as they are oxidized to the corresponding sulfoxides and sulfones:

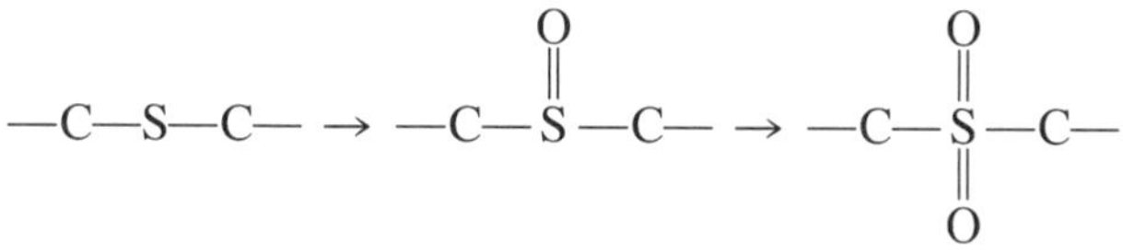

$$-C-S-C- \rightarrow -C-\overset{\overset{\displaystyle O}{\|}}{S}-C- \rightarrow -C-\overset{\overset{\displaystyle O}{\|}}{\underset{\underset{\displaystyle O}{\|}}{S}}-C-$$

Similar reactions occur in soil, presumably largely by microbial action, and in pure cultures of certain microorganisms. The chief toxicants in nature appear to be the sulfoxide, the sulfone, or both. Three compounds marketed as insecticides have been widely studied in this regard, namely, aldicarb (synonym Temik), phorate (synonym Thimet), and disulfoton. Aldicarb is also used as a miticide and nematicide. The structures of the

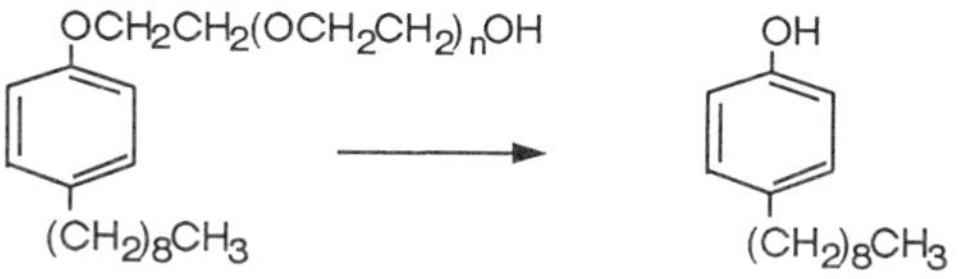

FIG. 5.6 Microbial conversion of nonylphenol polyethoxylates to 4-nonylphenol.

compounds are

aldicarb:

$$\underset{\underset{CH_3}{|}}{\overset{\overset{CH_3}{|}}{CH_3SCCH}}=NO\overset{\overset{O}{\|}}{C}NHCH_3$$

phorate:

$$CH_3CH_2SCH_2S\overset{\overset{S}{\|}}{P}(OCH_2CH_3)_2$$

disulfoton:

$$CH_3CH_2SCH_2CH_2S\overset{\overset{S}{\|}}{P}(OCH_2CH_3)_2$$

In each instance, the sulfur adjacent to the terminal CH_3 or CH_3CH_2 can be oxidized to the corresponding sulfoxide and sulfone. The formation of the toxic analogues occurs in soil treated with aldicarb (Richey *et al.*, 1977), phorate (Getzin and Shanks, 1970), and disulfoton (Clapp *et al.*, 1976). Microorganisms also form sulfoxides, sulfones, or both in cultures incubated with aldicarb (Jones, 1976) and phorate (Le Patourel and Wright, 1976). Other putative pesticides are also converted to the more toxic sulfoxide and sulfone derivatives in soil (Whitten and Bull, 1974).

Hydrolysis of Esters

Several esters marketed as herbicides are activated by hydrolysis to give the actual phytotoxin, which is the free acid (Hatzois and Penner, 1982; Kerr, 1989):

$$\overset{\overset{O}{\|}}{R}COR' \rightarrow \overset{\overset{O}{\|}}{R}COH$$

This reaction is carried out in soils amended with flamprop-methyl (Roberts and Standen, 1978), benzoylprop-ethyl (Beynon *et al.*, 1974), and dichlorfop-methyl (Gaynor, 1984). As the names of these pesticides indicate, R' is CH_3 or CH_2CH_3, respectively. The second product of the conversion is presumably the nontoxic alcohol HOR'.

Other Activations

A number of other microbial activities may lead to the formation of toxic products. Insofar as is presently known, these activations are not widespread and may be unique to the individual compound. An early example of activation is the conversion in soil, apparently by microorganisms, of 2,4-dichlorophenoxyethyl sulfate to 2,4-D. The former is innocuous, but the latter is a potent phytotoxin (Fig. 5.7). The reaction also is carried out by *Bacillus cereus* (Audus, 1952; Vlitos and King, 1953). The presumed intermediate is 2,4-dichlorophenoxyethanol. Although microorganisms cleave other sulfate esters, the products are not generally toxic.

Microorganisms are able to convert the fungicide benomyl (synonym Benlate) to benzimidazole carbamic acid methyl ester. Both the parent compound and the product are fungicidal (Clemons and Sisler, 1969). However, for some fungi, the transformation is an activation because they are more sensitive to the product (Felsot and Pedersen, 1991).

Chlorinated dibenzo-*p*-dioxins and dibenzofurans are among the most toxic substances known, especially 2,3,7,8-tetrachloro-*p*-dibenzodioxin (TCDD). These extremely hazardous compounds can be produced from 3,4,5- and 2,4,5-trichlorophenols by peroxidases (Öberg *et al.*, 1990; Svenson *et al.*, 1989). However, the biological formation of such toxicants in nature or by microorganisms has not been described.

Many chlorophenols are harmful, and many are persistent. It is possible that these may be produced microbiologically in nature in view of the finding that a fungal chloroperoxidase halogenates phenol to yield monochlorophenols, the latter to give dichlorophenols; the dichlorophenols to produce trichlorophenols; the trichloro compounds to give rise to tetrachlorophenols; and even the latter to generate pentachlorophenol (Wannstedt *et al.*, 1990). Fungal peroxidases may also dimerize 3,4-dichloroaniline to 3,4,3',4'-tetrachloroazobenzene, a compound similar in toxicity to TCDD (Pieper *et al.*, 1992). It is not known whether any such processes occur in nature, but the enzymatic precedent is cause for further inquiry.

Several other types of activation occur in microbial cultures, but their significance in nature is uncertain. For example, the fungus *Cunningham-*

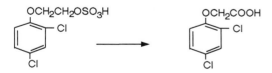

OCH₂CH₂OSO₃H → OCH₂COOH

FIG. 5.7 Activation of 2,4-dichlorophenoxyethyl sulfate.

ella elegans converts pyrene to 1,6- and 1,8-dihydroxypyrenes, which probably can be injurious to higher organisms (Cerniglia *et al.*, 1986). Several microorganisms convert the fungicide triadimefon to triadimenol, and since some fungi are resistant to the first but sensitive to the second compound, this reaction represents an activation (Deas *et al.*, 1986); the process involves the reaction

$$\underset{R'}{\overset{R}{\diagdown}}C{=}O \rightarrow \underset{R'}{\overset{R}{\diagdown}}CHOH$$

Some bacteria, in culture, are able to *O*-methylate chloroguaiacols:

$$ROH \rightarrow ROCH_3$$

The products of such a methylation are toxic to fish (Allard *et al.*, 1985).

Activation by microbial methylation is well known in the cases of mercury, arsenic, and tin. The methylation of mercury has received considerable attention because it has resulted in the contamination of fish and bans on the use of several bodies of water for commercial or sports fishing. The concentrations of methylmercury in aquatic animals may be several orders of magnitude greater than that in the ambient water because the methylation of cationic mercury results in bioconcentration of the element. The transformation occurs in aerobic and anaerobic environments, and both monomethyl- (CH_3Hg^+) and dimethylmercury (CH_3HgCH_3) may be formed (Bisogni and Lawrence, 1975). The process takes place in fresh water, estuarine and marine sediments, and soils, but the focus of interest has been in sediments because of the uptake of methylmercury by fish. A variety of bacteria are able to carry out this metabolic step in culture.

Inorganic arsenic is also subject to methylation. Although arsenite and arsenate are themselves toxic, and indeed are more toxic than some methylarsenic compounds, so that a true activation is not involved, methylation of this element is of special interest because some of the methylated species are volatile, and thus respiratory exposures may occur with harmful outcomes. Indeed, arsenic poisoning has occurred because of human inhalation of methylarsenic released in houses by the activity of microorganisms growing on As-containing wallpapers. Monomethylarsine (CH_3AsH_2), dimethylarsine (($CH_3)_2AsH$), and arsine (AsH_3) are formed in soil and appear as volatile metabolites (Cheng and Focht, 1979). In addition, the following compounds have been found in marine and fresh waters as well as in microbial cultures (Braman, 1975; Cox and Alexander, 1973):

$CH_3AsO(OH)_2$ methylarsonic acid
$(CH_3)_2AsO(OH)$ dimethylarsinic acid
$(CH_3)_3As$ (or the oxide)   trimethylarsine

Although inorganic tin is of little toxicological significance, methylated forms of this element are highly toxic. Trimethyltin, for example, at low concentrations produces irreversible neuronal damage and neuronal necrosis in the brain, and it is absorbed into the body from the stomach, intestine, and even the skin (Aldridge and Brown, 1988). Microorganisms in sediments and a number of species in pure culture are able to methylate inorganic tin and form mono-, di-, and trimethyltin (Gilmour *et al.*, 1987; Hallas *et al.*, 1982). The fact that sterilized sediments do not form methyltin compounds confirms that the process is microbial. Inasmuch as mono-, di-, and trimethyltin are found in natural waters, albeit at low concentrations (Byrd and Andreae, 1982), the process is not merely a laboratory phenomenon. Nevertheless, it is not clear whether the microbial transformation constitutes a substantive threat to human health.

The reverse reaction—demethylation—may also result in activation, witnessed by the fact that the demethylation of diphenamid (*N*,*N*-dimethyl-2,2-diphenylacetamide) to yield the monomethyl and the unmethylated 2,2-diphenylacetamide by *Trichoderma viride* and *Aspergillus candidus* converts a nontoxic precursor to two phytotoxins (Kesner and Ries, 1967); a similar conversion has been noted in soil treated with diphenamid (Golab *et al.*, 1968):

$$RN\begin{array}{c} {}^{CH_3} \\ {}_{CH_3} \end{array} \rightarrow RN\begin{array}{c} {}^{CH_3} \\ {}_{H} \end{array} \rightarrow RNH_2$$

Another activation to yield a phytotoxin from an innocuous precursor has been noted both in soil as well as in cultures of *Pseudomonas putrefaciens*. The reaction is the conversion of α-amino-2,6-dichlorobenzaldoxime to 2,6-dichlorobenzonitrile (Milbarrow, 1963).

Several other carcinogens are generated microbiologically. For example, the carcinogens 1,1- and 1,2-dimethylhydrazine are produced during anaerobic biodegradation of the explosive known as RDX (Fig. 5.8) (McCormick *et al.*, 1981).

In some instances, the microbial transformation itself does not create a toxicant. Nevertheless, the product can be of public health or ecological importance because it undergoes a nonbiological reaction that leads to a hazardous substance. This is particularly true of compounds that are chlorinated as a result of Cl_2 treatment of water for human consumption.

$$O_2N-N \overbrace{\qquad} N-NO_2$$

$$\longrightarrow \quad \begin{array}{c} CH_3 \\ \\ CH_3 \end{array} \hspace{-0.3em} \diagdown \hspace{-0.2em} N-NH_2 + CH_3NHNHCH_3$$

with NO$_2$ substituent

FIG. 5.8 Biodegradation of RDX.

An example of such a problem is the conversion by sewage bacteria of dodecylsulfate to acetoacetic acid; the latter is converted in high yield to chloroform, a toxicant, when the water is treated with Cl_2 (Itoh *et al.*, 1985).

The examples cited here represent activations in a strict sense, that is, the product is more toxic than the precursor. However, microorganisms may increase the exposure of sensitive organisms to toxicants not only because they form more harmful metabolites but also because they alter the *mobility* or *persistence* of toxic compounds. The microbial product is often more mobile and sometimes more persistent than the parent molecule, and the product but not the parent may be found in the groundwater underlying soil receiving the precursor or in soil long after the original chemical has disappeared. In this way, the exposure can be enhanced. Thus, dieldrin persists long after aldrin applied to soil has been metabolized and constitutes a long-term source of pollution. With the *N*-nitrosodimethylamine formed from dimethylamine, not only does the conversion result in the appearance of a highly potent carcinogen but the nitroso compound moves through soil into groundwaters, is carried with the flow of the groundwater, and is highly persistent; this contrasts with the lack of appreciable injury, poor mobility, and the frequently rapid biodegradation of the precursor amine.

DEFUSING

A compound that is potentially activated will pose a health or environmental hazard if it undergoes that type of reaction. However, if the microflora converts that substrate to a different metabolite that itself is both harmless and not subject to activation, the potential problem posed by the initial substrate does not arise. Thus, A is converted to C rather than to B:

By analogy to a bomb that only explodes after its fuse is lit, this phenomenon has been termed *defusing*. Just as a bomb loses its capacity to do
injury by having the fuse removed, the chemical is not activated by the
prior occurrence of the defusing reaction.

Defusing is best illustrated by some pesticides that undergo activation.
Among the phenoxy herbicides, 4-(2,4-DB) is activated when it undergoes
β-oxidation to yield 2,4-D. Hence, bacteria that cleave the molecule by
removing butyric acid from the side chain to release 2,4-dichlorophenol
in culture (MacRae *et al.*, 1963) are defusing the molecule (Fig. 5.9).
Diazinon is activated when the P=S is converted to P=O, but its cleavage
to 2-isopropyl-4-methyl-6-hydroxypyrimidine and diethylthiophosphate in
soil (Konrad *et al.*, 1967) and in culture (Sethunathan and Yoshida, 1973b)
similarly represents a defusing (Fig. 5.9). Defusing has also been reported
for a number of other insecticides that are activated by the conversion of
P=S to P=O, for example, when part or all of the parathion (Munnecke
and Hsieh, 1976) or malathion (Rosenberg and Alexander, 1979) is cleaved
in bacterial cultures or dimethoate is cleaved in soil (El Beit *et al.*, 1978)
before the chemical is activated.

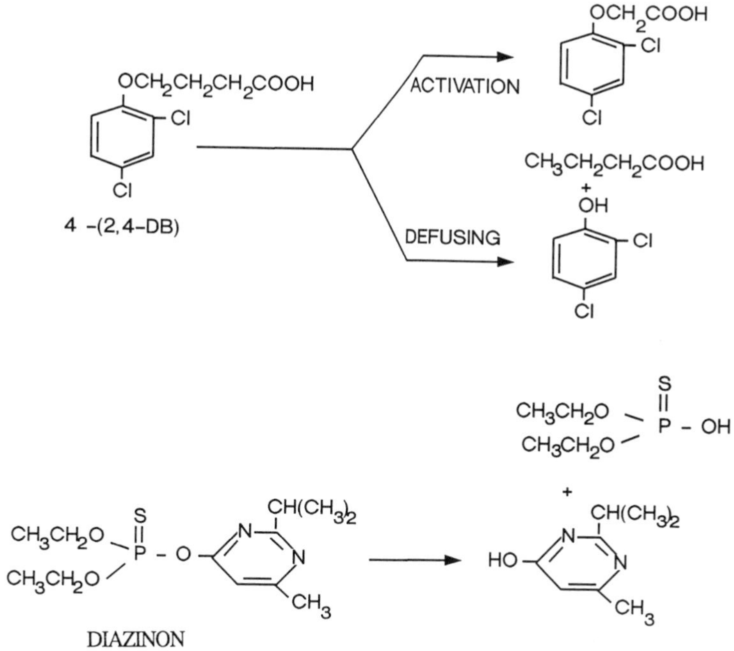

FIG. 5.9 Defusing of two pesticides.

CHANGE IN TOXICITY SPECTRUM

On occasion, a compound harmful to one group of organisms is converted to a molecule injurious to an entirely different group of organisms. Although such a change in spectrum of toxicity is not quite what is meant by activation in a strict sense, it still is an activation for the second group of organisms because for them the substance that was originally benign is converted to a form that causes harm.

One of the most dramatic instances of a change in spectrum of toxicity occurred in Japanese agriculture. Pentachlorobenzyl alcohol, a compound tolerated by many plant species even at greater than 2000 mg/kg, was introduced for the control of a fungal disease of the rice plant known as rice blast disease. However, two years after beginning its use, toxicity to tomato, cucumber, and melon plants was noted, especially in fields where rice straw had been added to the soil the year before. As a result of the marked crop losses, it was found that the pentachlorobenzyl alcohol was converted in the soil to penta-, 2,3,4,6- and/or 2,3,5,6-tetra-, and 2,4,6-trichlorobenzoic acid, all of which, even at extremely low concentrations, were toxic to a variety of plants. As a result of this transformation, the production and marketing of pentachlorobenzyl alcohol immediately ceased (Ishida, 1972).

Many other examples have been described in which the spectrum of toxicity has been changed, presumably as a result of microbial activity (Table 5.1). However, some of the conversions cited could be abiotic. Differing from the case of pentachlorobenzyl alcohol, moreover, the instances cited are not known to have resulted in harm to susceptible species in nature. With other compounds, both the parent molecule and one or

TABLE 5.1
Changes in Spectrum of Toxicity Arising from Chemical Transformations

Process	Reaction	Reference
Antifungal to carcinogen precursor	Thiram to dimethylamine	Odeyemi and Alexander (1977)
Insecticide to fish toxin	DDT to 1,1-dichloro-*bis*(4-chlorophenyl)ethane	Day (1991)
Insecticide to acaricide	DDT to 1,1-dichloro-*bis*(4-chlorophenyl)ethane	Matsumura *et al.* (1971)
Herbicide to genotoxic product	2,4,5-T to 2,4,5-trichlorophenol	George *et al.* (1992)
	Propanil to 3,3',4,4'-tetrachloroazobenzene	Prasad (1970)

more of its metabolic products are toxic to a single species. Thus, atrazine is deethylated in soil, and both parent and daughter are toxic to plants (Sirons *et al.*, 1973), and 2,6-dichlorobenzonitrile (the herbicide known as dichlobenil) is metabolized microbiologically in soil to yield 2,6-dichlorobenzamide, and both precursor and product are phytotoxic (Verloop, 1972). The microbial conversion of avermectin B_{2a} to its 23-keto derivative in soil represents the conversion of one nematicide to another and, as in some of the instances cited, the toxic product is more persistent than the original chemical (Gullo *et al.*, 1983). A variety of fungicides are converted microbiologically to substances that are themselves antifungal; for example, benomyl is converted to the antifungal benzimidazole carbamic acid methyl ester by *Saccharomyces cerevisiae* (Clemons and Sisler, 1969), several microorganisms convert triadimefon to the toxic triadimenol (Deas *et al.*, 1986), and *Neurospora crassa* converts the fungicide captan to the antifungal carbonyl sulfide (Somers *et al.*, 1967). These antifungal-to-antifungal conversions have been described in microbial cultures, but the biodegradation of the fungicide chlorothalonil to the fungicide 1,3-dicarbamoyl-2,4,5,6-tetrachlorobenzene takes place in soil (Rouchaud *et al.*, 1988).

RISKS FROM BIODEGRADATION

Clearly, many harmful metabolites are generated microbiologically in a variety of environments. These products may represent substantive threats to the health, growth, or vigor of humans and a variety of animals and plants. In view of such conversions, studies of breakdown pathways in natural environments are of particular importance. It is not sufficient merely to measure the persistence and disappearance of the parent substance. What microorganisms do to that chemical may be of as great or even greater importance to human health, agricultural productivity, or populations in natural ecosystems.

The view is often expressed, particularly by the purveyors of a technology, that the technology is not only useful but is also without risk. Whether this view is actually believed or is espoused solely to convince the unwary to use the technology is often difficult to assess.

Every new technology has a risk. That risk may be large or small, but it does exist. Recognizing the issues or factors coupled to the risk is a first step in reducing or avoiding the risk. In the examples of activation in this chapter, such risks have been discussed. These issues are not insubstantial, and by learning more about the dangers associated with microbial metabolites, it should be possible to establish approaches to avoid their occurrence or reduce their concentrations. The biologically active metabolite formed from a toxicant may not always be toxic. Sometimes, it may be stimulatory. The conversion of toxic compounds to stimulators has been shown particularly for phytotoxic parents. For example, the herbicides DNOC (Bruinsma, 1960), dinoseb (Crafts, 1949), and 2,4-D (Newman *et al.*, 1952) are each converted to breakdown products that actually stimulate plant growth, but the identities of the stimulatory metabolites have not been established.

REFERENCES

Aldridge, W. N., and Brown, A. W., in "The Biological Alkylation of Heavy Elements" (P. J. Craig and F. Glockling, eds.), pp. 147–163. Royal Society of Chemistry, London, 1988.

Allard, A. S., Remberger, M., and Neilson, A. H., *Appl. Environ. Microbiol.* **49**, 279–288 (1985).

Audus, L. J., *Nature (London)* **170**, 886–887 (1952).

Ayanaba, A., and Alexander, M., *Appl. Microbiol.* **25**, 862–868 (1973).

Ayanaba, A., and Alexander, M., *J. Environ. Qual.* **3**, 83–89 (1974).

Ayanaba, A., Verstraete, W., and Alexander, M., *J. Natl. Cancer Inst. (U.S.)* **50**, 811–813 (1973).

Beynon, K. I., Roberts, T. R., and Wright, A. N., *Pestic. Sci.* **5**, 451–463 (1974).

Bisogni, J. J., Jr., and Lawrence, A. W., *J. Water Pollut. Control Fed.* **47**, 135–152 (1975).

Braman, R. S., in "Arsenical Pesticides" (E. A. Woolson, ed.), pp. 108–123. American Chemical Society, Washington, DC, 1975.

Brewer, W. S., Draper, A. C., III, and Wey, S. S., *Environ. Pollut. Ser. B* **1**, 37–83 (1980).

Bruinsma, J., *Plant Soil* **12**, 249–258 (1960).

Byrd, J. T., and Andreae, M. O., *Science* **218**, 565–569 (1982).

Cerniglia, C. E., Kelly, D. W., Freeman, J. P., and Miller, D. W., *Chem.-Biol. Interact.* **57**, 203–216 (1986).

Cheng, C.-N., and Focht, D. D., *Appl. Environ. Microbiol.* **38**, 494–498 (1979).

Clapp, D. W., Naylor, D. V., and Lewis, G. C., *J. Environ. Qual.* **5**, 207–208 (1976).

Clemons, G. P., and Sisler, H. D., *Phytopathology* **59**, 705–706 (1969).

Collins-Thompson, D. L., Sen, N. P., Aris, B., and Schwinghamer, L., *Can. J. Microbiol.* **18**, 1968–1971 (1972).

Cox, D. P., and Alexander, M., *Bull. Environ. Contam. Toxicol.* **9**, 84–88 (1973).

Crafts, A. S., *Hilgardia* **19**, 159–169 (1949).

Day, K. E., in "Pesticide Transformation Products" (L. Somasundaram and J. R. Coats, eds.), pp. 217–241. American Chemical Society, Washington, DC, 1991.

Deas, A. H. B., Carter, G. A., Clark, T., Clifford, D. R., and James, C. S., *Pestic. Biochem. Physiol.* **26**, 10–21 (1986).

Duff, W. G., and Menzer, R. E., *Environ. Entomol.* **2**, 309–318 (1973).

Duffy, J. R., and Wong, N., *J. Agric. Food Chem.* **15**, 457–464 (1967).

El Beit, I. O. D., Wheelock, J. V., and Cotton, D. E., *Int. J. Environ. Stud.* **12**, 215–225 (1978).

Elsner, E., Bieniek, D., Klein, W., and Korte, F., *Chemosphere* **1**, 247–250 (1972).
Felsot, A. S., and Pedersen, W. L., *in* "Pesticide Transformation Products" (L. Somasundaram and J. R. Coats, eds.), pp. 172–187. American Chemical Society, Washington, DC, 1991.
Fujii, K., Kobayashi, M., and Takahashi, E., *J. Sci. Soil Manure Jpn.* **43**(5), 160–164 (1972).
Gaynor, J. D., *Can. J. Soil Sci.* **64**, 283–291 (1984).
George, S. E., Whitehouse, D. A., and Claxton, L. D., *Environ. Toxicol. Chem.* **11**, 733–740 (1992).
Getzin, L. W., and Shanks, C. H., Jr., *J. Econ. Entomol.* **63**, 52–58 (1970).
Giger, W., Brunner, P. H., and Schaffner, C., *Science* **225**, 623–625 (1984).
Gilmour, C. C., Tuttle, J. H., and Means, J. C., *Microb. Ecol.* **14**, 233–242 (1987).
Golab, T., Gramlich, J. V., and Probst, G. W., *Abstr. Pap., 155th Meet., Am. Chem. Soc.,* San Francisco, Abstr. No. A-50 (1968).
Greene, S., Alexander, M., and Leggett, D., *J. Environ. Qual.* **10**, 416–420 (1981).
Gullo, V. P., Kempf, A. J., MacConnell, J. G., Mrozik, H., Arison, B., and Putter, I., *Pestic. Sci.* **14**, 153–157 (1983).
Gutenmann, W. H., Loos, M. A., Alexander, M., and Lisk, D. J., *Soil Sci. Soc. Am. Proc.* **28**, 205–207 (1964).
Hallas, L. E., Means, J. C., and Cooney, J. J., *Science* **215**, 1505–1507 (1982).
Hatzois, K. K., and Penner, D., "Metabolism of Herbicides in Higher Plants." Burgess Publ. Co., Minneapolis, MN, 1982.
Hawksworth, G., and Hill, M. J., *Br. J. Cancer* **25**, 520–526 (1971).
Ishida, M., *in* "Environmental Toxicology of Pesticides" (F. M. Matsumura, G. M. Boush, and T. Misato, eds.), pp. 281–306. Academic Press, New York, 1972.
Itoh, S.-I., Naito, S., and Unemoto, T., *Water Res.* **19**, 1305–1309 (1985).
Jones, A. S., *J. Agric. Food Chem.* **24**, 115–117 (1976).
Kerr, M. W., *in* "Herbicides and Plant Metabolism" (A. D. Dodge, ed.), pp. 199–210. Cambridge Univ. Press, Cambridge, UK, 1989.
Kesner, C. D., and Ries, S. K., *Science* **155**, 210–211 (1967).
Konrad, J. G., Armstrong, D. E., and Chesters, G., *Agron. J.* **59**, 591–594 (1967).
Korte, F., Ludwig, G., and Vogel, J., *Justus Liebig's Ann. Chem.* **656**, 135–140 (1962).
Kozlowski, T. T., *Nature (London)* **205**, 104–105 (1965).
Kunisaki, N., and Hayashi, M., *Appl. Environ. Microbiol.* **37**, 279–282 (1979).
Le Patourel, G. N. J., and Wright, D. J., *Comp. Biochem. Physiol.* **53C**, 73–74 (1976).
Lichtenstein, E. P., and Schulz, K. R., *J. Econ. Entomol.* **53**, 192–197 (1960).
MacRae, I. C., Alexander, M., and Rovira, A. D., *J. Gen. Microbiol.* **32**, 69–76 (1963).
Mallik, M. A. B., Tesfai, K., and Pancholy, S. K., *Proc. Okla. Acad. Sci.* **61**, 31–35 (1981).
Matsumura, F., Patil, K. C., and Boush, G. M., *Nature (London)* **230**, 325–336 (1971).
McCormick, N. G., Cornell, J. H., and Kaplan, A. M., *Appl. Environ. Microbiol.* **42**, 817–823 (1981).
Milbarrow, B. V., *Biochem. J.* **87**, 255–258 (1963).
Mills, A. L., and Alexander, M., *J. Environ. Qual.* **5**, 437–440 (1976).
Munnecke, D. M., and Hsieh, D. P. H., *Appl. Environ. Microbiol.* **31**, 63–69 (1976).
Neurath, G. B., Dünger, M., Pein, F. G., Ambrosius, D., and Schreiber, O., *Food Cosmet. Toxicol.* **15**, 275–282 (1977).
Newman, A. S., Thomas, J. R., and Walker, R. L., *Soil Sci. Soc. Am. Proc.* **16**, 21–24 (1952).
Newman, L. M., and Wackett, L. P., *Appl. Environ. Microbiol.* **57**, 2399–2402 (1991).
Öberg, L. G., Glas, B., Swanson, S. E., Rappe, C., and Paul, C. G., *Arch. Environ. Toxicol. Chem.* **19**, 930–938 (1990).

O'Brien, R. D., "Toxic Phosphorus Esters." Academic Press, New York, 1960.

Odeyemi, O., and Alexander, M., *Appl. Environ. Microbiol.* **33**, 784–790 (1977).

Pieper, D. H., Winkler, R., and Sandermann, H., Jr., *Angew. Chem., Int. Ed. Engl.* **31**, 68–70 (1992).

Prasad, I., *Can. J. Microbiol.* **16**, 369–372 (1970).

Read, D. C., *J. Econ. Entomol.* **62**, 1328–1334 (1969).

Read, D. C., *J. Econ. Entomol.* **64**, 796–800 (1971a).

Read, D. C., *J. Econ. Entomol.* **64**, 800–804 (1971b).

Richardson, M. L., Webb, K. S., and Gough, T. A., *Ecotoxicol. Environ. Saf.* **4**, 207–212 (1980).

Richey, F. A., Jr., Bartley, W. J., and Sheets, K. P., *J. Agric. Food Chem.* **25**, 47–51 (1977).

Roberts, T. R., and Standen, M. E., *Pestic. Biochem. Physiol.* **9**, 322–333 (1978).

Rosenberg, A., and Alexander, M., *Appl. Environ. Microbiol.* **37**, 886–891 (1979).

Rouchaud, J., Roucourt, P., Vanachter, A., Benoit, F., and Ceustermans, N., *Rev. Agric. (Brussels)* **41**, 889–899 (1988).

Sander, J., Schweinsberg, E., Ladenstein, M., and Schweinsberg, F., *Zentralbl. Bakteriol., Mikrobiol. Hyg., Abt. I, Orig. A* **227**, 71–80 (1974).

Sethunathan, N., and Yoshida, T., *J. Agric. Food Chem.* **21**, 504–506 (1973a).

Sethunathan, N., and Yoshida, T., *Can. J. Microbiol.* **19**, 873–875 (1973b).

Singh, G., Kathpal, T. S., Spencer, W. F., and Dhankar, J. S., *Environ. Pollut.* **70**, 219–239 (1991).

Sirons, G. J., Frank, R., and Sawyer, T., *J. Agric. Food Chem.* **21**, 1016–1020 (1973).

Somers, E., Richmond, D. V., and Pickard, J. A., *Nature (London)* **215**, 214 (1967).

Svenson, A., Kjeller, L.-O., and Rappe, C., *Environ. Sci. Technol.* **23**, 900–902 (1989).

Tate, R. L., III, and Alexander, M., *Soil Sci.* **118**, 317–321 (1974).

Tate, R. L., III, and Alexander, M., *Appl. Environ. Microbiol.* **31**, 399–403 (1976).

Taylor, H. F., and Wain, R. L., *Proc. R. Soc. London, Ser. B.* **156**, 172–186 (1962).

Verloop, A., *Residue Rev.* **43**, 55–103 (1972).

Vlasenko, N. L., Zhuravleva, I. L., Terenina, M. B., Golovnya, R. V., and Ilnitskii, A. P., *Gig. Sanit.* **11**, 15–17 (1981).

Vlitos, A. J., and King, L. J., *Nature (London)* **171**, 523 (1953).

Wannstedt, C., Rotella, D., and Siuda, J. F., *Bull. Environ. Contam. Toxicol.* **44**, 282–287 (1990).

Whitten, C. J., and Bull, D. L., *J. Agric. Food Chem.* **22**, 234–238 (1974).

Wilson, B. H., Smith, G. B., and Rees, J. F., *Environ. Sci. Technol.* **20**, 997–1002 (1986).

Yordy, J. R., and Alexander, M., *J. Environ. Qual.* **10**, 266–270 (1981).

6 KINETICS*

Knowledge of the kinetics of biodegradation is essential to the evaluation of the persistence of organic pollutants and to assessing exposure of humans, animals, and plants. Once degradation of a chemical commences, the amount disappearing with time and the shape of the disappearance curve will be a function of the compound in question, its concentration, the organisms responsible, and a variety of environmental factors. Information on kinetics is extremely important because it characterizes the concentration of the chemical remaining at any time, permits prediction of the levels likely to be present at some future time, and allows assessment of whether the chemical will be eliminated before it is transported to a site at which susceptible humans, animals, or plants may be exposed. Such knowledge is thus essential for the assessment of the potential risk associated with exposure of susceptible individuals and species to the chemical.

Research on kinetics has focused on two topics. The first is assessing factors that affect the amounts of the compound transformed per unit time. In this regard, much information is available on the influence of temperature, pH, soil moisture, and other C sources on the rates of transformation. The second topic is determining the shapes of the curves that depict the transformation and evaluating which of the patterns of decomposition best fit the metabolism of given compounds in a microbial culture, in laboratory microcosms, or, occasionally, in the field. This second topic is the subject of this chapter.

In many instances, the information on kinetics comes only from evaluations of the loss of the parent molecule. This is warranted for toxicants whose inhibitory effects are totally destroyed as a result of the first enzymatic reaction or in metabolic sequences in which intermediates do not accumulate. It is not warranted if intermediates accumulate, especially if

*Significant portions of this chapter were published earlier (Reprinted with permission from Alexander and Scow, 1989).

they are toxic or their hazard has yet to be evaluated. In other instances, the knowledge comes from tests of mineralization of the compounds; although mineralization reflects detoxication, the pattern of disappearance of toxicity from the natural environment could be quite different from the patterns of conversion of the organic molecule to inorganic products. Unfortunately, essentially no attention has been given to the kinetics of the various steps in transformation in nature that result in product accumulation.

Many models have been proposed to represent the kinetics of biodegradation. An understanding of when to use these models and why they may fail to describe the data requires knowledge of the theoretical bases for these models. The models commonly used to fit data from evaluations of biodegradation are, in essence, either empirical or theoretical.

The study of kinetics of biodegradation in natural environments is often empirical, reflecting the rudimentary level of knowledge about microbial populations and activity in these environments. An example of an empirical approach is the power rate model (Hamaker, 1972)

$$-dC/dt = kC^n, \tag{1}$$

where C is substrate concentration, t is time, k is the rate constant for chemical disappearance, and n is a fitting parameter. This model can be fit to substrate-disappearance curves by varying n and k until a good fit is achieved. From this equation, it is evident that the rate is proportional to a power of the substrate concentration. The power-rate law provides a basis for comparison of different curves, but it gives no insight into the reasons for the shapes. Therefore, often it may have no predictive ability. Moreover, investigators interested in kinetics do not always state whether the model they are using has a theoretical basis or is simply empirical, and whether constants in an equation have physical meaning or are only fitting parameters (Bazin et al., 1976).

An appropriate introduction to the kinetics of biodegradation is to consider a pure culture of a single bacterial population that is growing on and degrading a single, soluble organic chemical, and to assume that no barriers exist between the substrate and the cells.

PROCESSES LINKED TO GROWTH

The biodegradation of a particular organic substrate may be carried out by microorganisms that are: (a) growing at the expense of that substrate and using it as a source of C, energy, or possibly another nutrient element needed for proliferation; (b) growing at the expense of another organic

nutrient that is used as a source of C, energy, or both but metabolizing
the substrate of interest (although not using it to supply building blocks
for cell synthesis); or (c) not growing as they metabolize the chemical of
concern.

Consider the case of a population of cells of one bacterial species using
a particular organic compound as its source of C and energy. To simplify
the illustration, assume that the organic molecule is water soluble and
nontoxic, that the organism is growing in well-aerated liquid media, and
that the inorganic nutrients and growth factors needed by the bacterium
are present in excess of the organism's need. At low concentrations of
the C and energy source, the growth rate of the organism is slow because
it is limited by the low level of the substrate. If the organism is inoculated
into flasks containing increasing concentrations of the C compound, its
growth rate will increase in proportion to the increase in concentration.
Above some moderately high level of substrate, the growth rate does not
rise as markedly with increasing concentrations. Ultimately, at a quite high
level, the growth rate does not increase with further rises in concentration.
Monod (1949) mathematically formulated this relationship. It is now commonly written as

$$\mu = \frac{\mu_{max} S}{K_s + S},$$ (2)

where μ is the specific growth rate of the bacterium, μ_{max} is its maximum
specific growth rate (which occurs at the higher range of substrate concentrations), S is the substrate concentration, and K_s is a constant that represents the substrate concentration at which the rate of growth is half the
maximum rate. The expression for the Monod equation is presented graph-
ically as the hyperbolic curve shown in Fig. 6.1. The value for K_s represents
the affinity of the organism for the growth-supporting organic nutrient;
the lower the value, the greater is the bacterium's affinity for that molecule.

The values for K_s extend over a considerable range. For a single bacterium, the K_s value differs with different substrates. For a single substrate,
the value will depend on the bacterium. Moreover, a single microorganism
may have one K_s value at low substrate concentrations and another at
high concentrations. Some values are given in Table 6.1. From the data
presented, it is immediately evident that the values vary enormously,
and no generalizations are evident at first glance. Nevertheless, bacteria
characteristic of nutrient-rich environments often have higher K_s values
than those obtained from habitats with low levels of organic constituents.
Also given in Table 6.1 are the K_s values for metabolism in natural waters,
the data probably representing the species in the water sample that most

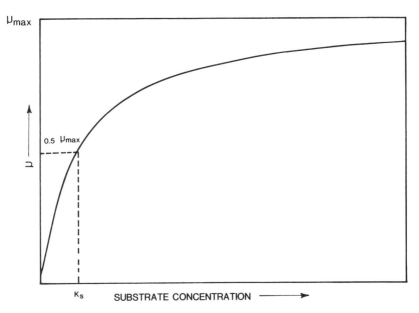

FIG. 6.1 Relationship between growth rate of a bacterium and the concentration of the C source supporting its growth.

rapidly metabolizes the added molecule. It is worth pointing out that the concentrations of the C source commonly included in culture media are far higher than the values recorded for K_s.

Two K_s values or affinity constants may be evident for a single substrate, both in pure cultures and in microbial communities. One value for pure

TABLE 6.1
Values for K_s for Bacteria and for Samples of Water

Substrate	Organism or sample	K_s value (mg/liter)	Reference
Glucose	*Flavobacterium* 1	0.0071	van der Kooij and Hijnen (1981)
	Flavobacterium 2	29, 1314	Ishida *et al.* (1982)
	River water	26	Larson (1980)
Glutamate	*Aeromonas* sp.	0.163, 1.3	Ishida *et al.* (1982)
Maltose	*Butyrivibrio fibrisolvens*	2.1	Russell and Baldwin (1979)
Xylose	*Butyrivibrio fibrisolvens*	55	Russell and Baldwin (1979)
m-Cresol	Natural water	0.0006–0.0018	Bartholomew and Pfaender (198?
Chlorobenzene	Natural water	0.0010–0.0051	Bartholomew and Pfaender (198?
NTA	Natural water	0.060–0.170	Bartholomew and Pfaender (198?
Phenol	Wastewater	1.3–270	Rozich *et al.* (1985)

cultures may be quiet low (i.e., a high affinity), in the range of less than 20 μg/liter; the other may be quite high (i.e., a low affinity) and in the range of 1 to more than 10 mg/liter (Ishida et al., 1982; Lewis et al., 1983). Two affinity constants have also been noted for microorganisms in lake water that degrade phenol, the concentrations at which the rates are half the maximum (0.5 μ_{max}) being ca. 5 and 400 μg/liter (Jones and Alexander, 1986). In samples of natural environments, however, the existence of more than one affinity constant may reflect the activity of a single population or, alternatively, different species with dissimilar affinities for the same organic molecule. Similar studies with suspensions of biofilms from river water also show that more than one K_s may characterize the degradation of 4-chlorophenol and glucose in a single environmental sample (Lewis et al., 1988).

Under certain circumstances, Monod kinetics may not be an adequate description of bacterial growth on soluble substrates in pure culture (Koch and Wong, 1982). Nevertheless, such anomalies are not widely known and may be assumed to be uncommon. However, as pointed out in the following, such kinetics may be inappropriate for substrates that are sorbed, insoluble, or toxic, even for pure cultures.

When a pure culture of bacteria is growing in media containing a C source at concentrations far in excess of the K_s, most of the period during the growth cycle occurs without the declining substrate level greatly affecting the growth rate. Therefore, the rate of degradation of the substrate is not markedly influenced by its concentration until nearly all is gone. During this period when the C source (and presumably other nutrients) is in excess, the time for one cell to divide to give two, two to divide to give four, four to give eight, etc., is constant. Thus, for cells that multiply by binary fission, the population density increases in a geometric progression with, in the case of nonlimiting nutrients, a constant time interval between each doubling. This relationship can be expressed mathematically as

$$N_0(2^n) = N, \tag{3}$$

where N_0 is the initial cell number and N is the number of cells after n divisions. This is a simple exponential (logarithmic) progression that can be expressed as

$$\ln N - \ln N_0 = kt \tag{4}$$

or in $\log_{10}$,

$$\log N - \log N_0 = kt/2.303, \tag{5}$$

where t is time and k is the specific growth rate constant. On rearrangement, the relationship is

$$\log N = kt/2.03 + \log N_0, \qquad (6)$$

and a plot of $\log N$ versus t yields a straight line with a slope of $k/2.303$ and an intercept of N_0. Thus, during this exponential phase of bacterial growth, a plot of cell number versus time gives a straight line (Neidhardt *et al.*, 1990).

If it is assumed that each cell metabolizes the same amount of organic substrate during this phase, a plot of the logarithm of the amount of *substrate metabolized* (or the amount of C recovered in cells plus all products) versus time should also give a straight line. However, this does not mean that a plot of the logarithm of *substrate remaining* versus time gives a straight line. The shape of the curve that reflects logarithmic growth is depicted in Fig. 6.2. Initially, little loss of the chemical is evident because the cell density is low, but then as the mass of cells becomes large and doubles exponentially, a rapid loss of chemical is evident. Indeed, many doublings in population size occur with little or no loss of C source being

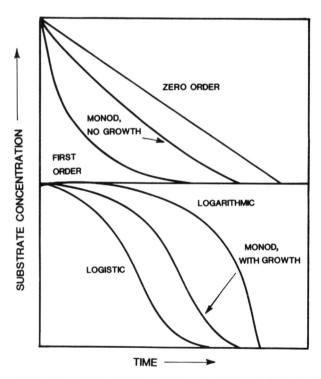

FIG. 6.2 Disappearance curves for chemicals that are metabolized by different kinetics (From Alexander, 1985. Reprinted with permission from the American Chemical Society.)

evident, and much of the disappearance can be attributed to the last few doublings in population size. Logarithmic kinetics of substrate disappearance can be expressed in differential form as

$$dS/dt = \mu_{max}(S_0 + X_0 - S) \tag{7}$$

and in integral form as

$$S = S_0 + X_0[1 - \exp(\mu_{max}t)], \tag{8}$$

where S_0 is initial substrate concentration, S is substrate concentration, and X_0 is the amount of the substrate required to produce the initial population (Simkins and Alexander, 1984). Conditions for logarithmic kinetics are satisfied when a single bacterial population is multiplying and S_0 is much above K_s ($S_0 \gg K_s$). Logarithmic kinetics have been verified for the disappearance of benzoate in cultures of a *Pseudomonas* sp. (Simkins and Alexander, 1984) and apparently for the formation of $^{14}CO_2$ from ^{14}C-labeled 2,4-D added at high concentrations to soil (Kunc and Rybarova, 1983).

The kinetics of growth at initial substrate concentrations much below K_s ($S_0 \ll K_s$) are quite different. From Fig. 6.1, it is evident that the growth rate declines at progressively lower substrate concentrations. Thus, when *Escherichia coli* is multiplying in media with low glucose levels, its growth rate is proportional to glucose concentration (Shehata and Marr, 1971). When the culture multiplies in a medium with such low glucose levels, the cells continue to increase in number, but because the progressively falling substrate concentration is causing a progressively slower growth rate, the period between each doubling becomes progressively longer. Thus, in contrast with the logarithmic phase, in which the doubling time is constant, growth at substrate concentrations below K_s is characterized by increasingly long doubling times even as the cell number is rising. This is known as logistic growth, and the kinetics of this pattern of multiplication will be mirrored in the kinetics of substrate disappearance if each cell metabolizes essentially the same amount of its C source.

The logistic growth curve has interested population ecologists for many years (Odum, 1971). The logistic curve has an S shape and is symmetrical about the point of inflection. This is evident in the plot depicting logistic kinetics of substrate disappearance (Fig. 6.2). Because of the low concentration of substrate, one can only have appreciable increases in cell numbers when the initial population size is small, so that both little substrate and few cells are necessary for this pattern of biodegradation. Logistic kinetics may be written in differential form as

$$-\frac{dS}{dt} = dS(S_0 + X_0 - S) \tag{9}$$

or an integral form as

$$S = \frac{S_0 + X_0}{1 + (X_0/S_0) \exp [k (S_0 + X_0)t]},$$ (10)

where $k = \mu_{max}/K_s$ (Simkins and Alexander, 1984). Logistic kinetics have been noted in the mineralization of 2.0 μg of phenol and 2.0 and 7.0 μg of 4-nitrophenol per liter in lake water (Jones and Alexander, 1986).

When bacterial growth follows logistic kinetics, the curve for the time in which the density rises from 10 to 90% of the maximum population density can be closely approximated by a straight line (Schmidt et al., 1985). Thus, when the initial substrate concentration is below K_s ($S_0 \ll K_s$) and few cells are present that are able to use that chemical as a C source for growth, the finding of what appears to be linear kinetics may, in fact, be a reflection of a logistic transformation.

When the initial substrate concentration is approximately the same as K_s ($S_0 \sim K_s$), neither logarithmic nor logistic kinetics apply. The situation is somewhat more complex because μ is not directly dependent on substrate concentration (as when $S_0 \ll K_s$) or largely independent of it (as when $S_0 \gg K_s$) (Fig. 6.1). The pattern of substrate disappearance in this range of concentrations is termed Monod-with-growth kinetics, and the shape of the curve of chemical loss is depicted in Fig. 6.2. Monod-with-growth kinetics can be expressed mathematically in differential form as

$$-\frac{dS}{dt} = \frac{\mu_{max} S(S_0 + X_0 - S)}{K_s + S}$$ (11)

and in integral form as

$$K_s \ln (S/S_0) = (S_0 + X_0 + K_s) \ln (X/X_0) - (S_0 + X_0)\mu_{max}t,$$ (12)

where X is the amount of substrate to produce the population density. These kinetics describe the metabolism of benzoate by *Pseudomonas* sp. at benzoate levels near K_s (Simkins and Alexander, 1984) and the mineralization of 4-nitrophenol in lake water (Jones and Alexander, 1986).

At high concentrations, many pollutants are toxic to the very microorganisms that use them as C sources. For these chemicals, the typical relationship between bacterial growth rate and concentration of its C source, as presented in Fig. 6.1, is not found. Although the growth rate increases with increasing but low substrate concentration, a concentration is reached above which the growth rate falls as the substrate level rises further. This decline is a result of the antimicrobial action of the chemical. Such a relationship between growth rate and concentration of a potentially toxic organic nutrient can be characterized by the Haldane modification of the Monod equation

$$\mu = \frac{\mu_{max} S}{K_s + S + (S^2/K_I)},$$ (13)

where K_I is an inhibition constant that reflects the suppression of the growth rate by the toxic substrate. This equation has been used for describing the kinetics of phenol and pentachlorophenol metabolism by microorganisms (Klecka and Maier, 1985; Rozich et al., 1985).

BIODEGRADATION BY NONGROWING ORGANISMS

For bacteria to grow appreciably, the amount of substrate must be sufficiently high relative to the number of cells active on that compound to permit several or many doublings. If the bacterial cell density is high relative to the substrate concentration, little or no increase in cell numbers is possible. Under these conditions, one can again consider three cases: $S_0 \gg K_s$, $S_0 \sim K_s$, and $S_0 \ll K_s$. Such kinetics resemble those of enzyme reactions because multiplication is not involved. The relationship between the rate of an enzyme reaction and the concentration of the substrate for that enzyme is often best expressed by the Michaelis–Menten equation

$$v = \frac{V_{max} S}{K_m + S},$$ (14)

where v is reaction rate, V_{max} is the maximum reaction rate, and K_m is a constant called the Michaelis constant. Equation (14) represents the same relationship as the Monod equation [Eq. (2)], differing only in the replacement of v, V_{max}, and K_m for μ, μ_{max}, and K_s, respectively. Moreover, a graphic representation of the equation would be the same as that shown in Fig. 6.1, with the substitution of v, V_{max}, and K_m for μ, μ_{max}, and K_s. The essential difference is that in Michaelis–Menten kinetics, the quantity of reactive material (enzyme) is constant, whereas in Monod kinetics, the amount of reactive material (cells) is increasing because of microbial proliferation. It is because Michaelis–Menten kinetics are formulated on the basis of constant reactive material that such kinetics are useful for nongrowing cells.

Referring to Fig. 6.1 and using v, V_{max}, and K_m in place of μ, μ_{max}, and K_s, the kinetics of biodegradation by nongrowing cells become immediately evident. At initial substrate concentrations much higher than K_m ($S_0 \gg K_m$), the rate does not fall appreciably as the cells transform the organic substrate; that is, moving from the extreme right of the curve to a point somewhat to the left does not greatly alter v. In other words, the rate is essentially constant as the concentration falls from high levels to concentrations that are lower but still above K_m. The rate is thus linear or, to use the term from chemical kinetics, it follows zero-order kinetics. Conversely, when the initial substrate concentration is much below K_m ($S_0 \ll K_m$) and metabolism of the compound further lowers the concentration, the rate falls in proportion to the decline in substrate concentration because the rate is a direct function of concentration. In this case, the rate is continuously falling as the substrate level falls, owing to microbial metabolism. This relationship is known as first-order kinetics.

For nongrowing cells, the kinetics when $S_0 \gg K_s$, $S_0 \sim K_s$, and $S_0 \ll K_s$ (referring back to the terms used for growing organisms) are thus termed zero-order, Monod-no-growth, and first-order kinetics. The kinetics are expressed mathematically as follows.

For first-order kinetics, the differential form is

$$-\frac{dS}{dt} = k_1 S \tag{15}$$

and the integral form is

$$S = S_0 \exp(-k_1 t), \tag{16}$$

where t = time, S = substrate concentration at time t, and $k_1 = \mu_{max}$ (X_0/K_s). The term k_1 is the first-order rate constant and is expressed in units of (time)$^{-1}$, that is, h^{-1}, days^{-1}, etc. If $k_1 = 0.01$ h^{-1}, the rate is 1% per hour. This is simply another way of saying a constant percentage is lost per unit time.

For Monod-no-growth kinetics, the differential form is

$$-\frac{dS}{dt} = \frac{k_2 S}{K_s + S} \tag{17}$$

and the integral form is

$$K_s \ln(S/S_0) + S - S_0 = -k_2 t, \tag{18}$$

where $k_2 = \mu_{max} X_0$.

For zero-order kinetics, the differential form is

$$-\frac{dS}{dt} = k_2 \tag{19}$$

and the integral form is

$$S = S_0 - k_2 t \tag{20}$$

(Simkins and Alexander, 1984). In zero-order kinetics, a *constant amount* is lost per unit time.

The patterns of chemical biodegradation that occur by zero-order, Monod-no-growth, and first-order kinetics are presented in Fig. 6.2.

To express Monod-no-growth as Michaelis–Menten kinetics, K_s is replaced by K_m, and k_2 is equal to $V_{max} B_0$ in both differential and integral equations. The term B_0 is the initial population density. Such Michaelis–Menten or nongrowing-cell kinetics have been reported to describe the kinetics of biodegradation of picloram in soil (Hamaker et al., 1968; Meikle et al., 1973) and the initial metabolism of 3,5-dichlorobenzoate and 4-amino-3,5-dichlorobenzoate by an anaerobic enrichment culture converting these molecules to methane (Suflita et al., 1983), and first-order, Monod-no-growth, and zero-order kinetics sometimes best describe the mineralization of phenol in lake water (Jones and Alexander, 1986).

The terms first and zero order come from chemical kinetics. In a first-order process, the rate is proportional to the concentration of a single reactant; for the purposes of this discussion, that reactant is the substrate. In a zero-order process, the rate is independent of the concentration of the reactants, that is, independent of the substrate concentration. When the concentration is plotted against time, as in Fig. 6.2, the concentration decreases at a constant rate in zero-order processes, but it falls quickly initially and then more slowly in first-order processes. As stated earlier, a constant amount is lost per unit time in zero-order reactions, and a constant percentage disappears per unit time in first-order reactions.

ZERO-ORDER KINETICS

Zero-order kinetics or linear biodegradation of organic substrates (or formation of organic products) has been observed frequently. According to the theory presented in the preceding section, such rates should be evident in processes effected by nongrowing cells when $S_0 \gg K_s$ (or $S_0 \gg K_m$), and they may seem to be evident when bacteria are growing logistically because $S_0 < K_s$. Linear transformation may also occur under the following conditions.

(a) The nutrient that limits the growth of the active population becomes available at a constant rate, but the rate does not fully meet the demand of the organisms. For example, several bacteria grow linearly in liquid media when O_2 enters the solution at a rate that limits their further multiplication (Volk and Myrvik, 1953; Brown et al., 1988). The O_2 limitation probably explains why some fungi grown in media with supplemental aeration enter a linear growth phase (Gillie, 1968). Such O_2 limitation to biodegradation is likely to occur at high substrate concentrations.

(b) The organisms use up the supply of some essential nutrient element or growth factor. For example, methane production by *Methanosarcina* switches from a logarithmic to a linear rate when phosphate in the medium becomes depleted (Archer, 1985). A possible reason is that the concentration of some enzyme or enzyme system essential for further multiplication is constant (Monod, 1949).

(c) The population size of organisms active on the organic compound has become large as a result of previous addition of the chemical, and a second increment of the compound is introduced. Under these conditions, the biomass of the already large population may not increase as it decomposes the second increment, presumably when the concentration of the second addition is above K_s. This has been observed when an anaerobic enrichment culture that had been acclimated to metabolize 3-chlorobenzoate received a second increment of the chemical (Suflita *et al.*, 1983).

(d) The population is growing on certain C compounds that have low water solubilities, and the amount in aqueous solution has been totally consumed. The reasons for this are not yet clear. Linear growth has been reported for pure cultures of bacteria, yeasts, and sometimes fungi growing on tri-, tetra-, penta-, hexa-, and octadecane (Lindley and Heydeman, 1986; Yoshida and Yamane, 1971; Thomas *et al.*, 1986), a material known as slack wax, which contains 70 to 90% of straight-chain solid paraffin (Lonsane *et al.*, 1979), cholesterol, β-sitosterol (Goswami *et al.*, 1983), phenanthrene (Stucki and Alexander, 1987), and crystalline or adsorbed polycyclic aromatic hydrocarbons (Volkering *et al.*, 1992). Such linear growth in pure culture frequently follows a period of logarithmic growth (Goswami *et al.*, 1983; Stucki and Alexander, 1987; Lonsane *et al.*, 1979), because the initially small population of microorganisms probably first grows unrestrictedly on the soluble chemical or other dissolved organic nutrients and then, when the supply of those is depleted, uses the chemical that initially is not in the aqueous phase.

Zero-order kinetics have been reported frequently for biodegradation (Table 6.2). Moreover, linear rates have been found at extremely low concentrations (102 pg/liter), which are undoubtedly below the K_s, to such high concentrations (10 g/liter) that they are undoubtedly far above K_s. In a few instances, the metabolic conversion is zero order at high concentrations and first order at low concentrations; for example, for the mineralization of maleic hydrazide in soil (Helweg, 1975) or the mineralization of glucose and linear alcohol ethoxylate in bay water (Vashon and Schwab, 1982).

TABLE 6.2

Organic Compounds That Are Metabolized by Zero-Order Kinetics in Samples from Natural Environments

Chemical	Concentration (per liter of water or kg of soil)	Environmental sample	Reference
Phenol	102 pg–10 g	Lake water	Subba-Rao et al. (1982)
2,4-D	1.5 ng	Lake water	Subba-Rao et al. (1982)
Aniline	5.7 ng–500 μg	Lake water	Subba-Rao et al. (1982)
Diethanolamine	21 ng	Stream water	Boethling and Alexander (1979)
Toluene	380, 3900 ng	Seawater	Button et al. (1981)
4,6-Dinitro-2-methylphenol	5–2500 μg	Soil	Hurle and Pfefferkorn (1972)
NTA	10, 200 μg	Estuarine water	Pfaender et al. (1985)
Benzylamine	20, 200 μg	Lake water	Subba-Rao et al. (1982)
Di(2-ethylhexyl) phthalate	21–200 μg	Lake water	Subba-Rao et al. (1982)
N-Nitrosodiethanolamine	54, 940 μg	Lake water	Yordy and Alexander (1980)
	940 μg	Sewage	Yordy and Alexander (1980)
Glucose	1, 10 mg	Bay water	Vashon and Schwab (1982)
2,4-D	25–100 mg	Aquatic	Hemmett and Faust (1969)
Glyphosate	90 mg	Soil	Torstensson and Aamisepp (1977)
Maleic hydrazide	120 mg	Soil	Helweg (1975)
Glucose	400 mg	Activated sludge	Gaudy et al. (1963)
Butyrate	1.15 g	Activated sludge	Mateles and Chian (1969)

FIRST-ORDER KINETICS

First-order biodegradation is to be expected when the chemical concentration is below K_s (or K_m) and the organisms are not increasing in abundance, possibly because there is not sufficient available C to support a doubling, possibly because some other limiting nutrient is lacking. A common way of presenting first-order kinetics is to plot the logarithm of the concentration of chemical remaining (or logarithm of S/S_0) as a function of time; if the reaction is first order, a straight line is obtained (Fig. 6.3). First-order kinetics are sometimes termed half-life kinetics because if half of the chemical is gone in time t, half of what is then left will remain at time $2t$, and half again at time $3t$. In other words, if the half-life is 20 days, the amount remaining at 20, 40, 60, and 80 days is 1/2, 1/4, 1/8, and 1/16 of the initial quantity.

First-order kinetics have been observed for the metabolism of glucose by *Salmonella typhimurium* at a concentration (0.4 μg/liter) below that

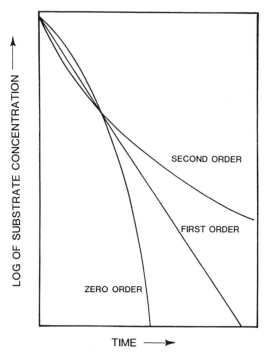

FIG. 6.3 A plot of logarithm of substrate remaining as a function of time, as is typical of first-, zero-, and second-order reactions (From Hamaker, 1966. Reprinted with permission from the American Chemical Society.)

supporting growth (Schmidt *et al.*, 1985) and for the biodegradation of 200 μg of methyl parathion per liter in seawater (Badawy and El-Dib, 1984), 5.3 μg of methyl parathion per liter in anaerobic sediments (Wolfe *et al.*, 1986), 4.0 mg of hexazinone per kilogram of soil (Bouchard *et al.*, 1985), 0.5 μg of phenol per liter in lake water (Jones and Alexander, 1986), 1,1,1-trichloroethane at concentrations up to 1 mg/liter in a biofilm reactor inoculated with methane-oxidizing bacteria (Arvin, 1991), 11 mg of dichlobenil per kilogram of soil (Montgomery *et al.*, 1972), 0.04 and 4.0 mg of chlorosulfuron per kilogram of soil (Walker and Brown, 1983), 0.85 to 140 μg of linear alcohol ethoxylate per liter in bay water (Vashon and Schwab, 1982), and a number of other chemicals. Many other claims of first-order transformations cannot be accepted, although they may be correct, because of too few data points to determine the appropriate kinetics, and many must be rejected because the data do not fit first-order plots well enough.

First-order kinetics are commonly used to describe biodegradation in environmental fate models because mathematically the expression can be easily incorporated into the models. Unfortunately, this compatibility with models often takes precedence over other evaluation criteria that are more important, and blind acceptance of this type of kinetics can lead to incorrect conclusions on the persistence of toxic chemicals. Many investigators grasp at first-order kinetics because of the ease of presenting and analyzing the data, the simplicity of plotting the logarithm of the chemical remaining versus time as a straight line regardless of the poorness of fit of the line to the points, and the ease of predicting future concentrations once the time is determined for loss of half of the chemical.

It is important to stress that a pollutant whose destruction follows first-order kinetics persists long after the first half-life is over because the level is falling at diminishing rates. This is in contrast with logarithmic or zero-order transformations, the former resulting in more and more being lost per unit time period, the latter resulting in a constant rate until all the chemical is gone.

In developing predictive kinetic models, use has been made of the fact that at substrate concentrations below K_s, the rates of substrate destruction are first order and the cells responsible for the destruction are not growing to any significant degree. In different environments, the first-order constants and the number of cells able to metabolize the substrate will differ. However, it has been proposed that for predictive purposes, special use can be made of the value obtained by dividing the first-order rate constant by the number of cells present in natural environments.

One may write the Monod equation as

$$-\frac{dS}{dt} = \frac{\mu_{max} B S}{Y(K_s + S)},$$ (21)

where B is the bacterial cell density and Y is the yield coefficient or the number of bacteria per milligram of substrate. When the substrate concentration is much lower than K_s, this expression can be approximated by

$$-\frac{dS}{dt} = k_b B S,$$ (22)

where $k_b = \mu_{max}/YK_s$ and is termed the "second-order" rate constant (Paris and Rogers, 1986; Paris et al., 1981). The use of the term "second-order" is unfortunate becaue it confuses such expressions with second-order kinetics of chemistry, and it will not be used here for that reason. An important value needed to determine this relationship is the number of cells actually degrading the substrate, but this value is difficult to obtain for populations in nature. Thus, the numbers used are the total bacterial counts, which makes the approach less valuable, especially because the percentage of the total cell number able to degrade a different chemical often will vary greatly in different ecosystems. Such kinetics have been used to characterize the metabolism of malathion in microbial cultures (Falco *et al.*, 1977), methyl parathion and diethyl phthalate by attached microbial growths (Lewis and Holm, 1981), the hydrolysis of an ester of 2,4-D by microorganisms growing on surfaces submerged in fresh water (Lewis *et al.*, 1983), and the transformation of several chemicals in lake water (Paris *et al.*, 1981).

The various orders of reaction, two of which were considered in the foregoing, are commonly summarized by

$$\text{Rate} = -\frac{dC}{dt} = kC^n, \tag{23}$$

where k is the rate constant, C is concentration, and n is the order of the reaction [this is given above as Eq. (1)]. From this equation, it is evident that the rate is proportional to a power of the chemical (or substrate, S, in the present context) concentration (Hamaker, 1972). In the first- and zero-order reactions, the equation is

$$-\frac{dC}{dt} = k \qquad \text{for a zero-order reaction}$$

and

$$-\frac{dC}{dt} = kC \qquad \text{for a first-order reaction.}$$

In a second-order reaction, the rate is proportional to the second power (i.e., the square) of the concentration of a single reactant molecule (rate $= kC^2$) or in other circumstances to the concentrations of two reactants (rate $= kC_1C_2$). In both instances, the concentrations of both compounds involved in the reaction change with time. It is in the sense of having two reactants that the term "second order" has been applied to biodegradation, one reactant being the substrate and the other being the microbial biomass. However, both reactants in abiotic processes typically decline in concentration in such transformations, whereas in biodegradation, the cells either increase or the numbers remain constant. Hence, such kinetics really represent simply first-order rates divided by cell number. In more classical chemical terms, a plot of the logarithm of the chemical concentration remaining, in the case of only a single reacting chemical, against time would give a line that is straight, concave down, or concave up for first, zero, or second (or higher) orders of reaction, respectively (Fig. 6.3).

Frequently, only two or three samples are taken prior to making predictions of the amount of a chemical that will remain at a site in the future. Given the same initial analytical values, the various kinetic models will predict vastly different amounts of chemical remaining at later times. Consider the case of a polluted site with 10 mg/liter initially and 9 mg/liter after 30 days. If predictions were made of the time for the concentration to fall to a regulatory standard of, say, 10 μg/liter, the predictions would be 33 days for the logarithmic model, 300 days for the zero-order model, more than 5 years for the first-order model, and possibly centuries for a model that appears to be initially first order and is followed by a second, slower phase of degradation. In this regard, it is important to note that the similarities in the x axes in Fig. 6.2 could be misleading; the time periods for some models are short and those for other models are very long.

The preceding discussion considered kinetics in relation to density of bacteria active on the substrate and substrate concentration. The relationship can be depicted in a simple fashion (Fig. 6.4). For the purpose of this presentation, it is considered that either (a) the cells are not growing (the three sectors above the diagonal line) because the substrate concentration is not sufficiently high to support even a single doubling or (b) they are growing because enough C is present to permit the population to increase in size (the three sectors below the line). A key feature of the illustration is K_s; it can be any value but, for purposes of the example, it is arbitrarily chosen as 1.0 mg/liter and is represented by a vertical line. The values for bacterial density are given per milliliter because that is the convention among microbiologists. For the purposes of this illustration, it is assumed that 1.0 pg of substrate is required to form one bacterial cell, a figure that is often approximated in nature. The precise positions of the vertical lines other than K_s are not certain, but they are placed at positions to indicate that (a) the growth rate, μ, is maintained at about μ_{max} until nearly all the substrate is degraded (the vertical line to the right of K_s), and (b) the rate of degradation per cell varies greatly with substrate concentration (the vertical line to the left of K_s). Thus, the two sectors each at the left, middle, and right denote $S_0 \ll K_s$, $S_0 \sim K_s$, and $S_0 \gg K_s$, and the three sectors above and the three below the diagonal denote nongrowing and growing cells, respectively. From this diagram, the ranges of relative cell densities and substrate concentrations corresponding to each of the six models are evident.

Scow et al. (1986) found that the models of the Monod family did not provide good fits to curves depicting the mineralization of low concentra-

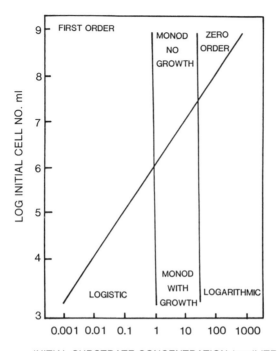

INITIAL SUBSTRATE CONCENTRATION (mg/LITER)

FIG. 6.4 Kinetics of biodegradation as a function of substrate concentration and number of bacteria able to metabolize that substrate. Patterns below the diagonal line represent kinetics of biodegradation by bacteria using the substrate as a source of C and energy for growth. (From Simkins and Alexander, 1984. Reprinted with permission from the American Society for Microbiology.)

tions of phenol, 4-nitrophenol, aniline, 2,4-dichlorophenol, benzylamine, NTA, and cyclohexylamine. The models of the Monod family also did not fit satisfactorily to curves of atrazine, linuron, and picloram disappearance (Hance and McKone, 1971). Other studies have shown that first-order kinetics fit substrate-disappearance curves; however, when biodegradation of more than one concentration of the organic compound was tested, the estimated rate constant often varied with initial concentration, which is not consistent with first-order kinetics.

Natural environments are highly complex, both physically and chemically, the composition of their microbial communities is quite heterogeneous, and the abiotic constituents are commonly reactive. Hence, the application of existing models to biodegradation kinetics in many natural ecosystems often is questionable.

The following factors often confound the facile extrapolation of the kinetics described here to natural circumstances.

(a) Diffusional barriers may limit or prevent contact between microbial cells and their organic substrates.
(b) Many organic molecules sorb to clay or humus constituents of soils and sediments, and the kinetics of decomposition of sorbed substrates may be quite different from the same compounds free in solution.
(c) The presence of other organic molecules that can be metabolized by the biodegrading species may repress or enhance use of the test chemicals.
(d) The supply of inorganic nutrients, O_2, or growth factors may govern the rate of transformation, and the process will then be regulated by the diffusion of those nutrients or the rate of their formation or regeneration by other inhabitants of the community.
(e) Many species may be metabolizing the same organic compounds simultaneously, and these organisms may have different K_s and K_m values for the substrate.
(f) Protozoa or possibly species parasitizing the biodegrading populations may govern the growth, size, or activity of the populations responsible for the biodegradation.
(g) Many synthetic chemicals or pollutants have exceedingly low solubilities in water, and the kinetics of their transformation may be wholly dissimilar from compounds that are in the aqueous phase.
(h) The cells of the active population may be sorbed or develop in microcolonies, and the kinetics of processes effected by sorbed bacteria or microcolonies are as yet unresolved.
(i) Many organic compounds disappear only after an acclimation period, and methods do not now exist that can predict the length of this period or anticipate the percentage of the time between introduction of the chemical and its total destruction. Hence, nearly all the available kinetics models ignore the acclimation period.

DIFFUSION AND ADSORPTION

Because an important variable in models of biodegradation kinetics is the concentration of the substrate, a process that significantly lowers the concentration should affect the rate of biodegradation. Both physical and chemical processes, for example, diffusion to unavailable sites and adsorption, may be involved. A lack of consideration of the kinetics of diffusion and adsorption may contribute to the common failure of environmental-fate models in simulating laboratory measurements (van Genuchten et al., 1974; Davidson and Chang, 1972). Sorption is usually treated as a rapid-equilibrium, reversible process, but kinetic studies have shown that sorption is better represented as a two-phase process with an initial fast stage (<1 h) followed by a longer slow phase (days), and that diffusion of the solute to internal adsorption sites controls the second rate (Cameron and Klute, 1977; Karickhoff, 1980; McCall and Agin, 1985). The rates of adsorption and diffusion to inaccessible sites may be similar to many rates of biodegradation, and so these abiotic processes may be effectively competing with microorganisms for the substrate.

Sorption of a chemical has a major impact on the biodegradation of that compound. Nevertheless, surprisingly little attention has been given to the kinetics of biodegradation of sorbed molecules. A model was proposed by Mihelcic and Luthy (1988), however, that considered diffusion as a controlling factor, and it was based on the assumption that only the compound in the aqueous phase was acted on by microorganisms. A sorption-retarded radial diffusion model that took into account the effect of the size of soil aggregates was developed to describe the degradation of hexachlorocyclohexane, and this model also considered the diffusion of the substrates within the soil particles (Rijnaarts et al., 1990). In the latter study, moreover, the initial rates of biodegradation were greater than the initial rates of desorption, so that the rate of spontaneous desorption may not be an appropriate parameter for the kinetics of biodegradation. A different model was proposed by Miller and Alexander (1991) for sorbed organic substrates that are readily desorbed from the surface of solids, and this model gave a good fit to biodegradation of benzylamine that was initially sorbed to montmorillonite clay.

It is likely that diffusion also controls the availability of many organic substrates to microorganisms and influences the rate of degradation of these chemicals. Rovira and Greacen (1957) suggested that much of the native organic matter of soil was protected from microbial attack by its being sequestered within small pores. Other studies have provided evidence that the persistence of 1,2-dibromoethane may result from its entrapment in soil micropores, which makes it unavailable for microbial degradation (Steinberg et al., 1987; Pignatello et al., 1987). Soil consists of pores of different sizes, and measurements of moisture tension in a silt loam show that as much as 50% of the total pore volume consists of pores with radii estimated to be <1 μm (Cary and Hayden, 1973). Casida (1971) found that most soil bacteria range in size from 0.5 and 0.8 μm, and studies indicate that the mean diameter of soil pores occupied by bacteria is even larger, approximately 2 μm (Kilbertus, 1980). These findings suggest that in many soils, a significant portion of the soil solution retained in pores is inaccessible to most bacteria. Hence, diffusion of organic compounds into and out of these pores may be an important factor in controlling the rate of mineralization of the compounds.

The availability of many hydrophobic pollutants is markedly affected if the molecule is in a nonaqueous phase liquid (NAPL) at the site of pollution. That nonaqueous phase liquid may be oil from a marine spill, petroleum from a gasoline storage tank, a solvent from a leaking storage tank that is placed within the soil, or a mixture of solvents at a hazardous waste site. Models for the kinetics of biodegradation of substrates within NAPLs have not been devised. The rate of degradation is undoubtedly affected by the interfacial area, that is, the area of the surface between the NAPL and the aqueous phase, and consideration must be given to the kinetics of microbial growth at the interface, in the aqueous phase only, or at both the interface and in aqueous solution.

An approach to more complex kinetics is exemplified by two-compartment models. In such kinetics models, it is assumed that the substrate exists in two compartments, the identities of which usually are not known. The chemical in one compartment may be unavailable for microbial use, and that in the other compartment may be the form in which the chemical is transformed. In an environment containing particulate matter, a solution phase, and possible air-filled pores such as characterizes soils, one compartment could represent the substrate freely available to microorganisms and subject to rapid mineralization. The second compartment might then be substrate that is not readily available because it is adsorbed to colloidal surfaces or deposited in inaccessible micropores. After the supply of substrate in the first compartment is depleted, the subsequent rate of biodegradation would be governed by the rate of desorption or diffusion of the substrate from the inaccessible micropores to sites containing the active microorganisms. The rates of mass transfer of the substrate between the two compartments may be designated k_1 and k_2, and the rate of microbial transformation of the substrate in the labile compartment to product may be designated

$$S_1 \underset{k_2}{\overset{k_1}{\rightleftharpoons}} S_2 \xrightarrow{k_3} \text{products.}$$

S_1 and S_2 are the quantities of substrate in the unavailable and available compartment, respectively (Hamaker and Goring, 1976). The substrate in both compartments may be available to some extent, in which case the two-compartment kinetics model may be written as

$$S_1 \underset{k_2}{\overset{k_1}{\rightleftharpoons}} S_2$$

$$k_3 \searrow \swarrow k_4$$

products.

As an example, it is assumed that the substrate is at a concentration too low to support growth and the mass transfer of substrate between compartments follows first-order kinetics; therefore, k_1, k_2, k_3, and k_4 are first-order rate constants. In many cases, for example, when growth occurs, the process will be more complex. The simplified form of the model may be described mathematically by the two differential equations (Scow et al., 1986)

$$\frac{dS_1}{dt} = -(k_1 + k_3)S_1 + k_2 S_2 \tag{24}$$

$$\frac{dS_2}{dt} = k_1 S_1 - (k_2 + k_4) S_2. \tag{25}$$

Of several models tested, the two-compartment model provides the best fit for the mineralization in soil of low concentrations of NTA and phenol (Scow et al., 1986). It also fits the mineralization of aniline at concentrations ranging from 0.3 μg to 500 mg per kilogram. The two-compartment model also gives good fits for the decomposition of monocrotophos in soil; in this instance, the unavailable compartment is assumed to contain herbicide that is adsorbed by soil constituents, and the microflora is assumed to degrade the chemical that is free in soil solution as the available compartment (Furmidge and Osgerby, 1967).

Two-compartment models have been used to describe the biodegradation of organic compounds added to soil. Hamaker and Goring (1976) fit a two-compartment model to curves of degradation of triclopyr in soil. Two first-order curves provided the best fit to the patterns of disappearance in soil of three dinitroaniline herbicides and of metribuzin at 30°C (Hyzak and Zimdahl, 1974; Zimdahl and Gwynn, 1977). At low temperatures, however, first-order kinetics provided the best fit to the data. Parker and Doxtader (1983) fit two first-order functions to curves depicting the metabolism of 2,4-D in soil, but the second rate was faster than the first, possibly a result of the growth of the 2,4-D-metabolizing population. The metabolism of dodecane by the fungus *Cladosporium resinae* appears to occur in two linear phases (Lindley and Heydeman, 1986).

Diffusion may also control the rate of biodegradation at high concentrations of the organic chemical. The rate of diffusion of O_2 or inorganic nutrients may be limiting, and such limitations may be especially prominent for bacteria growing in microcolonies (Brunner and Focht, 1984). The rate of diffusion of toxic products away from the active organisms may also control their growth and metabolism.

It is difficult to study the effect of diffusion on biodegradation in natural soil given the difficulties in eliminating other potential variables in such a complex system. However, in a defined system in which *Pseudomonas* sp. metabolized glutamate in a gel exclusion–bead matrix, evidence of a role for diffusion in controlling biodegradation was obtained. In this instance, increasing the volume of solution retained inside the beads, which excludes bacteria but not substrate, results in increasingly slower initial rates of mineralization, lower final extents of mineralization, and greater rates in the second, or tail, phase (Scow and Alexander, 1992). Such a defined system enables explicit definition mathematically of the physical and biological processes involved and independent determination, experimentally, of the rates of transfer and degradation.

Two-compartment kinetics also apply to certain circumstances in which products accumulate for some time before they are converted to CO_2. In tests for biodegradation involving determinations of the formation of $^{14}CO_2$ from ^{14}C-labeled substrates, the labeled substrate is essentially one compartment, and the labeled product that temporarily accumulates is the second. Such kinetics also might be evident when two different populations carry out a process that has two separate steps in conversion of the parent chemical to CO_2 (Scow *et al.*, 1986; Simkins *et al.*, 1986).

METABOLISM OF ONE SUBSTRATE DURING GROWTH ON ANOTHER

In the three growth models presented in the preceding section, the cells are multiplying at the expense of the chemical whose biodegradation is being determined. However, the bacteria may be growing at the expense of a different organic compound. The compound whose biodegradation is being measured may still be a substrate, but it is not contributing substantially to the C supply that the cell is using to make more biomass. Such metabolism without providing C to the metabolizing organisms may occur because the concentration of the substance of interest is below the threshold needed to sustain growth or because it is only acted upon by cometabolism.

Mathematical formulations have been developed for the kinetics of biodegradation of one organic chemical when the transformations reflect both the metabolism of that substrate and the simultaneous growth of bacteria on a second compound. The formulations are based on coupling of Monod growth kinetics and Michaelis–Menten kinetics, which were presented earlier. Nine models have thus been advanced, the nine reflecting linear, logistic, and exponential growth on one substrate and concentrations of the second substrate (whose biodegradation is of interest) that are below, at about, or much above K_m (Table 6.3). The models are arbitrarily numbered I to IX, and differential and integral forms of the equations for each have been published (Schmidt et al., 1985). The shapes of the curves depicting the kinetics of disappearance are presented in Fig. 6.5.

To illustrate the applicability of such models, consider the case of the transformation of a low concentration of a chemical that does not support growth but is acted on by a reasonably large number of cells; its disappearance will be first order. However, if a second organic compound is present at a level that does support growth, the metabolism of the first molecule will reflect both the kinetics of growth on this second compound as well as the kinetics that would normally apply to the enzyme system catalyzing the metabolism of the first.

Tests have been conducted to determine the applicability of the models of Table 6.3. Thus, the breakdown of low concentrations of phenol or glucose by two bacteria growing on other C sources is best fit by Models I and IV (Schmidt et al., 1985), and degradation of 4-nitrophenol by a strain of *Pseudomonas* in the presence of glucose is best fit by Model V (Schmidt et al., 1987). The kinetics of mineralization in soil of 5 μg of 4-nitrophenol per kilogram in the presence of increasing concentrations of phenol change from nongrowth kinetics to kinetics that reflect growth of the 4-nitrophenol-mineralizing population on phenol (Scow et al., 1989).

TABLE 6.3
Models for the Kinetics of Biodegradation of Substrates That Do Not Support Growth But Are Metabolized by Populations Growing on Other Organic Substrates

Type of growth	Concentration of test compound		
	$S \ll K_m$	$S \sim K_m$	$S \gg K_m$
Logistic	I	II	III
Logarithmic	IV	V	VI
Linear	VII	VIII	IX

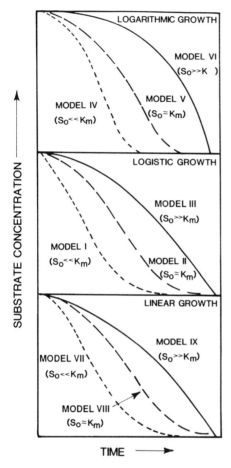

FIG. 6.5 Shapes of substrate disappearance curves for nine kinetics models based on growth on one organic substrate and simultaneous metabolism of a second (From Schmidt *et al.*, 1985. Reprinted with permission from the American Society for Microbiology.)

Model IV best fits the curve of 4-nitrophenol mineralization in the presence of 5 μg of phenol per kilogram, and Model I provides the best fit in the presence of 10 and 250 mg of phenol per kilogram of soil. Bacterial counts confirm that the population of microorganisms capable of using 4-nitrophenol does indeed multiply. Model IV also gives good fits to the biodegradation of 0.2–20 μg of phenol per liter and comparable levels of 4-nitrophenol in lake water, and Model I also gives good fits to the mineralization of carbonyl-labeled carbofuran (Scow *et al.*, 1990b) and of 1.0 and 7.0 μg of phenol per liter of lake water (Jones and Alexander, 1986). The biodegradation of sodium dodecyl sulfate in 16 of 19 samples of river water is also best fit by Model IV (Anderson *et al.*, 1990).

The models in Table 6.3 are probably appropriate for compounds acted on by cometabolism, although the possibility has not yet been evaluated experimentally. On the other hand, a different model was used to describe the kinetics of cometabolism of trichloroethylene by methane-oxidizing bacteria in the absence of methane (Alvarez-Cohen and McCarty, 1991a), and a somewhat different model has been developed by Criddle (1993). Competitive inhibition was considered in a model for the cometabolism of chlorinated compounds by methane-oxidizing bacteria (Alvarez-Cohen and McCarty, 1991b). In some instances, even first- or zero-order kinetics may give good approximations of the kinetics, as in the transformation of propachlor in fresh water and sewage (Novick and Alexander, 1985). It is unlikely that the three models in the right-hand column of Table 6.3 (Models III, VI, and IX) would be realistic for test compounds that could support growth because (a) if S_0 is sufficiently high (the models consider $S_0 \gg K_m$), the organisms would then multiply on the test chemical, or (b) one of two compounds that both support growth would probably repress metabolism of the second; in both cases, these formulations of kinetics would not apply.

THREE-HALF-ORDER KINETICS

Brunner and Focht (1984) proposed the three-half-order model for the mineralization of organic compounds in soil. There are two forms of the model, one assuming linear growth and the other exponential growth. The linear growth form of the three-half-order model can be used when little or no growth occurs. Those kinetics can be represented as

$$P = S_0\{1 - \exp[-k_1 t - (k_2 t^2)/2]\} + k_0 t, \qquad (26)$$

where P is concentration of product, k_1 is a proportionality constant per unit time, k_2 is a constant in units of reciprocal time squared, and k_0 is a zero-order rate constant. The exponential growth model has the form

$$P = S_0\{1 - \exp[k_1 t - \frac{E_0}{\mu}(\exp(\mu t) - 1)]\} + k_0 t. \qquad (27)$$

Some of the advantages of the three-half-order model are that it fits a set of data containing an acclimation phase, it can be applied to various microbial growth conditions, and it can fit the slow phase of mineralization often observed in the latter portion of mineralization curves. The linear growth form of the three-half-order model has been fit to curves of biphenyl mineralization in soil (Focht and Brunner, 1985) and to curves of mineralization of low concentrations of 4-nitrophenol and benzylamine (Scow *et al.*, 1986). The exponential form of the model has been fit to curves of CO_2 evolution from glucose that was added to soil that was gamma-irradiated to reduce the population density (Brunner and Focht, 1984).

KINETICS OF FUNGAL PROCESSES

A virtually unexplored area in biodegradation kinetics is the contribution of fungi to the rate of metabolism of organic compounds. In many soils, fungi appear to be more important than bacteria in community respiration (Anderson and Domsch, 1973). On the other hand, the theory underlying growth-linked kinetics of biodegradation is derived from the expectations for multiplication of bacteria. To the extent that the biomass of fungi may increase logarithmically, the kinetics of their metabolism may be assumed to resemble those of bacteria. However, the biomass of fungi increases through hyphal lengthening and branching (not by binary fission) and the organisms undergo morphological changes during the life cycle, so there is no reason to expect that Monod kinetics would be applicable to fungi. Logarithmatic kinetics have been reported to describe the unrestricted growth of fungi in liquid culture (Righelato, 1975). Cubic kinetics, in which a plot of the cube root of the dry weight or respiratory activity of fungi versus time yields a straight line, have been observed in shake cultures for a number of species of fungi and actinomycetes (Marshall and Alexander, 1960). This pattern of biomass increase probably reflects the fact that the growth of many species is largely restricted to the end of the filaments and those filaments develop linearly at a constant rate. If this linear extension then takes place in three dimensions, the hyphal mass of fungi developing as pellets may be viewed as spheres that have radii that increase at a constant rate. For the many fungi that do not form such pellets in mixed, liquid culture, cube-root kinetics presumably do not apply (Mandels, 1965), and they would not apply once unrestricted growth comes to an end. Cubic kinetics may also characterize increases in the biomass of other filamentous microorganisms, such as some actinomycetes (Marshall and Alexander, 1960). The growth rate on the surface of agar media, on the other hand, is linear (Trinci, 1970), and it is likely that in a three-dimensional porous matrix such as soil, the kinetics would be different yet.

Direct tests of the kinetics of mineralization in sand or liquid culture of low concentrations of phenol by a strain of *Penicillium* and glucose by species of *Fusarium* and *Rhizoctonia* showed that, of the models of the Monod family, the best fit was obtained with the logistic model (Scow *et al.*, 1990a).

OVERVIEW

A problem in kinetics analysis is to distinguish among models when the theoretical curves depicting the pattern of decomposition are quite similar. It is not always possible to distinguish among models, even with nonlinear regression techniques, so that the final choices among models are arbitrary. This is especially true for investigations yielding data that are not highly precise or do not result in many points. In instances in which models cannot be distinguished, it is critical to evaluate the models by comparing estimated parameters to independent experimental measurements (Schmidt *et al.*, 1987).

The models formulated to date often are based on sound microbiological and biochemical principles, but they rely on concepts derived from studies of single populations or single enzymes. The fact that the models often adequately characterize the data may be a reflection that a single species dominates the process or that the rate is governed by the kinetics of a single enzymatic step. Should several species be involved, predators or parasites act on the biodegrading species, or the species carrying out the transformation require the activities of neighboring populations, more complex models may be necessary. Also, if the substrate is insoluble or retained by abiotic components of the environment, physical and chemical processes may have to be considered in the models. Given the array of chemicals, the complexity of some environments, and the variety of microorganisms that may bring about biodegradation, it is unlikely that a single model or equation would be useful for the description of rates of biodegradation of all organic substrates in all environments.

REFERENCES

Alexander, M., *Environ. Sci. Technol.* **19**, 106–111 (1985).
Alexander, M., and Scow, K. M., *in* "Reactions and Movement of Organic Chemicals in Soils" (B. L. Sawhney and K. Brown, eds.), pp. 243–269. Soil Science Society of America, Madison, WI, 1989.
Alvarez-Cohen, L., and McCarty, P. L., *Environ. Sci. Technol.* **25**, 1381–1387 (1991a).
Alvarez-Cohen, L., and McCarty, P. L., *Appl. Environ. Microbiol.* **57**, 1031–1037 (1991b).
Anderson, D. J., Day, M. J., Russell, N. J., and White, G. F., *Appl. Environ. Microbiol.* **56**, 758–763 (1990).

Anderson, J. P. E., and Domsch, K. H., *Arch. Mikrobiol.* **93,** 113–127 (1973).

Archer, D. B., *Appl. Environ. Microbiol.* **50,** 1233–1237 (1985).

Arvin, E., *Water Res.* **25,** 873–881 (1991).

Badawy, M. I., and El-Dib, M. A., *Bull. Environ. Contam. Toxicol.* **33,** 40–49 (1984).

Bartholomew, G. W., and Pfaender, F. K., *Appl. Environ. Microbiol.* **45,** 103–109 (1983).

Bazin, M. J., Saunders, P. T., and Prosser, J. I., *CRC Crit. Rev. Microbiol.* **4,** 463–498 (1976).

Boethling, R. S., and Alexander, M., *Environ. Sci. Technol.* **13,** 989–991 (1979).

Bouchard, D. C., Lavy, T. L., and Lawson, E. R., *J. Environ. Qual.* **14,** 229–233 (1985).

Brown, D. E., Gaddum, R. N., and McEvoy, A., *Biotechnol. Lett.* **10,** 525–530 (1988).

Brunner, W., and Focht, D. D., *Appl. Environ. Microbiol.* **47,** 167–172 (1984).

Button, D. K., Schell, D. M., and Robertson, B. R., *Appl. Environ. Microbiol.* **41,** 936–941 (1981).

Cameron, D. R., and Klute, A., *Water Resour. Res.* **13,** 183–188 (1977).

Cary, J. W., and Hayden, C. W., *Geoderma* **9,** 249–256 (1973).

Casida, L. E., Jr., *Appl. Environ. Microbiol.* **21,** 1040–1045 (1971).

Criddle, C. S., *Biotechnol. Bioeng.* **41,** 1048–1056 (1993).

Davidson, J. M., and Chang, R. K., *Soil Sci. Soc. Am. Proc.* **36,** 257–261 (1972).

Falco, J. W., Sampson, K. T., and Carsel, R. F., *Dev. Ind. Microbiol.* **18,** 193–202 (1977).

Focht, D. D., and Brunner, W., *Appl. Environ. Microbiol.* **50,** 1058–1063 (1985).

Furmidge, C. G. L., and Osgerby, J. M., *J. Sci. Food Agric.* **18,** 269–273 (1967).

Gaudy, A. F., Jr., Komolrit, K., and Bhatla, M. N., *J. Water Pollut. Control Fed.* **35,** 903–922 (1963).

Gillie, O. J., *J. Gen. Microbiol.* **51,** 179–184 (1968).

Goswami, P. C., Singh, H. D., Bhagat, S. D., and Baruah, J. N., *Biotechnol. Bioeng.* **25,** 2929–2943 (1983).

Hamaker, J. W., *in* "Organic Pesticides in the Environment" (A. A. Rosen and H. F. Kraybill, eds.), pp. 122–131. American Chemical Society, Washington, DC, 1966.

Hamaker, J. W., *in* "Organic Chemicals in the Soil Environment" (C. A. I. Goring and J. W. Hamaker, eds.), pp. 253–340. Dekker, New York, 1972.

Hamaker, J. W., and Goring, C. A. I., *in* "Bound and Conjugated Pesticide Residues" (D. D. Kaufman, G. G. Still, D. D. Paulson, and S. K. Bandal, eds.), pp. 219–243. American Chemical Society, Washington, DC, 1976.

Hamaker, J. W., Youngson, C. R., and Goring, C. A. I., *Weed Res.* **8,** 46–57 (1968).

Hance, R. J., and McKone, C. E., *Pestic. Sci.* **2,** 31–34 (1971).

Helweg, A., *Weed Res.* **15,** 53–58 (1975).

Hemmett, R. B., Jr., and Faust, S. D., *Residue Rev.* **29,** 191–207 (1969).

Hurle, K., and Pfefferkorn, V., *Proc. Br. Weed Control Conf., 11th, 1972,* Vol. 2, pp. 806–810 (1972).

Hyzack, D. L., and Zimdahl, R. L., *Weed Sci.* **22,** 75–79 (1974).

Ishida, Y., Imai, I., Miyagaki, T., and Kadota, H., *Microb. Ecol.* **8,** 23–32 (1982).

Jones, S. H., and Alexander, M., *Appl. Environ. Microbiol.* **51,** 891–897 (1986).

Karickhoff, S. W., *in* "Contaminants and Sediments" (R. A. Baker, ed.), Vol. 2, pp. 193–205. Ann Arbor Sci. Publ., Ann Arbor, MI, 1980.

Kilbertus, G., *Rev. Ecol. Biol. Sol* **17,** 543–557 (1980).

Klecka, G. M., and Maier, W. J., *Appl. Environ. Microbiol.* **49,** 46–53 (1985).

Koch, A. L., and Wong, C. H., *Arch. Microbiol.* **131,** 36–42 (1982).

Kunc, F., and Rybarova, J., *Soil Biol. Biochem.* **15,** 141–144 (1983).

Larson, R. J., *in* "Biotransformation and Fate of Chemicals in the Aquatic Environment" (A. W. Maki, K. L. Dickson, and J. Cairns, Jr., eds.), pp. 67–86. American Society for Microbiology, Washington, DC, 1980.

Lewis, D. L., and Holm, H. W., *Appl. Environ. Microbiol.* **42,** 698–703 (1981).

Lewis, D. L., Kollig, H. P., and Hall, T. L., *Appl. Environ. Microbiol.* **46,** 146–151 (1983).

Lewis, D. L., Hodson, R. E., and Hwang, H.-M., *Appl. Environ. Microbiol.* **54,** 2054–2057 (1988).

Lindley, N. D., and Heydeman, M. T., *Appl. Microbiol. Biotechnol.* **23,** 384–388 (1986).

Lonsane, B. K., Singh, H. D., Nigam, J. N., and Baruah, J. N., *Indian J. Exp. Biol.* **17,** 1263–1264 (1979).

Mandels, G. R., in "The Fungi" (G. C. Ainsworth and A. S. Sussman, eds.), Vol. 1, pp. 599–612. Academic Press, New York, 1965.

Marshall, K. C., and Alexander, M., *J. Bacteriol.* **80,** 412–416 (1960).

Mateles, R. I., and Chian, S. K., *Environ. Sci. Technol.* **3,** 569–574 (1969).

McCall, P. J., and Agin, G. L., *Environ. Toxicol. Chem.* **4,** 37–44 (1985).

Meikle, R. W., Youngson, C. R., Hedlund, R. T., Goring, C. A. I., Hamaker, J. W., and Addington, W. W., *Weed Sci.* **21,** 549–555 (1973).

Mihelcic, J. M., and Luthy, R. G., *Pap., Int. Conf. Physiochem. Biol. Detox. Hazard. Wastes,* Vol. 2. pp. 708–721. Technomic Publishing Co., Lancaster, PA, 1988.

Miller, M. E., and Alexander, M., *Environ. Sci. Technol.* **25,** 240–245 (1991).

Monod, J., *Annu. Rev. Microbiol.* **3,** 371–394 (1949).

Montgomery, M., Yu, T. C., and Freed, V. H., *Weed Res.* **12,** 31–36 (1972).

Neidhardt, F. C., Ingraham, J. L., and Schaechter, M., "Physiology of the Bacterial Cell." Sinauer Associates, Sunderland, MA, 1990.

Novick, N. J., and Alexander, M., *Appl. Environ. Microbiol.* **49,** 737–743 (1985).

Odum, E. P., "Fundamentals of Ecology." Saunders, Philadelphia (1971).

Paris, D. F., and Rogers, J. E., *Appl. Environ. Microbiol.* **51,** 221–225 (1986).

Paris, D. F., Steen, W. C., Baughman, G. L., and Barnett, J. T., Jr., *Appl. Environ. Microbiol.* **41,** 603–609 (1981).

Parker, L. W., and Doxtader, K. G., *J. Environ. Qual.* **12,** 553–558 (1983).

Pfaender, F. K., Shimp, R. J., and Larson, R. J., *Environ. Toxicol. Chem.* **4,** 587–593 (1985).

Pignatello, J. J., Sawhney, B. L, and Frink, C. R., *Science* **236,** 898 (1987).

Righelato, R. C., in "The Filamentous Fungi" (J. E. Smith and D. R. Berry, eds.), Vol. 1, pp. 79–103. Edward Arnold, London, 1975.

Rijnaarts, H. H. M., Bachmann, A., Jumelet, J. C., and Zehnader, A. J. B., *Environ. Sci. Technol.* **24,** 1349–1354 (1990).

Rovira, A. D., and Greacen, E. L., *Aust. J. Agric. Res.* **8,** 659–673 (1957).

Rozich, A. F., Gaudy, A. F., Jr., and D'Adamo, P. C., *Water Res.* **19,** 481–490 (1985).

Russell, J. B., and Baldwin, R. L., *Appl. Environ. Microbiol.* **37,** 531–536 (1979).

Schmidt, S. K., Simkins, S., and Alexander, M., *Appl. Environ. Microbiol.* **50,** 323–331 (1985).

Schmidt, S. K., Scow, K. M., and Alexander, M., *Appl. Environ. Microbiol.* **53,** 2617–2623 (1987).

Scow, K. M., and Alexander, M., *Soil Sci. Soc. Am. J.* **56,** 128–134 (1992).

Scow, K. M., Simkins, S., and Alexander, M., *Appl. Environ. Microbiol.* **51,** 1028–1035 (1986).

Scow, K. M., Schmidt, S. K., and Alexander, M., *Soil Biol. Biochem.* **21,** 703–708 (1989).

Scow, K. M., Li, D., Manilal, V., and Alexander, M., *Mycol. Res.* **94,** 793–798 (1990a).

Scow, K. M., Merica, R. R., and Alexander, M., *J. Agric. Food Chem.* **38,** 908–912 (1990b).

Shehata, T. E., and Marr, A. G., *J. Bacteriol.* **107,** 210–216 (1971).

Simkins, S., and Alexander, M., *Appl. Environ. Microbiol.* **47,** 1299–1306 (1984).

Simkins, S., Mukherjee, R., and Alexander, M., *Appl. Environ. Microbiol.* **51,** 1153–1160 (1986).

Steinberg, S. M., Pignatello, J. J., and Sawhney, B. L., *Environ. Sci. Technol.* **21,** 1201–1208 (1987).

Stucki, G., and Alexander, M., *Appl. Environ. Microbiol.* **53,** 292–297 (1987).

Subba-Rao, R. V., Rubin, H. E., and Alexander, M., *Appl. Environ. Microbiol.* **43,** 1139–1150 (1982).

Suflita, J. M., Robinson, J. A., and Tiedje, J. M., *Appl. Environ. Microbiol.* **45,** 1466–1473 (1983).

Thomas, J. M., Yordy, J. R., Amador, J. A., and Alexander, M., *Appl. Environ. Microbiol.* **52,** 290–296 (1986).

Torstensson, N. T. L., and Aamisepp, A., *Weed Res.* **17,** 209–212 (1977).

Trinci, A. P. J., *Arch. Mikrobiol.* **73,** 353–367 (1970).

van der Kooij, D., and Hijnen, W. A. M., *Appl. Environ. Microbiol.* **41,** 216–221 (1981).

van Genuchten, M. T., Davidson, J. M., and Wierenga, P. J., *Soil Sci. Soc. Am. Proc.* **38,** 29–35 (1974).

Vashon, R. D., and Schwab, B. S., *Environ. Sci. Technol.* **16,** 433–436 (1982).

Volk, W. A., and Myrvik, Q. N., *J. Bacteriol.* **66,** 386–388 (1953).

Volkering, F., Breure, A. M., Sterkenburg, A., and van Endel, J. G., *Appl. Microbiol. Biotechnol.* **36,** 548–552 (1992).

Walker, A., and Brown, P. A., *Bull. Environ. Contam. Toxicol.* **30,** 365–372 (1983).

Wolfe, N. L., Kitchens, B. E., Macalady, D. L., and Grundl, T. J., *Environ. Toxicol. Chem.* **5,** 1019–1026 (1986).

Yordy, J. R., and Alexander, M., *Appl. Environ. Microbiol.* **39,** 559–565 (1980).

Yoshida, F., and Yamane, T., *Biotechnol. Bioeng.* **13,** 691–695 (1971).

Zimdahl, R. L., and Gwynn, S. M., *Weed Sci.* **25,** 247–251 (1977).

7 THRESHOLD

Organic pollutants in many surface and groundwaters, soils, and sediments are present at low concentrations. Even at these trace levels, they may be of concern. Among the reasons that they are of practical importance are the following: (a) Risk analyses suggest that many of the chronic toxicants will be injurious to a small portion of the human population consuming waters or foods containing them. Chronic toxicants include a diversity of carcinogens, mutagens, and teratogens. (b) Some of the chemicals at these low concentrations (e.g., micrograms-per-liter levels) are acutely toxic to aquatic organisms. (c) Some are subject to bioconcentration within tissues of organisms in natural food chains and ultimately reach levels that are injurious to species at higher trophic levels in these food chains. (d) Regulatory agencies of national or local governments have established levels of many chemicals that are deemed to be safe, especially for public health, and the concentrations given by these regulatory guidelines or standards are often quite low. The standards for drinking water are often set based on the risk analyses.

The public health and ecological concerns with low chemical concentrations have fostered interest in the biodegradative processes affecting trace concentrations of organic chemicals. In the past, microbiologists have not paid attention to the problem because it was deemed far easier to grow organisms at the high substrate concentrations that would give large cell yields. However, as the interest grew, previously unanticipated phenomena became apparent. One such phenomenon is the existence of a threshold, or, more specifically, a concentration of a nutrient source below which microorganisms cannot grow.

To maintain its viability, every organism must expend energy. In animals and humans, the energy used is reflected in basal metabolism. In microorganisms, the amount of energy to permit the organism to remain alive is designated *maintenance energy*. For heterotrophs, this energy is derived from the oxidation of organic compounds. When the concentration of the carbon source for growth is high, diffusion of the substrate from solution

102

to the cell surface and the subsequent transfer of the molecule across the surface into the cell provide enough of the substrate to satisfy the needs for maintenance energy and for processes that lead to increases in cell size, growth, and multiplication. The same is not the case at low substrate levels. Considering only diffusion of the molecule from the liquid to the cell surface (and ignoring transfer across the membrane, which cannot exceed the rate that molecules reach the surface of the organism), as a low substrate concentration is reduced to a still lower level, the energy for maintenance represents an ever higher percentage of substrate-C that reaches the organism by diffusion, and an ever smaller percentage is used for growth and multiplication. At some lower value, all the energy in the form of C that reaches (and enters) the cell is used simply to keep the cell alive, and none is used for growth. At this concentration, therefore, although the substrate is being metabolized, the cells are not growing, and the population size and biomass are not increasing. This concentration represents the threshold.

Moreover, if the population size initially is so small that biodegradation is inconsequential, undetectable, or both, that absence of multiplication will be reflected in the absence of significant or detectable biodegradation, even though the organisms are metabolizing part of the substrate pool to maintain themselves. The threshold is the lowest concentration that sustains growth and represents the level below which a species (that needs to proliferate to cause a detectable change) brings about little or no chemical destruction.

The possible existence of a threshold was first postulated because of the presence of relatively constant levels of dissolved organic C in the oceans. This C, presumably because of its low concentration, was not available to support microbial proliferation and hence mineralization of the C (Jannasch, 1967). The level of such dissolved organic C is approximately 1 mg/liter in marine waters and is commonly less than 5 mg/liter in oligotrophic fresh waters. Moreover, if significant decomposition of this organic matter were occurring, the concentration should fall at increasing distances away from the water's surface, where the organic matter is being generated photosynthetically by the phytoplankton. Because no such marked decline is evident with depth, it was hypothesized that mineralization must be slow. However, this line of evidence in support of the existence of a threshold for growth is weak because (a) much of the organic matter, when concentrated, is intrinsically resistant to microbial degradation (Barber, 1968) and (b) the concentration of some aquatic constituents may represent a steady state, that is, a balance between the continuous formation and continuous mineralization. For example, the concentration of individual amino acids at a site in the Pacific Ocean ranges from less than 0.05 to 3 µg/liter; however, these amino acids are continuously being destroyed, so that they also must be constantly formed to maintain the quantities that are found (Williams *et al.*, 1976).

More convincing evidence has come from studies of biodegradable synthetic compounds in waters and soils. Because these compounds are not formed biologically, their presence at reasonably constant levels or their persistence at low levels indicates that the biodegradation one might expect is not occurring. These studies indicate that no biodegradation occurs in the test period below a certain concentration or that the rate is less than that which might be expected from the rates observed at higher levels (if it is assumed that the rate is proportional to concentration) (Fig. 7.1). Typical data for fresh waters and soils are shown in Table 7.1. It is evident that the threshold is sometimes very low and sometimes reasonably high. Nevertheless, the data suggest that the threshold is often at about 0.1 to 5 µg per liter of water or per kilogram of soil. In instances in which water

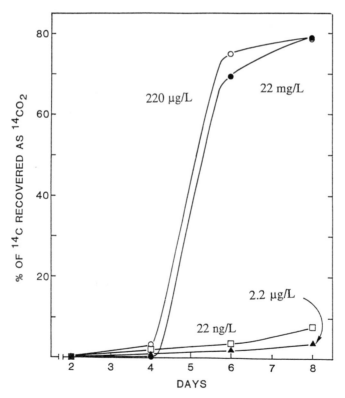

FIG. 7.1 Mineralization of 2,4-D added at several concentrations to river water. (From Boethling and Alexander, 1979a. Reprinted with permission from the American Society for Microbiology.)

ABLE 7.1

hemical Concentrations at Which Biodegradation Does Not Occur or Is Slower
han Predicted

Chemical	Environmental source	Concentration (μg per liter of water or kg of soil)	Reference
4-D	Stream	2.2	Boethling and Alexander (1979a)
vin	Stream	3.0	Boethling and Alexander (1979a)
niline	Lake	0.1	Hoover et al. (1986)
Nitrophenol	Lake	1.0	Hoover et al. (1986)
4-Dichlorophenol	Lake	2.0	Hoover et al. (1986)
yrene	Lake	2.5	Fu and Alexander (1992)
enol	Lake	0.0015	Rubin and Alexander (1983)
arbofuran	Soil	10, 100[a]	Chapman et al. (1986)
4,5-T	Soil	100	McCall et al. (1981)
2-, 1,3-, and 1,4-Dichlorobenzenes	Biofilm on glass	0.2–7.1[b]	Bouwer and McCarty (1982)

oncentration that did not result in a population increase to rapidly destroy a second addition of
rbofuran.
oncentration in effluent from a column containing glass beads supporting microorganisms degrading
e chemicals.

containing the chemical is passing through a solid (e.g., glass beads or soil)
on which the microorganisms reside, the concentration for the apparent
threshold may be anomalous because the chemical may escape in the
liquid emerging from the bottom of the column of particles before all
the chemical, which is being slowly destroyed at these trace levels, is
degraded.

The experimental values recorded in Table 7.1 are not surprising in
view of monitoring data obtained from analyses of samples taken from
natural ecosystems. For example, in natural waters of Canada receiving
NTA, the average level of this chelating agent was 5 μg/liter (International
Joint Commission, 1978). Certain nonylphenoxycarboxylic acids persist
in river waters at concentrations of 2 to 116 μg/liter (Ahel et al., 1987).
Similarly, a great variety of synthetic chemicals are present in surface
waters at low concentrations (Meijers and van der Leer, 1976), and groundwater accidentally contaminated with 2,4-D and 2,4-dichlorophenol still
showed their presence at low levels years after an inadvertent release
(Faust and Aly, 1964).

Analogous observations have been made when wastewaters are passed
through soil as a means of destroying a harmful chemical by microbial

action. In experimental trials, the concentrations of many compounds fell
to undetectable levels as solutions containing them passed through soil
columns, but a reasonable percentage of the 1,2-, 1,3-, and 1,4-dichloro-
benzenes and diisobutyl phthalate in the influent water was still present
in the effluent, and a readily biodegradable molecule like di(2-ethylhexyl)
phthalate at 70 ng/liter did not disappear at all as a result of passage
through soil (Bouwer et al., 1981). Benzophenone and diethyl and dibutyl
phthalate also have been reported to not disappear when passed at low
concentration through soil columns set up to simulate the rapid infiltration
of contaminated waters through soil, and waters moving out and away
from land-infiltration sites in the field have been found to contain 0.02 μg
of toluene, 0.05 to 1.14 μg of xylenes, 0.07 to 0.50 μg of naphthalene,
0.05 to 2.1 μg of benzophenone, and 0.01 to 2.4 μg of individual phthalate
esters per liter (Hutchins et al., 1983). Each of these chemicals is mineralized at higher concentrations. Similarly, passing a solution of 1,2-dichloro-
benzene through a column of sand reduced the concentration from 25 μg/
liter but only to a concentration of 0.1 μg/liter (van der Meer et al., 1987).
In a plume of contaminated water derived from secondary sewage effluent
subjected to rapid infiltration, a number of compounds were found to have
persisted at low concentration in the aquifer for more than 30 years; the
average concentrations in the groundwater were 20–70 ng of 2,3-dimethyl-
2-butanol, 2-methyl-2-hexanol, ethylbenzene, and propylbenzene isomers
(Barber et al., 1988), compounds that are probably all metabolized at
higher concentrations.

In some instances in which microbial colonization on glass beads is
promoted to give biofilms, the minimum concentration below which there
is no growth of the biofilm is quite high; for example, 100 to 1000 μg/liter
(Rittmann, 1985).

Determining the existence of a threshold concentration for growth of
bacteria in pure culture is complicated by the ability of many species to
grow in media to which no C source is deliberately added. The liquid or
inorganic salts used to formulate the medium, the air in the gas phase
above the liquid medium in the flask, or both typically contain sufficient
organic matter to support the multiplication of these species, which may
reach densities of 10^4 to 10^5 cells per milliliter in such allegedly C-free
media. Inasmuch as a population of 10^5 cells per milliliter would probably
consume 100 ng of substrate per milliliter (or 100 μg per liter) and the
threshold is usually below 100 μg/liter, it is difficult with such species to
show that they cannot grow in solutions with little added C; that is, the
growth is nearly entirely at the expense of the uncharacterized, contaminating substrates rather than the test compound. This experimental difficulty is a result of the artificial conditions in pure cultures because most

species in nature that actively destroy synthetic organic molecules are probably not effective competitors with their neighboring species for naturally occurring chemicals. Nevertheless, this procedural obstacle in pure culture can be overcome by using species that grow little, if at all, on the contaminating C.

Investigations of pure cultures of bacteria clearly show the existence of a threshold concentration of the C source below which multiplication does not occur. This value is about 18 μg/liter for *Escherichia coli* and *Pseudomonas* sp. growing on glucose (Shehata and Marr, 1971; Boethling and Alexander, 1979b), 180 μg/liter for *Aeromonas hydrophila* growing on starch (van der Kooij *et al.*, 1980), 210 μg/liter for a coryneform bacterium using glucose (Law and Button, 1977), approximately 300 μg/liter for a strain of *Pseudomonas* growing at the expense of 2,4-dichlorophenol (Goldstein *et al.*, 1985), about 5 μg/liter for *Salmonella typhimurium* provided with glucose (Schmidt and Alexander, 1985), and 2 μg/liter for a bacterium mineralizing quinoline (Brockman *et al.*, 1989). Such information as well as individual studies of a variety of marine bacteria, for which threshold concentrations of 0.15 to greater than 100 mg/liter were found (Jannasch, 1967), demonstrate that the threshold concentrations below which individual bacterial species are unable to multiply vary enormously. Some species have surprisingly high thresholds, but others are able to grow down to about 2 μg/liter but no lower. These values are of special significance for biodegradation if the population is initially small so that multiplication is essential for appreciable destruction of the substrate. Indeed, it has been noted that the indigenous population of 2,4-dinitrophenol-metabolizing bacteria in soil could not be maintained and would not multiply if the concentration was 0.1 mg/kg, although the bacteria multiplied at higher concentrations (Schmidt and Gier, 1989).

A model has been developed for estimating, on theoretical grounds, the threshold concentration of an organic compound required to support the multiplication of a bacterium. Below the value so calculated, the size of the population should not increase. The model is formulated on the basis of (a) the maximum rate that an organism can acquire energy at a particular concentration of substrate and (b) the rate it uses energy just to maintain its viability. It predicts that a threshold exists when the organisms' need for C to supply the energy for maintenance is just equal to the rate of diffusion of the chemical to the cell surface. Below this concentration, too little energy is available to the cell to allow it to be maintained, and thus it will die. The equation for the relationship is

$$\tau = \frac{1/Y_{max}(R_d^2 - R_b^2)/2}{D_{AB}C_b/\rho - (m/\ln 2)(R_d^2 - R_b^2)/2},$$

where τ is the maximum doubling time for the cells, Y_{max} is the yield coefficient, R_b and R_d are the radii of the cell at its first appearance and at the time of its cell division, respectively, D_{AB} is the diffusivity of the chemical, C_b is the chemical concentration, ρ is the dry weight density of the cell (i.e., dry weight divided by the volume of the cell), and m is the maintenance coefficient. Diffusion constants for most organic pollutants are about 10^{-5} cm²/sec. Common values found for bacteria are 0.55 g dry wt/g of substrate for Y_{max}, 0.31 g dry wt/cm³ for ρ, and 15 mg of substrate/g dry wt/h for m. Using such common values and assuming that the radius of the cell as it first appears after cell division is 0.50 μm (R_b) and that the radius before the next cell division is 0.63 μm (R_d), the maximum doubling times for the cells at 10, 1.0, 0.5, and 0.20 μg of substrate per liter are 1.71, 21, 57, and infinite hours, respectively, that is, the threshold for such cells would be 0.20 μg/liter (Schmidt et al., 1985). Obviously, however, the actual threshold must be somewhat higher than those suggested by the model because the cell needs energy to grow, not every molecule that reaches the cell surface passes through the membrane, and not every penetrating molecule is utilized. Nevertheless, the model does provide a minimum value. Moreover, because the requirements for maintenance energy differ appreciably among species, so too will the thresholds. Similar assumptions are used to calculate the threshold concentration below which a biofilm of bacteria would not be maintained (Rittmann and McCarty, 1980; Rittmann et al., 1980).

The threshold is lowered if the bacteria carrying out the transformation have certain alternative C sources available to them. In continuous culture of a marine bacterium, for example, the threshold concentration for glucose utilization was reported to be 0.48 mg/liter if the sugar was the sole C source, but it was lowered to 8 μg/liter in the presence of arginine and reduced even further in the presence of a mixture of amino acids (Law and Button, 1977). Similarly, the lower-than-predicted rate of mineralization of 0.39 to 1.5 ng of phenol per liter by a mixture of lake water bacteria was increased to the expected rate if much higher concentrations of arginine were added (Rubin and Alexander, 1983). Too few observations have been made to determine the frequency that the threshold can be changed by second C sources, and the effect of alternative organic substrates is likely to be expressed only when the specific population carrying out the biodegradation is able to compete effectively for the second nutrient in communities containing many other species.

A threshold may also exist for the acclimation of microbial communities. Thus, a freshwater microbial community became acclimated to the mineralization of 4-nitrophenol at levels above but not below 10 μg/liter (Spain and Van Veld, 1983); because such acclimation probably is merely an indication of the time for the cells to become sufficiently numerous to cause a detectable loss of the chemical, the threshold may merely reflect that which characterizes growth per se. On the other hand, the induction of metabolic activity in bacterial cells may have a threshold even in the absence of growth, witness the reported induction of 3- and 4-chlorobenzoate degradation by *Acinetobacter calcoaceticus* at concentrations above 160 μg/liter but not below (Reber, 1982).

The threshold phenomenon may not be restricted to C sources, and no growth may take place at concentrations of other nutrients below some threshold value, for example, P (Button, 1985). At this time, however, the occurrence of thresholds for other nutrients and their significance for biodegradation have scarcely been explored.

The fact that the biodegradation of some compounds, both in pure culture and in nature, does not occur below some measurable concentration does not mean that thresholds always exist—or at least at concentrations measurable by currently available methods. Quite the contrary: many organic chemicals are mineralized in natural environments (or in samples collected from these environments and tested in the laboratory) at levels below which organic substrates fail to support growth. In stream water, for example, glucose is mineralized at 1.8 ng/liter and dimethylamine and diethanolamine at less than 10 ng/liter (Boethling and Alexander, 1979b), and a linear alcohol ethoxylate is mineralized in estuarine water at 33 ng/liter (Larson *et al.*, 1983). In lake water, mineralization of benzylamine is evident at less than 1 ng/liter, of phenol at 0.10 ng/liter, of aniline at 5.7 ng/liter, of 2,4-D at 1.5 ng/liter, and of di(2-ethylhexyl) phthalate at 21 ng/liter (Rubin *et al.*, 1982; Subba-Rao *et al.*, 1982) (Fig. 7.2). In many of these instances, the rate, as expected, is directly correlated with concentration, so that the mineralization rate is 10-fold less at 10-fold lower levels. Similarly, mineralization of 0.32 μg of phenol, 0.30 μg of aniline, and 1.0 μg of 4-nitrophenol per kilogram occurs in soil (Scow *et al.*, 1986), and several compounds are mineralized in sediment at 0.5 μg/kg (Ursin, 1985). Mineralization at substrate levels below the threshold for replication may be carried out either by nongrowing cells or by cells growing at the expense of other organic compounds present at levels above the threshold. For nongrowing cells to cause a significant change in nature, however, their biomass must be large.

Many environments contain levels of organic C in excess of that needed to support growth, or the levels may be regenerated constantly by excretions of other organisms (e.g., phytoplankton) or by new additions. Under these conditions, the energy needs for maintenance and growth of the populations degrading the compounds of interest may be met by use of these other organic molecules. As will be discussed elsewhere, microor-

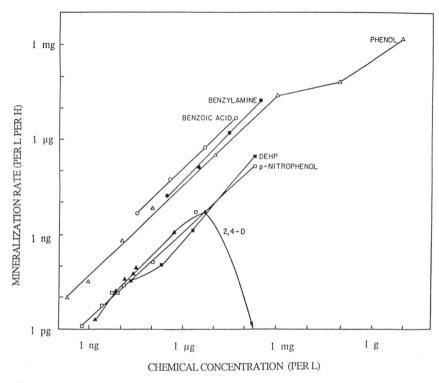

CHEMICAL CONCENTRATION (PER L)

FIG. 7.2 Rates of biodegradation in lake water of several organic compounds added at a wide range of concentrations. (From Rubin *et al.*, 1982. Reprinted with permission from the American Society for Microbiology.)

ganisms may metabolize two, or sometimes more, organic substrates simultaneously provided that their concentrations are not excessively high. The compound sustaining growth and that is present at levels above the threshold has been called the *primary substrate,* and the compound that is below the threshold but is still catabolized has been designated the *secondary substrate* (Rittmann, 1985). For example, *Salmonella typhimurium,* which has a threshold for growth on glucose of slightly below 5 µg/ liter, is able to destroy that sugar at 0.5 µg/liter if the bacterium is multiplying at the expense of arabinose that is initially present at 5.0 mg/liter (Fig. 7.3). Similarly, *Aeromonas hydrophila* destroys starch at concentrations too low to allow for proliferation of the bacterium if it is provided with glucose (van der Kooij *et al.,* 1980). A biofilm composed of a microbial mixture that colonized the surfaces of glass beads, in like fashion, is capable of destroying alanine at 30 µg/liter if the amino acid is a secondary

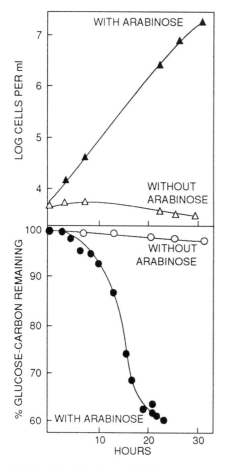

FIG. 7.3 Numbers of *Salmonella typhimurium* and mineralization of ¹⁴C-labeled glucose (0.5 µg/liter) in media with and without arabinose at 5 mg/liter. Only glucose mineralization was determined (From Schmidt and Alexander, 1985. Reprinted with permission from the American Society for Microbiology.)

substrate, although the threshold for maintenance of the biofilm is 200 µg/liter when it is the sole substrate (Namkung *et al.*, 1983). The microbial destruction of organic molecules present in trace amounts may, in fact, frequently occur because of the presence of primary substrates for the active species.

The apparent existence in natural waters and wastewaters, and possibly in other environments, of traces of potentially degradable organic pollutants may thus be attributable to the thresholds below which growth does not occur. A microorganism whose sole selective advantage in these environments is its ability to grow by using particular novel substrates therefore may not increase in abundance, and the substrate may then not disappear. Moreover, the fact that thresholds exist points to the danger of drawing conclusions about what will happen at low chemical concentrations in nature based on laboratory tests with solutions containing much higher concentrations of the substrate. Nevertheless, it is not presently possible to predict which biodegradable chemicals in what environments will persist because of the threshold and which will be destroyed because of the ability of the responsible populations to function at still lower levels of the substrate.

REFERENCES

Ahel, M., Conrad, T., and Giger, W., Environ. Sci. Technol. 21, 697–703 (1987).
Barber, L. B., II, Thurman, E. M., Schroeder, M. P., and LeBlanc, D. R., Environ. Sci. Technol. 22, 205–211 (1988).
Barber, R. T., Nature (London) 220, 274–275 (1968).
Boethling, R. S., and Alexander, M., Appl. Environ. Microbiol. 37, 1211–1216 (1979a).
Boethling, R. S., and Alexander, M., Environ. Sci. Technol. 13, 989–991 (1979b).
Bouwer, E. J., and McCarty, P. L., Environ. Sci. Technol. 16, 836–843 (1982).
Bouwer, E. J., McCarty, P. L., and Lance, J. C., Water Res. 15, 151–159 (1981).
Brockman, F. J., Denovan, B. A., Hicks, R. J., and Fredrickson, J. F., Appl. Environ. Microbiol. 55, 1029–1032 (1989).
Button, D. K., Microbiol. Rev. 49, 270–297 (1985).
Chapman, R. A., Harris, C. R., and Harris, C., J. Environ. Sci. Health, Part B B21, 125–141 (1986).
Faust, S. D., and Aly, O. M., J. Am. Water Works Assoc. 56, 267–279 (1964).
Fu, M. H., and Alexander, M., Environ. Sci. Technol. 26, 1540–1544 (1992).
Goldstein, R. M., Mallory, L. M., and Alexander, M., Appl. Environ. Microbiol. 50, 977–983 (1985).
Hoover, D. G., Borgonovi, G. E., Jones, S. H., and Alexander, M., Appl. Environ. Microbiol. 51, 226–232 (1986).
Hutchins, S. R., Tomson, M. B., and Ward, C. H., Environ. Toxicol. Chem. 2, 195–216 (1983).
International Joint Commission, 1978. Cited by J. M. Tiedje, in "Biotransformation and Fate of Chemicals in the Aquatic Environment" (A. W. Maki, K. L. Dickson, and J. Cairns, Jr., eds.), pp. 114–119. American Society for Microbiology, Washington, DC, 1980.
Jannasch, H. W., Limnol. Oceanogr. 12, 264–271 (1967).
Larson, R. J., Vashon, R. D., and Games, L. M., in "Biodeterioration 5" (T. A. Oxley and S. Barry, eds.), pp. 235–245. Wiley, Chichester, 1983.
Law, A. T., and Button, D. K., J. Bacteriol. 129, 115–123 (1977).
McCall, P. J., Vrona, S. A., and Kelley, S. S., J. Agric. Food Chem. 29, 100–107 (1981).
Meijers, A. P., and van der Leer, R. C., Water Res. 10, 597–604 (1976).
Namkung, E., Stratton, R. G., and Rittmann, B. E., J. Water Pollut. Control. Fed. 55, 1366–1372 (1983).

Reber, H. H., *Eur. J. Appl. Microbiol. Biotechnol.* **15,** 138–140 (1982).

Rittmann, B. E., *Sci. Total Environ.* **47,** 99–113 (1985).

Rittmann, B. E., and McCarty, P. L., *Biotechnol. Bioeng.* **22,** 2343–2357 (1980).

Rittmann, B. E., McCarty, P. L., and Roberts, P. V., *Ground Water* **18,** 236–243 (1980).

Rubin, H. E., and Alexander, M., *Environ. Sci. Technol.* **17,** 104–107 (1983).

Rubin, H. E., Subba-Rao, R. V., and Alexander, M., *Appl. Environ. Microbiol.* **43,** 1133–1138 (1982).

Schmidt, S. K., and Alexander, M., *Appl. Environ. Microbiol.* **49,** 822–827 (1985).

Schmidt, S. K., and Gier, M. K., *Microb. Ecol.* **18,** 285–296 (1989).

Schmidt, S. K., Alexander, M., and Schuler, M. L., *J. Theor. Biol.* **114,** 1–8 (1985).

Scow, K. M., Simkins, S., and Alexander, M., *Appl. Environ. Microbiol.* **51,** 1028–1035 (1986).

Shehata, T. E., and Marr, A. G., *J. Bacteriol.* **107,** 210–216 (1971).

Spain, J. C., and Van Veld, P. A., *Appl. Environ. Microbiol.* **45,** 428–435 (1983).

Subba-Rao, R. V., Rubin, H. E., and Alexander, M., *Appl. Environ. Microbiol.* **43,** 1139–1150 (1982).

Ursin, C., *Chemosphere* **14,** 1539–1550 (1985).

van der Kooij, D., Visser, A., and Hijnen, W. A. M., *Appl. Environ. Microbiol.* **39,** 1198–1204 (1980).

van der Meer, I. R., Roelofsen, W., Schraa, G., and Zehnder, A. J. B., *FEMS Microbiol. Ecol.* **45,** 333–341 (1987).

Williams, P. J. Le B., Berman, T., and Holm-Hansen, O., *Mar. Biol. (Berlin)* **35,** 41–47 (1976).

8 SORPTION

Some substances appear to be nonbiodegradable under all circumstances, for example, various synthetic polymers. Many compounds that are potentially subject to microbial attack, however, are not destroyed. It is thus essential to distinguish between a molecule that is biodegrad-ABLE and one that, under particular circumstances, is not biodegradED. The former term indicates a susceptibility to destruction, the latter describes what actually occurs in a particular set of conditions.

Several reasons can be suggested for the lack of biodegradation of a molecule that is biodegradable. (a) The concentration of toxic substances may be so high at the site that microbial proliferation and metabolism is precluded. (b) One or more nutrients needed for microbial growth are at levels too low to permit appreciable growth. (c) The substrate itself may be at a concentration too low to allow for replication of the organisms containing the catabolic enzymes. (d) The substrate may not be in a form that is readily available for the microorganisms. *Bioavailability* is of extreme importance because it frequently accounts for the persistence of compounds that are biodegradable and that might otherwise be assumed to be readily decomposed. It may also limit attempts to bioremediate polluted sites.

The unavailability of an organic molecule could result from its sorption to solids in the environment, its presence in nonaqueous phase liquids (NAPLs), or its entrapment within the physical matrix of the soil, sediment, or aquifer. These topics will be considered in detail.

The solid surfaces in many environments may dramatically affect the activity of indigenous microorganisms. These surfaces may alter the availability of organic chemicals, change the levels of various organic and inorganic nutrients, modify the pH or O_2 relationships, render inhibitors less toxic, retain microorganisms, or depress the activity of extracellular enzymes. The active surfaces may be clay minerals, the organic fraction (or humic substances) of soils or sediments, other complex carbonaceous matter, or sometimes amorphous Fe or Al oxides or hydroxides. The solid

114

surfaces often act by *adsorption,* which refers to the retention of solutes originally present in solution by the surfaces of a solid material. *Absorption* may also be prominent in certain circumstances, this term referring to the retention of the solute within the mass of the solid rather than on its surfaces. The term *sorption* is used to include both adsorption and absorption. The zone in which sorption occurs represents a microenvironment immediately adjacent to the solid material, but this microenvironment is so different from the surrounding solution and is so important that much attention has been given to its understanding.

A wide array of organic compounds are sorbed by constituents of soils, sediments, wastewaters, subsoil materials, and other natural ecosystems. Included in this array are amino compounds, organic phosphates and phosphonates, nitrogen heterocycles, alkylbenzenesulfonates, cationic surfactants, and certain high-molecular-weight materials, to mention only a few classes of compounds. Certain organic molecules are sorbed more to the clay minerals, and others are bound largely or entirely to the organic matter. Not only are many of the organic substrates sorbed but so too are many of the inorganic nutrients needed by microorganisms.

A number of factors influence sorption of organic compounds. These include the type and concentration of solutes in the surrounding solution, the type and quantity of clay minerals, the amount of organic matter in the soil or sediment, pH, temperature, and the specific compound involved. The type of cation that is saturating the clay (e.g., whether the clay is saturated with Fe, Ca, or H ions) and the exchange capacity and specific surface area of clays also are of importance. Many of the major processes of concern to sorption occur at the surfaces of the clay minerals and humic materials, and these may have large areas per unit of mass; for example, a gram of clay may have a surface area of 20 to 80 m^2.

CHEMISTRY OF SORPTION

Much attention has been given to the sorption of organic compounds both to the clay and to the organic matter of soils and sediments, and it is important to consider the chemistry of sorption by these major constituents of soils and aquatic sediments.

The major clay minerals are of two main types, one in which the Si and Al layers are assembled in a 1 : 1 ratio of Si and Al (–Si·Al·Si·Al·Si·Al–), the second in which the layers are in a 2 : 1 ratio (–Si·Al·Si·Si·Al·Si–). In a 1 : 1 clay, such as kaolinite, the layers are tightly held together. Such nonexpanding clays have smaller surface areas and lesser capacities for sorption than do the 2 : 1 clays. Molecules are adsorbed on the outer surfaces of the 1:1 clays. In a 2:1 clay such as montmorillonite, the lattice structure of the clay can expand, and such clays may sorb organic compounds on both the external and the internal surfaces. With these expanding clays, organic molecules, inorganic nutrients, and water may penetrate between the layers of the mineral crystal. The availability of both the outer and internal surfaces of some clay minerals for sorption is often of great importance.

Adsorption may involve physical or van der Waals forces, hydrogen bonding, ion exchange, or chemisorption. Large molecules may be retained on clay surfaces by hydrogen bonding, but for the low-molecular-weight organic compounds of importance as pollutants, of particular significance is ion exchange, in which an ion in solution of one type is exchanged for an ion of another type that is on the solid sorbing material. Clay minerals and colloidal organic materials have a net negative charge, and therefore they attract cations. A clay particle may have H, Ca, K, or Mg ions on its surface, but a positively charged organic molecule may displace another cation already on the surface of the clay and thus become retained by the mineral. Positively charged compounds may also be adsorbed to the organic fraction of soils and sediments since the humic substances also bear negative charges. The surfaces of both clay and humus colloids may retain organic cations by such means. Anionic organic molecules, in contrast, are generally repelled because of the negative charge on the surface (Morrill *et al.,* 1982). As a consequence, it is the molecules that are positively charged at pH values prevailing in nature that are chiefly retained by the negatively charged surfaces.

The capacity of clays to affect biodegradation differs according to clay type. This may be related in many instances to the *cation-exchange capacity* of the clays because the effect is related to the cationic properties of the substrate (or, for large molecules as substrates, the cationic properties of an extracellular enzyme). Montmorillonite, for example, frequently sorbs potential microbial substrates because of its high cation-exchange capacity and its expanding lattice structure. Many organic substrates can enter between the silicate sheets that make up this clay, and they thereby become protected. Typically, the effects of clay on sorption, if effects occur, are marked with montmorillonite and are less prominent with kaolinite and illite.

The organic fraction of soils and sediments is responsible for the sorption of many compounds, particularly those that are hydrophobic. Many polycyclic aromatic hydrocarbons and other nonpolar pollutants are sorbed chiefly by the native organic matter rather than the clay constituents of soils and sediments. The extent of this retention is directly correlated with the octanol–water partition coefficients, which is expressed as the

K_{ow} value (a measure of hydrophobicity of chemicals), and the percentage of organic C in the soil or sediment; the more organic matter present in the solid phase, the more the hydrophobic molecule is sorbed. Two views exist on how hydrophobic molecules are retained by the organic matter. One view maintains that the process is physical sorption by the organic matter, in which physical binding of the solute to the organic solids occurs (Calvet, 1989). The other view holds that the hydrophobic molecule exists in the organic matter because it diffuses and partitions into the solid organic matter, much as a hydrophobic compound will partition from aqueous solution into an organic solvent in which it is highly soluble, that is, the compound is within the physical matrix of the solid organic phase that is the sorbent (Chiou, 1989). One view is thus that the molecule is concentrated on the outer surface or within the pores of a solid, where it is sorbed by physical or chemical forces; the other is that the molecule is distributed in the entire volume of the organic matter. These two concepts have markedly different implications for the potential availability of organic molecules to microorganisms.

Ion exchange associated with the native organic matter of soil may account for adsorption of such cationic compounds as the herbicide paraquat, and also of compounds that, following protonation, acquire a positive charge (e.g., the triazine herbicides), and these positively charged molecules may be retained by negatively charged groups (e.g., $R-COO^-$) of the complex organic material. The retention of cationic organic compounds because of ion exchange associated with the native organic matter of soils is especially important at neutral to slightly alkaline pH.

Hence, the properties of the sorbent and the chemical of concern determine whether sorption to the inorganic or to the organic surfaces, if both are present, is more important. Cationic organic molecules may be sorbed to the cation-exchange sites of clay minerals, humic surfaces, or both. Anionic compounds are poorly sorbed by clay minerals, but they may be moderately retained by organic surfaces. The complex organic matter of soil and other environments, on the other hand, is often the chief sorbent for nonionic organic compounds.

In some instances in which organic compounds interact with the organic matter of soils or sediments, or even the soluble humic materials of natural waters, the interaction is not really sorption in the sense used for clays or even for the hydrophobic binding discussed earlier for nonionic organic compounds. Instead, the change involves the formation of stable linkages. These linkages may sometimes be covalent linkages between the low-molecular-weight chemical and the complex natural humic substances. The result is a new chemical species. These types of interactions are discussed in Chapter 10.

DIMINISHED AVAILABILITY OF
SORBED SUBSTRATES

Sorption of the organic substrates of microorganisms has a major impact on their growth and activity. The effect varies with the specific compound, the mechanism by which that compound is bound, the strength of retention if organic cations are the substrates, and the capacity of the microorganisms at the site to use the sorbed compound. The last factor is not well understood, but it is clear that some species are able to use sorbed compounds whereas others can metabolize the same molecules only when they are in aqueous solution.

In some instances, a biodegradable compound that is sorbed becomes completely resistant to microbial attack. For example, when all of the herbicide diquat is sorbed by montmorillonite clay, a mixture of soil microorganisms able to metabolize the organic cation free in solution no longer can mineralize the compound (Fig. 8.1). A complete inhibition as a result of sorption may also occur with compounds sorbed hydrophobically, as shown by the inability of microorganisms in some soils to metabolize certain polycyclic aromatic hydrocarbons (PAHs) that are sorbed (Weissenfels *et al.*, 1992). Similarly, EDB that is freshly added to soil and that can still be easily desorbed is readily metabolized, whereas the same compound that has been in the soil for many years and is scarcely desorbed is likewise almost entirely resistant to microbial degradation (Steinberg *et al.*, 1987).

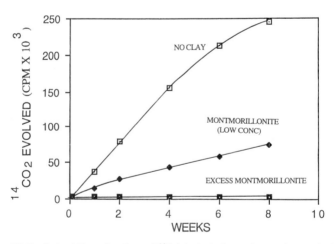

FIG. 8.1 Mineralization of ^{14}C-labeled diquat in nutrient solutions with or without clay. (Reprinted with permission from Weber and Coble, 1968.)

Reducing the availability of a molecule by sorption sometimes diminishes the rate or extent of transformation but does not wholly abolish the conversion. For example, dodecylbenzenesulfonates with linear or branched side chains are readily adsorbed by colloidal components of soil, and such binding reduces their degradation by a mixture of microorganisms (Inoue *et al.*, 1978). Similarly, the rate of mineralization of glyphosate in several soils declines as the amount of the herbicide bound by the soil colloids increases (Nomura and Hilton, 1977). Tests with diallate added to mixtures of activated charcoal and soil to give varying percentages of binding of the herbicide demonstrated again a direct relationship between the amount of chemical sorbed and the rate of degradation (Anderson, 1981).

The effect of clays in suppressing decomposition has been verified with many organic molecules. Thus, the extent of mineralization of low concentrations (20 ng to 200 μg per liter) of benzylamine in lake water is reduced by montmorillonite, apparently because the clay reduces the amount of substrate available to the lake water microflora by binding the aromatic amine (Subba-Rao and Alexander, 1982). Alkylamines such as *n*-heptyl-, *n*-octyl-, and *n*-decylamines are also protected from attack by amine-utilizing bacteria if the compounds are bound to montmorillonite (Wszolek and Alexander, 1979).

The role of sorption to the organic fraction, rather than to the clay constituents of soil, in reducing biodegradation is suggested by findings that herbicides known to be sorbed by soil organic matter are more persistent in soils rich in organic matter than those having low humus levels, for example, pyrazon (Smith and Meggitt, 1967), simazine, and atrazine (Briška *et al.*, 1974). The decrease with time in the rate of breakdown of N'-(4-chlorophenoxy)phenyl-N,N-dimethylurea in a humus-rich soil was attributed to the slow rate of desorption of this herbicide from the natural organic complexes (Geissbühler *et al.*, 1963). Humic acid, a major component of humic materials, may increase the acclimation period and also retard the mineralization of benzylamine, and the extent of benzylamine mineralization declines as the percentage of the amine initially bound increases (Amador and Alexander, 1988). In the latter instance, however, the binding of an amine to humic acid probably is not simple sorption, but rather involves the formation of a complex with stable chemical bonds.

Soils, clay, and organic matter also protect substrates against hydrolysis by individual enzymes; thus, although alkaline phosphatase hydrolyzes Guthion and parathion and acid phosphatase hydrolyzes parathion and pirimiphos-methyl, none is cleaved if these organophosphate insecticides are sorbed to soil (Heuer *et al.*, 1976). In addition, proteins that complex with soil organic matter are resistant to hydrolysis by a proteolytic enzyme of *Streptomyces griseus* (Burns *et al.*, 1972), and complexing of proteins with lignin makes them resistant to hydrolysis by chymotrypsin or by several bacteria (Estermann *et al.*, 1959). Proteins adsorbed on kaolinite or montmorillonite clays are protected to some degree from hydrolysis by chymotrypsin, although the sorbed proteins are still slowly attacked by the proteolytic enzymes (Ensminger and Gieseking, 1942; McLaren, 1954). Complexes of dextrans or hydroxyethylcellulose with montmorillonite are less readily attacked by microorganisms than are the free polysaccharides (Lynch and Cotnoir, 1956; Olness and Clapp, 1972); in these instances, microorganisms probably act on the polysaccharides by extracellular enzymes. Cationic compounds, such as paraquat, that enter into the clay lattice are resistant to biodegradation (Burns and Audus, 1970). Similarly, purines, amino acids, and peptides that enter the interlayer region of expanding lattice clays like montmorillonite may become protected from microbial attack (Greaves and Wilson, 1973), although extracellular enzymes excreted by microorganisms may slowly metabolize chemicals sorbed between lattices of such clays (Estermann *et al.*, 1959). It should be borne in mind, however, that low-molecular-weight compounds are generally believed to be degraded by intra- and not extracellular enzymes. Nevertheless, these studies have relevancy to approaches to bioremediation involving immobilized enzymes. It is highly likely that compounds degraded solely by intracellular enzymes would be resistant to biodegradation if located within the lattices of expanding-lattice clay minerals because of the inaccessibility of such chemicals to the microbial cell.

Several reasons can be advanced to explain the diminished rate of biodegradation because of the presence of surfaces. The diminished transformation often results from the fact that the substrate becomes less available or wholly unavailable in the sorbed state because biodegradation requires the compound to enter the cell to be acted on by intracellular enzymes. The chemical bound to a solid surface is not free to be transported across the outer surface of the cell to be transformed by the catalysts within the confines of the cell. Although most, and often all, of the enzymatic steps in the metabolism of low-molecular-weight compounds are intracellular, some steps in the microbial metabolism of low-molecular-weight substrates are extracellular, and the initial phases in the decomposition of high-molecular-weight molecules are also extracellular. Reactions catalyzed by such extracellular enzymes are markedly affected by reactive surfaces because the enzymes are subject to sorption, and they may then lose catalytic activity; for example, pronase in the presence of montmorillonite (Griffith and Thomas, 1979), uricase sorbed on montmorillonite (Durand, 1964), acid phosphatase on montmorillonite, illite, and kaolinite (Makboul and Ottow, 1979), a protease, amylases, and cellulase on allophane, montmorillonite, and halloysite (Aomine and Kobayashi, 1964), and arylsulfatase on montmorillonite and kaolinite (Hughes and Simpson, 1978). In addition, availability of a substrate acted on by extracellular enzymes may be diminished or become negligible because it is less accessible to the enzyme, possibly because the chemical is shielded from the catalyst or because of steric effects preventing the formation of the substrate–enzyme complex necessary for the enzyme to catalyze the alteration of the substrate. As indicated earlier, however, studies of sorbed substrates acted on by individual enzymes have little relevancy for most environmental pollutants, for which it is generally believed that metabolism is entirely intracellular.

Sorption may affect biodegradation in a number of ways in addition to removing the organic substrate from solution or binding extracellular enzymes. (a) Inorganic nutrients and growth factors are also sorbed, and the removal from solution of such essentials for microbial replication may reduce the rate or extent of growth. (b) The microenvironment around the surface may be less favorable for the transformation than the surrounding solution because of the frequently lower pH immediately around negatively charged surfaces (because these surfaces attract and concentrate H^+ from the solution). (c) Conversely, sorption concentrates the nutrients at the surface of the adsorbent, so that growth of organisms near the surface may be enhanced and the biodegradation may be stimulated, especially if the surrounding solution has a low concentration of nutrients. (d) Moreover, the microorganisms themselves are sorbed, and frequently most bacterial cells in ecosystems with much clay and particulate organic matter are associated with the solids rather than the free liquid. It is likely, indeed, that most bacteria active in degradation in soil, sediments, and aquifers are retained by the solids, a view supported by the finding that, as naphthalene is being metabolized in soil, the population of naphthalene degraders that are sorbed is two orders of magnitude greater than those present in the water phase (Di Grazio et al., 1990). Some bacteria that become attached may adhere reversibly, but some adhere irreversibly and are not released to the ambient solution (Kefford et al., 1982). At low cell densities, nearly all bacteria in the liquid may become adsorbed, but a small percentage may be retained at high cell densities (Gordon et al., 1983). Once the cells become attached to the surface, their physiological activity may alter, and their metabolic activity may be greater than, less than, or sometimes not different from the cells free in solution.

UTILIZATION OF SORBED COMPOUNDS

As indicated in the foregoing, although sorption often reduces the rate and extent of biodegradation, it does not necessarily prevent it. Many sorbed molecules can be used by microorganisms as sources of C, energy, N, and probably other elements, and the compounds are thereby transformed, frequently slowly, but sometimes at reasonable rates. Biodegradation is evident even when all the chemical is sorbed (Fig. 8.2). Such utilization occurs with compounds sorbed by clays as well as those retained by hydrophobic mechanisms. For example, certain amino acids or peptides that are bound to clays and are not spontaneously desorbed may serve as sources of C, N, or both for individual species of bacteria (Dashman and Stotzky, 1986). Many chemicals apparently adsorbed to the external surfaces of clay minerals can be metabolized, as is adenine present at the exposed edges of montmorillonite (Greaves and Wilson, 1973) and protein adsorbed on the outer surface of the same clay, especially if present on the clay in more than one layer (Pinck *et al.*, 1954). Some proteins adsorbed in monolayers on montmorillonite or kaolinite can be hydrolyzed by pure cultures of bacteria or by a purified proteolytic enzyme (Estermann *et al.*, 1959). The susceptibility to microbial attack of proteins complexed with clay varies with the type of protein, the site on the clay to

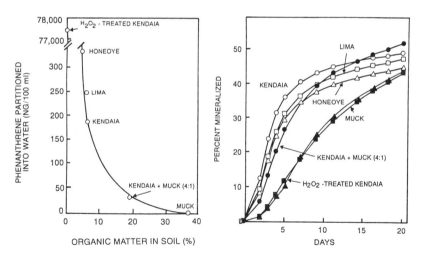

FIG. 8.2 Sorption (left panel) and mineralization (right panel) of phenanthrene in three mineral soils (Honeoye, Lima, and Kendaia), a muck soil, a soil–muck mixture, and soil treated with H_2O_2 to remove organic matter. (From Manilal and Alexander, 1991. Reprinted with permission from Springer-Verlag.)

which the protein is adsorbed, and the cation saturating the clay (Stotzky, 1986).

The availability of sorbed hydrophobic chemicals is evident in a study of biphenyl sorbed to polyvinylstyrene beads. On the one hand, a strain of *Pseudomonas* and microbial populations of a sediment not preexposed to biphenyl metabolized that compound in aqueous solution but not when sorbed to polyvinylstyrene beads. On the other hand, incubation of sediment with biphenyl for 14 days resulted in the proliferation of microorganisms able to mineralize the substrate sorbed to the beads (Fig. 8.3). These

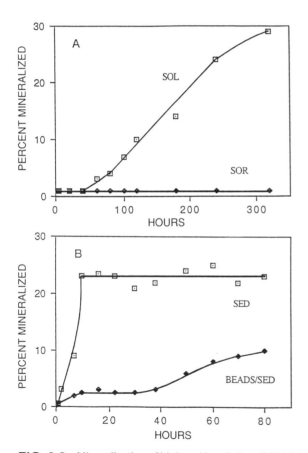

FIG. 8.3 Mineralization of biphenyl in solution (SOL) (100 mg/liter) and of an equivalent amount of biphenyl sorbed to polyvinylstyrene beads (SOR) by a nonacclimated community of sediment microorganisms (A) and of biphenyl sorbed to sediment (SED) or to beads in sediment (BEAD/SED) by sediment microorganisms acclimated to the sorbed compound (From R. Araujo and M. Alexander, unpublished data.)

observations suggest that bacteria acting on a sorbed molecule may need to have two physiological traits—the necessary catabolic enzymes and the capacity to make the sorbed molecule available. The first trait alone is not sufficient. This view is supported by a report that only one of two naphthalene-utilizing bacteria was able to degrade naphthalene sorbed by soil (Guerin and Boyd, 1992).

A methodological problem exists in verifying that an organism is actually using a sorbed compound: at equilibrium, even though most of the substrate may be retained by the solids, some is in aqueous solution. Then, as microorganisms grow and use up the substrate in the aqueous phase, more of the compound desorbs, and it may be the molecules initially present in solution and those subsequently desorbing and entering the liquid that sustain growth and that are being metabolized rather than those molecules that are sorbed (Fig. 8.4). However, even when little or no desorbed phenanthrene is detectable, it is readily mineralized in organic soils (Aronstein *et al.*, 1991; Manilal and Alexander, 1991), suggesting that the sorbed molecule is being degraded. Phenol and dialkyl quaternary ammonium compounds sorbed to sediments (Shimp and Young, 1988) and 4-nitrophenol and phenol sorbed to granular activated carbon (Speitel *et al.*, 1989a,b) also appear to be biodegraded, presumably by microorganisms acting on the molecules that are actually sorbed rather than those that are in solution. Organic molecules localized on other types of surfaces may be utilized, as shown by the ability of bacteria to use stearic acid or palmitic acid coated on glass (Hermansson and Marshall, 1985; Thomas and Alexander, 1987). Nevertheless, many sorbed substrates are not

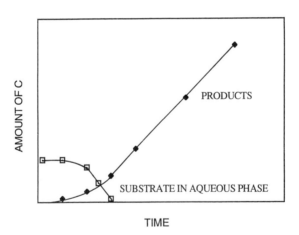

FIG. 8.4 Microbial utilization of a compound in solution prior to metabolism of a substrate that was initially sorbed.

readily used, or are not degraded at all, and any biodegradation of these compounds that occurs is at the expense of chemical in solution. Should the substrate be irreversibly sorbed so that there is no desorption and should there be no organism able to make use of the sorbed substrate, that molecule would be rendered resistant to degradation.

It is still not clear how sorbed molecules become available to microorganisms. To be used for growth and to be degraded, the chemical must be assimilated into the cell. This may require that the molecule be separated from the surface of the sorbent and enter the solution phase, from which it can pass through the cell membrane. If the chemical is thus used only when present in the solution phase, two hypotheses can be advanced to explain the mechanism of utilization. (a) The organism uses the chemical that is initially in solution, and it also metabolizes the compound that enters the aqueous phase as a result of spontaneous desorption from the solid. At the outset, an equilibrium exists between the chemical that is retained on the surface and that which is in the ambient liquid:

$$\text{Sorbed chemical} \underset{k_2}{\overset{k_1}{\rightleftharpoons}} \text{Chemical in solution.}$$

Once the chemical in solution is consumed, which will occur if the degrading population is initially large or when the cell density becomes high as the organisms grow at the expense of the substrate, the subsequent rate of metabolism will be governed by the rate at which additional amounts of the substrate enter the soluble phase, that is, the desorption rate (k_1). Thus, when the concentration of the compound in solution is effectively zero because all that was initially present and that which subsequently enters solution by desorption is consumed, the subsequent rate of degradation is governed by the rate of desorption. Figure 8.5 depicts the rate of spontaneous desorption and the attainment of equilibrium in the absence of microorganisms (A); desorption-limited biodegradation is also shown (B). In other words, desorption is the rate-limiting step. Studies of the mineralization in lake water of benzylamine sorbed to montmorillonite suggest that once the amine in solution is degraded, the subsequent mineralization is limited by desorption (Subba-Rao and Alexander, 1982). At very low cell densities, the rate of transformation is low, and the organisms may be subsisting on the chemical that is originally in solution. However, as the cell density increases, the microbial demand for organic nutrients increases concomitantly until none is left in solution; at this point, the controlling role of desorption becomes evident.

(b) According to the second mechanism of utilization, microorganisms excrete metabolites that facilitate desorption so that the rate of biodegrada-

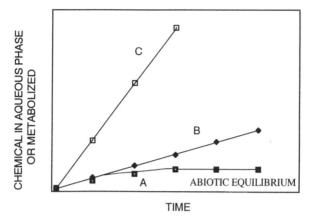

FIG. 8.5 (A) Spontaneous desorption and abiotic equilibrium in the aqueous phase of a reversibly sorbed compound, (B) biodegradation of a compound that is limited by the rate of abiotic desorption, and (C) biodegradation of a compound that is used more rapidly than abiotic desorption.

tion is actually greater than the rate of spontaneous desorption as determined in the absence of microorganisms. This is depicted in Fig. 8.5 by the plot labeled C. A study of decylamine–montmorillonite complexes showed that the rate of mineralization of the amine at high bacterial densities exceeds the spontaneous desorption rate and suggested that bacteria facilitate removal of the chemical from the clay (Wszolek and Alexander, 1979). Similarly, the initial rate of biodegradation of α-hexachlorocyclohexane in soil suspensions is often faster than the initial rate of desorption, also suggesting that the microflora enhances the desorption (Rijnaarts et al., 1990). In some instances, spontaneous desorption is so slow as to be undetectable, yet biodegradation is rapid, as with the mineralization of biphenyl sorbed to polyvinylstyrene beads (Y. M. Calvillo and M. Alexander, unpublished data). Under such circumstances, the rate of biodegradation is not limited by the spontaneous desorption rate. Although the mechanism of facilitated desorption is unresolved, it could involve the elaboration of surfactants that release the chemical from the surface, cations that displace charged compounds, or extracellular enzymes, or it may be associated with a microbiologically induced change in the pH at the surface of the charged material.

Surfactants desorb a number of hydrophobic compounds sorbed to soil, for example, anthracene, phenanthrene, pyrene (Liu et al., 1991), and PCBs (McDermott et al., 1989). However, high concentrations of most synthetic surfactants are needed both to desorb the hydrocarbons and to bring into aqueous solution these compounds of low water solubility. On the other hand, low concentrations of two nonionic alcohol ethoxylate surfactants markedly enhance the rate and/or extent of mineralization of phenanthrene and biphenyl sorbed to soils or aquifer solids, even though little of the compounds is desorbed (Aronstein et al., 1991; Aronstein and Alexander, 1992). Hence, it is plausible to believe that surfactants produced by microorganisms, which are typically excreted in low concentrations, may facilitate the use of bound compounds by these organisms.

It is also possible that some sorbed organic compounds are directly utilized by microorganisms that adhere to the same surfaces. The organism may come into direct contact with the compound, which may then penetrate into the cell without entering the surrounding liquid. This would be comparable to the utilization of water-insoluble organic compounds by bacteria that adhere to the substance undergoing attack (Thomas and Alexander, 1987).

KINETICS

A number of models have been proposed for the biodegradation of compounds that are initially sorbed. Typically, these models assume that the substrate must be in aqueous solution for it to be metabolized. In an early model, for example, use was made of rate constants for desorption (k_1), adsorption (k_2), and biodegradation (k_3) (Furmidge and Osgerby, 1967):

$$\text{Sorbed chemical} \underset{k_2}{\overset{k_1}{\rightleftharpoons}} \text{Chemical in solution}$$

$$\Big\downarrow k_3$$

$$\text{Products of biodegradation.}$$

The desorption term in such a formulation is usually considered to reflect spontaneous desorption, and facilitated desorption, direct use of the sorbed compound, or changes in the rate constants as the degrading population multiplies are not considered. If one makes some simplifying assumptions, assumes first-order kinetics of biodegradation, and introduces a partition coefficient (k_p) to reflect the affinity of the chemical for the sorbent, one might expect that the half-life of the chemical would be equal to k_p/k_3 (Briggs, 1976). The formulation in the preceding equation has been extended to consider the special cases in which the products of biodegradation themselves are sorbed by introducing rate constants for the sorption (k_4) and desorption (k_5) of these products (Holm et al., 1980).

The model of Dao and Lavy (1987) also uses rate constants for adsorption and desorption, as well as a rate constant for biodegradation, but their model assumes that only the substrate that is in aqueous solution is metabolized and that adsorption and desorption are essentially instantaneous. The model of Miller and Alexander (1991) was designed for sorbed molecules that are readily desorbed and that are used by nongrowing populations, and their model gives good fits to the biodegradation by a pseudomonad of benzylamine sorbed to montmorillonite. Several approaches are based on radial diffusion models for kinetics of sorption and desorption, in which the compounds are assumed to diffuse into and out of small pores within soil aggregates (Rijnaarts et al., 1990), and again it is commonly assumed that only the compound in solution is used (Mihelcic and Luthy, 1991). Several other models have been proposed for describing the rate of biodegradation of sorbed compounds (Guerin and Boyd, 1992; Scow and Hutson, 1992). However, the applicability of these models to a variety of substrate–sorbent systems and how well they describe the kinetics have yet to be determined.

STIMULATORY EFFECTS

The presence of particulate material sometimes is stimulatory to microorganisms, in contrast to the frequent findings that surfaces only reduce activity. This enhancement may be evident as an increase in the growth rate of marine oligotrophs by glass beads (Carlucci et al., 1986) or the stimulation of specific activities, such as starch decomposition in the presence of montmorillonite (Filip, 1973), the degradation of uric acid by Pseudomonas sp. in the presence of bentonite (Durand, 1964), the degradation of hexamethylenediamine by Bacillus subtilis in the presence of several clay minerals (Garbara and Rotmistrov, 1982), the mineralization of several aldehydes in soils to which montmorillonite is added (Kunc and Stotzky, 1977), the mineralization of glucose and growth of several actinomycetes upon the addition of montmorillonite to liquid media (Martin et al., 1976), or the hydrolysis of proteins by Pseudomonas sp. in the presence of kaolinite (Estermann and McLaren, 1959). The extent of mineralization may also be increased in the presence of clay minerals, and the enhancement in the rate or extent of mineralization may occur at one but not another substrate concentration (Subba-Rao and Alexander, 1982).

Definitive evidence to support one or another explanation for the stimulation is not at hand. Because the enhancement is frequently evident at low concentrations of organic chemicals, it has been proposed that the stimulation by charged surfaces, such as those of clays, results from the surface acting to concentrate the substrate, thereby promoting the metabolism of compounds present in solution at concentrations below K_m or K_s. The stimulation may result from the surface acting to reduce the concentration of a toxin in solution or by moderating the pH changes that may occur in poorly buffered environments.

From the scientific, engineering, and predictive viewpoints, one can only view current knowledge of the biodegradation of sorbed compounds with considerable unease. Some molecules are totally resistant, some are partially degraded, and some are fully available to microorganisms. At times, sorption is stimulatory but often it retards biodegradation. It is unclear how microorganisms make use of sorbed substrates. The models for the kinetics of biodegradation of such molecules have been applied only to an occasional compound. Hopefully, future research will provide meaningful information to fill this vast void of knowledge.

REFERENCES

Amador, J. A., and Alexander, M., Soil Biol. Biochem. 20, 185–191 (1988).
Anderson, J. P. E., Soil Biol. Biochem. 13, 155–161 (1981).
Aomine, S., and Kobayashi, Y., Trans. Int. Congr. Soil Sci., 8th, 1964, Vol. 3, pp. 697–703 (1964).
Aronstein, B. N., and Alexander, M., Environ. Toxicol. Chem. 11, 1227–1233 (1992).
Aronstein, B. N., Calvillo, Y. M., and Alexander, M., Environ. Sci. Technol. 25, 1728–1731 (1991).
Briggs, G. G., in "The Persistence of Insecticides and Herbicides." Monogr.—Br. Crop Prot. Counc. 17, 41–54 (1976).
Briška, A., Cencelj, J., Hočevar, J., Maček, J, and Sišakovič, V., Agrohemija 1/2, 37–43 (1973); cited in Weed Abstr. 23, 1391 (1974).
Burns, R. G., and Audus, L. J., Weed Res. 10, 49–58 (1970).
Burns, R. G., Pukite, A. H., and McLaren, A. D., Soil Sci. Soc. Am. Proc. 36, 308–311 (1972).
Calvet, R., Environ. Health Perspect. 83, 145–177 (1989).
Carlucci, A. F., Shimp, S. L., and Craven, D. B., FEMS Microb. Ecol. 38, 1–10 (1986).
Chiou, C. T., in "Reactions and Movement of Organic Chemicals in Soils" (B. L. Sawhney and K. Brown, eds.), pp. 1–29. Soil Science Society of America, Madison, WI, 1989.
Dao, T. H., and Lavy, T. L., Soil Sci. 143, 66–72 (1987).
Dashman, T., and Stotzky, G., Soil Biol. Biochem. 18, 5–14 (1986).
Di Grazio, P. M., Blackburn, J. W., Bienkowski, P. R., Hilton, B., Reed, G. D., King, J. M. H., and Sayler, G. S., Appl. Biochem. Biotechnol. 24/25, 237–252 (1990).
Durand, G., Ann. Inst. Pasteur, Paris 107, Suppl. 3, 136–147 (1964).
Ensminger, L. E., and Gieseking, J. E., Soil Sci. 53, 205–209 (1942).
Estermann, E. F., and McLaren, A. D., J. Soil Sci. 10, 64–78 (1959).
Estermann, E. F., Peterson, G. H., and McLaren, A. D., Soil Sci. Soc. Am. Proc. 23, 31–36 (1959).
Filip, Z., Folia Microbiol. (Prague) 18, 56–74 (1973).

Furmidge, C. G. L., and Osgerby, J. M., *J. Sci. Food Agric.* **18**, 269–273 (1967).

Garbara, S. V., and Rotmistrov, M. N., *Mikrobiologiya* **51**, 332–335 (1982).

Geissbühler, H., Haselbach, C., Aebi, H., and Ebner, L., *Weed Res.* **3**, 277–297 (1963).

Gordon, A. S., Gerchakov, S. M., and Millero, F. J., *Appl. Environ. Microbiol.* **45**, 411–417 (1983).

Greaves, M. P., and Wilson, M. J., *Soil Biol. Biochem.* **5**, 275–276 (1973).

Griffith, S. M., and Thomas, R. L., *Soil Sci. Soc. Am. J.* **43**, 1138–1140 (1979).

Guerin, W. F., and Boyd, S. A., *Appl. Environ. Microbiol.* **58**, 1142–1152 (1992).

Hermansson, M., and Marshall, K. C., *Microb. Ecol.* **11**, 91–105 (1985).

Heuer, B., Birk, Y., and Yaron, B., *J. Agric. Food Chem.* **24**, 611–614 (1976).

Holm, T. R., Anderson, M. A., Stanforth, R. R., and Iverson, D. G., *Limnol. Oceanogr.* **25**, 23–30 (1980).

Hughes, J. D., and Simpson, G. H., *Aust. J. Soil Res.* **16**, 35–40 (1978).

Inoue, K., Kaneko, K., and Yoshida, M., *Soil Sci. Plant Nutr.* **24**, 91–102 (1978).

Kefford, B., Kjelleberg, S., and Marshall, K. C., *Arch. Microbiol.* **133**, 257–260 (1982).

Kunc, F., and Stotzky, G., *Soil Sci.* **124**, 167–172 (1977).

Liu, Z., Laha, S., and Luthy, R. G., *Water Sci. Technol.* **23**, 475–485 (1991).

Lynch, D. L., and Cotnoir, L. J., Jr., *Soil Sci. Soc. Am. Proc.* **20**, 367–370 (1956).

Makboul, H. E., and Ottow, J. C. G., *Microb. Ecol.* **5**, 207–213 (1979).

Manilal, V. B., and Alexander, M., *Appl. Microbiol. Biotechnol.* **35**, 401–405 (1991).

Martin, J. P., Filip, Z., and Haider, K., *Soil Biol. Biochem.* **8**, 409–413 (1976).

McDermott, J. B., Unterman, R., Brennan, M. J., Brooks, R. E., Mobley, D. P., Schwartz, C. C., and Dietrich, D. K., *Environ. Prog.* **8**, 46–55 (1989).

McLaren, A. D., *Soil Sci. Soc. Am. Proc.* **18**, 170–174 (1954).

Mihelcic, J. R., and Luthy, R. G., *Environ. Sci. Technol.* **25**, 169–177 (1991).

Miller, M. E., and Alexander, M., *Environ. Sci. Technol.* **25**, 240–245 (1991).

Morrill, L. G., Mahilum, B., and Mohiuddin, S. H., "Organic Compounds in Soils: Sorption, Degradation and Persistence." Ann Arbor Sci. Publ., Ann Arbor, MI, 1982.

Nomura, N. S., and Hilton, H. W., *Weed Res.* **17**, 113–121 (1977).

Olness, A., and Clapp, C. E., *Soil Sci. Soc. Am. Proc.* **36**, 179–181 (1972).

Pinck, L. A., Dyal, R. S., and Allison, F. E., *Soil Sci.* **78**, 109–118 (1954).

Rijnaarts, H. H. M., Bachmann, A., Jumelet, J. C., and Zehnder, A. J. B., *Environ. Sci. Technol.* **24**, 1349–1354 (1990).

Scow, K. M., and Hutson, J., *Soil Sci. Soc. Am. J.* **56**, 119–127 (1992).

Shimp, R. J., and Young, R. L., *Ecotoxicol. Environ. Saf.* **15**, 31–45 (1988).

Smith, D. T., and Meggitt, W. F., *Weed Sci.* **18**, 260–264 (1967).

Speitel, G. E., Jr., Lu, C.-J., and Zhu, X.-J., *Environ. Sci. Technol.* **23**, 68–74 (1989a).

Speitel, G. E., Jr., Turahia, M. H., and Lu, C.-J., *J. Am. Water Works Assoc.* **81**(4), 168–176 (1989b).

Steinberg, S. M., Pignatello, J. J., and Sawhney, B. L., *Environ. Sci. Technol.* **21**, 1201–1208 (1987).

Stotzky, G., *in* "Interactions of Soil Minerals with Natural Organics and Microbes" (P. M. Huang and M. Schnitzer, eds.), pp. 305–428. Soil Science Society of America, Madison, WI, 1986.

Subba-Rao, R. V., and Alexander, M., *Appl. Environ. Microbiol.* **44**, 659–668 (1982).

Thomas, J. M., and Alexander, M., *Microb. Ecol.* **14**, 75–80 (1987).

Weber, J. B., and Coble, H. D., *J. Agric. Food Chem.* **16**, 475–478 (1968).

Weissenfels, W. D., Klewer, H.-J., and Langhoff, J., *Appl. Microbiol. Biotechnol.* **36**, 689–696 (1992).

Wszolek, P. C., and Alexander, M., *J. Agric. Food Chem.* **27**, 410–414 (1979).

9

NAPLs AND COMPOUNDS WITH LOW WATER SOLUBILITY

Many pollutants exist at contaminated sites not in the aqueous phase or sorbed to solids but rather in liquids that are immiscible with water. As such, the availability of the contaminants to biodegradation and bioremediation may be drastically reduced. These nonaqueous phase liquids (NAPLs) containing the environmental contaminants are present in aquifers, subsoils, sediments, soils, and at the top of the water column of marine, estuarine, and fresh waters. The NAPLs are most widely known because of spills or leakages from oil tankers, and these crude-oil NAPLs have contaminated surface waters, marine sediments, and coastal beaches of the Atlantic Ocean, Caribbean Sea, and Prince William Sound, Alaska. Much oil that is inadvertently discharged in marine waters eventually sinks to the bottom and persists in the sediments. Comparable spills of oil or petroleum products have occurred on land as a result of tank car spills or oil pipeline leaks. Gasoline, petroleum products, or industrial solvents have contaminated aquifers and groundwaters at uncounted numbers of sites because of underground storage tanks that, after many years of burial, corrode and begin to leak their contents. In addition, many hazardous waste sites contain industrial solvents, and not uncommonly, these organic solvents move from the site and enter adjacent groundwaters, frequently making nonpotable what was previously a safe water supply. If the spilled material in subsoils is dense, it will move downward and come to rest and remain as a pool at the bottom of the aquifer; these are called dense NAPLs (DNAPLs).

The NAPLs usually contain an array of organic molecules, although a spill of solvent from an industrial source or a leak from an underground storage tank may contain a single chemical. Typically, these NAPLs are composed of molecules that have low water solubilities and high solubilities in organic solvents, and the concentration in the water phase is thus quite low. However, the NAPL represents a long-term source of water pollution because the contaminants will continue to enter the water phase to replace that which is transported away from the site, is biodegraded, or is removed by some remediation technology.

Because nearly all the research on microbial physiology and the metabolism of organic substrates has been centered on molecules that are in aqueous solution, it is often assumed that the portion of a compound that is not dissolved in water is not readily accessible for utilization. The chemical must presumably make contact with the cell surface, so that it can enter the organism to be acted on by intracellular enzymes. If a pollutant is present initially in a NAPL, rapid uptake of that compound by the cell from the aqueous phase—and hence rapid growth on that compound as a C source—cannot occur since little will be present in the water, unless a special mechanism exists for uptake and assimilation of the substrate by the cell containing the degradative enzymes.

If a NAPL is a pure solvent (i.e., a single chemical), the water solubility of that solvent is of special importance. In this regard, it is worth considering not only liquids that are composed of a single type of molecule but also solids. Solids composed of a single compound are not environmental pollutants, but they are considered here because research on them has helped lay a foundation for our understanding of how chemicals of low water solubilities are metabolized.

The solubility of several chemicals in water is given in Table 9.1. Different values are reported by different investigators, but the data are typical of those that have been published. The values may be somewhat different in seawater or waters containing humus constituents than in distilled water (Boehm and Quinn, 1973; Sutton and Calder, 1974), but the differences are not of sufficient magnitude to convert a compound that otherwise has a very low solubility into one of appreciably greater solubility.

Biodegradation of the aliphatic hydrocarbons, whose solubilities decline as the number of carbons in the molecule increases (Table 9.1), has been extensively studied. Much of this research has been prompted by their presence in oil spills, but some of the interest has been a result of curiosity about how such poorly soluble substrates are assimilated. Alkanes up to at least a molecular weight of 618 ($C_{44}H_{90}$) can be mineralized by microorganisms (Haines and Alexander, 1974), even in instances in which the solubility apparently is less than 1 ng/liter. The quantity of such aliphatic hydrocarbons in water would allow for the growth of fewer than one bacterial cell per milliliter, so that clearly the organisms must be able to use the insoluble phase when extensive utilization occurs.

TABLE 9.1
Solubility of Several Organic Compounds in Water

Group	Chemical	mg/liter
Aliphatic hydrocarbons[a]	Heptane (C_7H_{16})	2.9
	Octane (C_8H_{18})	0.66
	Nonane (C_9H_{20})	0.22
	Decane ($C_{10}H_{22}$)	0.052
	Hexadecane ($C_{16}H_{34}$)	0.000020
	Eicosane ($C_{20}H_{42}$)	0.00000011
Aromatic hydrocarbons[b]	Naphthalene	31
	Biphenyl	7.2
	Acenaphthene	4.3
	Anthracene	0.050
	Phenanthrene	1.1
	Pyrene	0.13
	Chrysene	0.0020
	1,2-Benzpyrene	0.0053
Others[c]	4-Chlorobiphenyl	0.96
	Palmitic acid	0.0035
	DEHP	0.29

[a]Coates *et al.* (1985).
[b]Yalkowsky *et al.* (1983).
[c]Thomas *et al.* (1986).

Aromatic hydrocarbons with such low solubilities in water are also decomposed microbiologically. For example, anthracene, phenanthrene, pyrene, 1,2-benzpyrene, and chrysene are destroyed microbiologically in soil (Bossert and Bartha, 1986), and anthracene and naphthalene are mineralized by the microorganisms of sediments (Bauer and Capone, 1985). In pure culture, PAHs such as naphthalene, anthracene, and phenanthrene serve as carbon sources for bacterial growth (Stucki and Alexander, 1987; Wodzinski and Bertolini, 1972). The concentration of several of these aromatic compounds in water is so low that the finding of extensive bacterial growth, chemical loss, or mineralization shows that at least some, if not all, of the chemical that is not in the water phase is being utilized.

If a NAPL is not a pure solvent but instead contains two or—as is common in oil spills, leakages of underground storage tanks containing gasoline, or hazardous waste sites—a multitude of hydrophobic compounds, a critical factor is the amount of the compound present in the aqueous phase that is in equilibrium with the NAPL phase. The chemical partitions in the NAPL and in the water on the basis of its relative solubilities in these two phases, the more hydrophobic compounds being at higher concentrations in hydrophobic NAPLs and the more hydrophilic ones existing at higher concentrations in the water. This relative partitioning is usually expressed by determining the amounts present in an arbitrarily chosen organic solvent (*n*-octanol) and in water at equilibrium and is designated as the octanol–water partition coefficient K_{ow}, where K_{ow} is the ratio of the concentration of the test chemical (or solute) in octanol to the concentration present in water. Because the values for K_{ow} often are very high, the values are usually expressed logarithmically, as log K_{ow}. A compound with a high value is hydrophobic, and little would exist in water that is in equilibrium with octanol or with NAPLs. One with a low value would exist at higher concentrations in water. Values of log K_{ow} for a number of compounds are given in Table 9.2. Because the extent of sorption to the native organic matter of soils and sediments is correlated with the hydrophobicity of the molecule being retained, such log K_{ow} values are good predictors of the extent of binding to the humic fraction of these environments.

Field observations show that NAPLs are often extremely persistent. This is quite evident in sediments and in subterranean sites. Soils on which were built factories to convert coal to flammable gases that were used in cities for heat and lighting still contain the tarry materials that were deposited more than 100 years ago. The compounds that are present within these NAPLs obviously are also persistent. Under more defined conditions in the laboratory, the resistance of compounds present in NAPLs can be demonstrated, even otherwise readily metabolized compounds. For

TABLE 9.2
Octanol–Water Partition
Coefficients of Organic
Compounds[a]

Chemical	Log K_{ow}
Dioxane	−1.1
Acetone	0.23
Pyridine	0.71
Benzene	2.0
Toluene	2.4
Xylene	3.1
Diethyl phthalate	3.3
Diphenyl ether	4.3
Decane	5.6
Tetradecane	7.6
Dioctyl phthalate	8.8

[a]From Laane *et al.* (1987a).

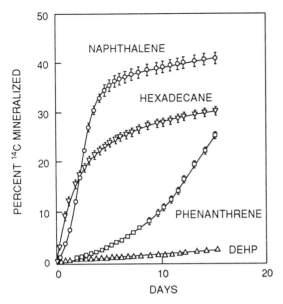

FIG. 9.1 Mineralization in soil of four hydrophobic compounds present in 2,2,4,4,6,8,8-heptamethylnonane. [Reprinted from *Environ. Toxicol. Chem.*, in press. Efroymson and Alexander (1994), with kind permission from Pergamon Press Ltd., Headington Hill Hall, Oxford OX3 OBW, U.K.]

example, in a soil in which ^{14}C-labeled naphthalene, hexadecane, phenanthrene, or di(2-ethylhexyl) phthalate (DEHP) is added with 2,2,4,4,6,8,8-heptamethylnonane as NAPL, two of the compounds are readily mineralized, phenanthrene is mineralized only after an acclimation, and little degradation of DEHP occurs (Fig. 9.1). Heptamethylnonane was chosen for these experiments because of its very low toxicity to microorganisms and its persistence. Similar tests confirm that some NAPLs reduce the availability of their hydrophobic constituents in subsoils (R. A. Efroymson and M. Alexander, unpublished data, 1994). The degree of protection is rarely known, but the finding of individual compounds in NAPLs introduced many years ago into subsoils or aquifers is ample proof of a high degree of unavailability of the constituents that remain.

The same compound in different NAPLs will be metabolized at considerably different rates. This is well known from measurements of the disappearance of individual alkanes in oils of different composition following spills of crude oil or under experimental conditions (Atlas, 1981). Similar observations have been made with soil into which DEHP in several NAPLs was introduced; the phthalate is metabolized quickly in the absence of a

NAPL and slowly with two of the NAPLs, but almost no destruction is observed with three of the NAPLs (Fig. 9.2). The data show not only a reduced availability of substrates within NAPLs but also that biodegradation does take place. This is also evident in studies of individual bacteria. Thus, *Corynebacterium equi* oxidizes 1- and 2-tetradecanol present initially in isooctane (Takazawa *et al.*, 1984), a strain of *Arthrobacter* mineralizes naphthalene and hexadecane dissolved in heptamethylnonane (Efroymson and Alexander, 1991), and several bacterial species utilize individual hydrocarbons present in crude oil (Foght *et al.*, 1990).

Many of the compounds present in a NAPL may be degraded simultaneously, or some may be metabolized initially and others may be destroyed only after the more susceptible molecules are transformed. Simultaneous metabolism is readily evident during the degradation of crude oils, which are composed of a highly heterogeneous mixture of aliphatic and aromatic hydrocarbons, heterocyclic compounds, and other chemical classes. Gas chromatograms depicting oil before and after incubation with a mixture of marine organisms are shown in Fig. 9.3. It is evident that many of the alkanes are destroyed even in this short period under laboratory condi-

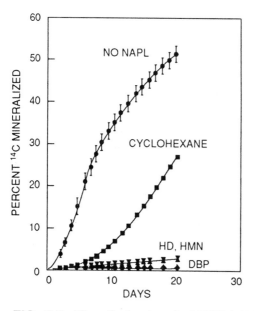

FIG. 9.2 Mineralization in soil of DEHP initially dissolved in heptamethylnonane (HMN), cyclohexane, hexadecane (HD), or dibutyl phthalate (DBP) or added with no NAPL. [Reprinted from *Environ. Toxicol. Chem.*, in press. Efroymson and Alexander (1994), with kind permission from Pergamon Press Ltd., Headington Hill Hall, Oxford OX3 OBW, U.K.]

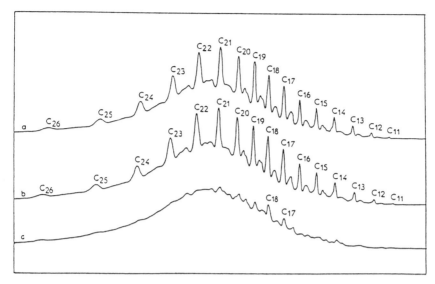

FIG. 9.3   Gas chromatograms of alkanes in oil in a medium at (a) 0 h and after (b) 2 and (c) 5 days incubation with a mixture of marine microorganisms. The solution was supplemented with N and P. (From LePetit and Barthelemy, 1968. Reprinted with permission from Elsevier Science Publishers.)

tions, the bacteria using the aliphatic hydrocarbons essentially at the same time. One must not extrapolate from such data to nutrient-poor or cool aquatic environments, in which the conversions may occur quite slowly, and especially not to NAPLs not spread out in the thin films common to surface waters. Because the exposed surface area is a key factor determining the rate of biodegradation, components in a NAPL existing as a thin film at the surface of the water column are far more likely to be biodegraded quickly than constituents of a NAPL that is thick and therefore has a lower surface-to-volume ratio. In contrast with the nearly simultaneous utilization of some components of heterogeneous NAPLs, other components are used only after some constituents have largely been destroyed (Oberbremer and Müller-Hurtig, 1989).

A major factor determining the biodegradation of compounds present in a NAPL is its toxicity. This toxicity results from the single solvent itself if the NAPL is a single compound, or it may be the result of the major solvent or of one or more minor components of a heterogeneous NAPL. Many organic solvents suppress microbial proliferation and metabolism. As a rule, organic solvents with high values for log K_{ow} (4.0 or greater) do not suppress microbial activity, whereas those with low values for log K_{ow} (often 2.0 or lower) are highly toxic (Inoue and Horikoshi, 1991; Laane *et al.*, 1987b). Nevertheless, some bacteria are tolerant of solvents with low log K_{ow} values, and a NAPL with a high log K_{ow} value may still prevent biodegradation because it contains a highly toxic component.

As indicated in the foregoing, the exposed area of the NAPL is of great significance; the greater the interfacial area between the NAPL and the water, the more rapid will be the degradation. Presumably those molecules that are at or near the surface of the NAPL will be initially utilized, so that the more surface of NAPL that is exposed to the water or the microorganisms, the faster it or its components will be metabolized. The best evidence for the key role of interfacial area comes from studies of individual compounds that exist as liquids and are sparingly soluble in water. For example, for aliphatic hydrocarbons present as droplets in water, the smaller the droplet, the larger is the interfacial area; as the surface area exposed is thus increased, so too does the growth rate of the organisms and hence the rate of their biodegradation (Moo-Young *et al.*, 1971). However, increasing the surface area may not always cause such a stimulatory response (Fogel *et al.*, 1985). Surfactants may also stimulate the degradation of hydrocarbons (Nakahara *et al.*, 1981), presumably by making more of the surface of the chemical available, and emulsification to increase the surface area of PCBs also promotes their destruction by a strain of *Pseudomonas* (Liu, 1980).

Measurements have been made of the kinetics of growth and sometimes mineralization by pure cultures of microorganisms acting on a number of pure hydrocarbons, both liquids and solids, that have low solubilities in water. Sometimes, the process is logarithmic, as in the mineralization of biphenyl by species of *Moraxella* and *Pseudomonas* (Stucki and Alexander, 1987), of palmitic acid by *Pseudomonas pseudoflava* (Thomas and Alexander, 1987), or the mineralization of palmitic acid, DEHP, and Sevin by a mixture of bacteria (Thomas *et al.*, 1986). Often linear kinetics are evident, as in the mineralization of octadecane by a mixture of bacteria (Thomas *et al.*, 1986), the decomposition of tri-, tetra-, penta-, hexa-, and octadecane by the fungus *Cladosporium resinae* (Lindley and Heydeman, 1986), the growth of *Torulopsis* sp. on aliphatic hydrocarbons (McLee and Davies, 1972), and the multiplication of three pseudomonads on a mixture of solid hydrocarbons known as slack wax (Amin *et al.*, 1973). Such linear growth or biodegradation might reflect a linear rate of partitioning of the organic substrate from the insoluble to the soluble phase. On the other hand, the growth of other species on poorly soluble chemicals is initially logarithmic, but the logarithmic phase is followed by a period of linear growth, as in the growth of *Flavobacterium* sp. and *Beijerinckia* sp. on phenanthrene (Stucki and Alexander, 1987), *Arthrobacter* sp. on a sterol (Goswami *et al.*, 1983), and *Candida tropicalis* on hexadecane (Blanch and Einsele, 1973). The kinetics of biodegradation of NAPLs in nature have received scant attention, although it has been noted that for phenanthrene dissolved in hexadecane, its mineralization in soil is linear (R. A. Efroymson and M. Alexander, unpublished data, 1994).

Studies of the utilization of individual alkanes by pure cultures have led to the formulation of a series of models to describe the kinetics of microbial growth on alkanes. These models are based on several assumptions, including the assumptions that the hydrocarbon is solubilized by the organisms before it is assimilated (Goma and Ribot, 1978), that growth takes place initially on the soluble fraction but subsequently by a microbial dissolution of the chemical (Chakravarty *et al.*, 1975), and that the cells obtain the C they need by attachment to or direct contact with the insoluble fraction (Moo-Young *et al.*, 1971; Mallee and Blanch, 1977). The validity of models of these sorts has yet to be tested for individual solvents or heterogeneous NAPLs in natural or polluted environments.

MECHANISMS OF UTILIZATION

Many scientists have been intrigued by the mechanism by which microorganisms make use of organic substrates that are not in the aqueous phase, through which cells get much of their nutrient supply. Because the literature on microbial metabolism is largely derived from studies of water-soluble substrates, a molecule not in solution appears to be an anomaly: how does the organism assimilate it into the cell, wherein the enzymes bring about its transformation? In instances in which the substrate is acted on by a hydrolytic enzyme excreted by the microorganism, the mechanism is straightforward and involves the cleavage by the enzyme of small, soluble fragments from the insoluble molecule. However, such hydrolytic enzymes, although involved in initiation of the transformation of such polymers as cellulose and chitin, are apparently not responsible for the initial phase of utilization of the low-molecular-weight compounds that are important pollutants. For these low-molecular-weight, water-insoluble substrates, the mechanism of initial uptake by the cell is quite different.

Three mechanisms may be postulated to explain how microorganisms utilize compounds in NAPLs or metabolize organic solvents having low solubilities in water. These mechanisms focus on how the chemical is transferred from the environment surrounding the organism to the cell surface, from which point it is transported through the membrane and to intracellular sites of enzymatic activity. (a) Only the chemical in the water phase is used. Once that supply is assimilated into the cell, the organism can only use molecules that enter the aqueous phase by spontaneous partitioning, and hence further degradation would depend on the rate of spontaneous partitioning into the water phase. (b) The microorganism excretes products that convert the substrate into droplets with sizes less than 1 μm, and these are then assimilated by the organism. Because of the small sizes of the droplets or particles, this process is sometimes called pseudosolubilization. The process commonly involves the excretion by the microorganism of surfactants or emulsifiers that facilitate the pseudo-solubilization. (c) The cells come into direct contact with the NAPL, on the surfaces of which the population develops, and the chemical at or near the point of contact with the organism passes through the cell surface into the cytoplasm.

Microbial utilization of only that portion of the compound in a NAPL (or of a sparingly soluble chemical) that is in the aqueous phase is likely with substrates that have low log K_{ow} values or compounds that dissolve rapidly in water as the supply of organic solutes in the aqueous phase is depleted by degradation. Such an organism would multiply as it uses the chemical in solution, but as its population size or biomass increases, its subsequent growth rate—and hence biodegradation—would often be limited by the rate of dissolution. In Fig. 9.4, which is identical to Fig. 8.5, A designates the rate of partitioning from the NAPL to the water

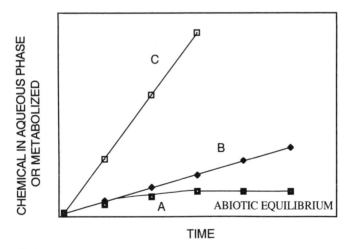

TIME

FIG. 9.4 (A) Rate of partitioning of a chemical from a NAPL to the surrounding water and the equilibrium reached in the absence of microorganisms, (B) rates of biodegradation by microorganisms using only substrate entering the aqueous phase, and (C) biodegradation using the substrate by some other mechanism.

phase, and the abiotic equilibrium reflects the partition coefficient, that is, how much of the compound would be in the water when equilibrium is reached between chemical in the NAPL and in the water. The rate of degradation of a substrate used by this mechanism is then shown by the line B.

One indication that some bacteria use only the substrate in the aqueous phase comes from studies showing that the growth rates of certain pure cultures of hydrocarbon-utilizing bacteria increase on aromatic hydrocarbons with increasing water solubilities—growing at a more rapid rate on naphthalene than on phenanthrene and more rapidly on phenanthrene than on anthracene (Wodzinski and Johnson, 1968). Since the concentrations in solution are probably far below the organism's presumed K_s for those substrates, the growth rate should increase with more chemical in solution. Further evidence comes from studies demonstrating that the rate of growth of certain bacteria on phenanthrene (or, with other bacteria, on naphthalene, bibenzyl, or diphenylmethane) is the same if the medium contains only the dissolved hydrocarbon or if it contains both dissolved and the insoluble chemical (Wodzinski and Bertolini, 1972; Wodzinski and Coyle, 1974; Wodzinski and Larocca, 1977).

Another indication of the importance of the water-soluble phase comes from a study in which bacteria were found to grow readily on the portion of naphthalene and 4-chlorobiphenyl that was in solution in culture media, the rate of dissolution of the chemicals exceeding the rate of microbial degradation as long as the biomass was small. However, when the concentration of organic substrate in solution became indetectable, the bacterial growth rate fell dramatically (Thomas et al., 1986). In considering the exclusive use of the chemical present in the water phase, both the concentration initially in the water and the spontaneous dissolution rate must be considered. If the organisms grow slowly, the rate of dissolution may be sufficiently rapid that soluble substrate is always available; the dissolution rate would then exceed the degradation rate. However, as the biomass increases, the microbial demand for the C source may ultimately exceed the dissolution rate, and subsequent activity will be limited by the rate of partitioning of the chemical from the NAPL (or the solid) to the water. In the case of phenanthrene biodegradation by *Flavobacterium* sp. and *Beijerinckia* sp., for example, the degradation appears to be limited by the rate of spontaneous dissolution of the PAH (Stucki and Alexander, 1987). Because studies of this mechanism have largely dealt with pure compounds—either liquids or solids—it is important to stress a key difference between a NAPL that is a single compound and the heterogeneous NAPLs that are more common as environmental pollutants: the compound of interest in the former case represents 100% of the NAPL phase, whereas the compound may represent 1% or far less of the NAPL phase in the latter. In addition, the partitioning from NAPL to water would be affected by other constituents of the NAPL (e.g., possible cosolvent effects on equilibrum partitioning), and the utilization by microorganisms of other constituents would lead to a competition among the active species for O_2, N, P, and other inorganic nutrients present in the water near the interface with the NAPL.

The production of emulsifiers and surfactants that facilitate the partitioning of the chemical from the NAPL to the water phase, thus resulting in enhanced biodegradation, has received considerable attention in the case of NAPLs that are composed of single compounds (specifically, the liquid alkanes), oils, or oil products. Similar studies have been conducted with some compounds or materials that are solids at ambient temperatures. The substance excreted by the microorganisms, commonly termed an emulsifier or a surfactant, brings about the conversion of the NAPL to small droplets. In contrast with a true solution, in which two or more substances are mixed homogeneously at the molecular or ionic level, an emulsion is an immiscible liquid that is dispersed intimately in another immiscible liquid as droplets that have diameters usually greater than 0.1 μm. In the present context, the first immiscible liquid is the NAPL and the second is water. If the droplets in suspension in the water are very small, that suspension has many of the properties of a solution. A typical emulsion is homogenized milk, in which water-insoluble fat is suspended as droplets so small that they are stable for long periods in the water phase. The smaller the droplets, the more stable is the emulsion. The emulsifying agent, that is, the substance that confers stability on these emulsions, dramatically increases the surface of the interface between the NAPL and water; witness the million-fold increase in area of this interfacial surface if 10 ml of oil is converted to droplets with a diameter of 0.2 μm. A surface-active agent, usually called a surfactant, may improve the stability of the emulsion and thus acts as an emulsifier, the surfactant reducing the surface or interfacial tension of the emulsion. Most emulsifying agents are surfactants, but not all surfactants are emulsifying agents (Becher, 1965). The surfactant, by definition, lowers the surface tension of a liquid, but the emulsifier may or may not do so; that is, the emulsifier may act by reducing interfacial tension or by some dissimilar mechanism. Microbial emulsifying agents are typically surfactants.

A surfactant molecule has a hydrophobic and a hydrophilic portion. At low concentrations, surfactants are fully soluble in water. If the concentration is increased, the molecules of the surfactant associate to form extremely small aggregates that are called micelles. The lowest concentration at which micelles begin to form is known as the critical micelle concentration (CMC). In water, the hydrophobic ends of the surfactant molecules are clustered in the center of the micelle, and the hydrophilic ends are on the outside toward the water phase (Fig. 9.5). A hydrophobic substrate derived from a NAPL would then be incorporated, presumably with some of the NAPL, in the inner region of the micelle and appear to be dissolved in the water, although it is really in a quasi-soluble form because it is entrapped within these small micelles. As a rule, the pseudosolubilization or apparent solubility becomes evident only at a surfactant concentration above but not below the CMC, but some surfactants may increase the water pseudosolubility of hydrophobic molecules below the CMC (Kile and Chiou, 1989).

Many species growing on and degrading NAPLs that are pure alkanes or oils excrete surface-active or emulsifying agents, and these convert the NAPL to droplets or particles with diameters of 0.1 to 1.0 μm (Einsele et al., 1975) that lead to the apparent solubility or "pseudosolubility" of molecules originally present in the NAPL. The rate of apparent solubilization caused by the excretions is often great enough to account for utilization of the substance by the microorganisms and their consequent growth (Cameotra et al., 1983; Goswami and Singh, 1991). These excretions often are capable of emulsifying various types of oil and mixtures of hydrocarbons (Rosenberg et al., 1979). In general, the smaller the droplet produced by the emulsifier, the faster the microorganism grows and degrades its organic substrate (Singer and Finnerty, 1984). The linear growth and degradation so often observed in pure cultures may result from the fact that

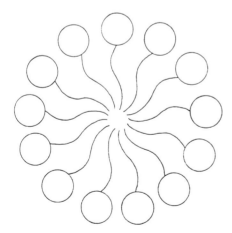

FIG. 9.5 Schematic representation of a micelle (From Robotham and Gill, 1989. Reprinted with permission from Elsevier Science Publishers.)

the exposed surface area of these droplets is limiting the rate of decomposition. The microbial emulsifiers that have been characterized chemically are polysaccharides, polysaccharide–protein complexes, or glycolipids (Rosenberg, 1986).

The role of emulsifiers or surfactants in degradation of NAPLs other than pure aliphatic hydrocarbons remains uncertain. Nevertheless, evidence exists for their importance in the degradation of sterols by an *Arthrobacter* sp. (Goswami *et al.*, 1983) and phthalate esters by strains of *Mycobacterium* and *Nocardia* (Gibbons and Alexander, 1989).

According to the third mechanism, the cells carrying out the biodegradation become attached directly to the NAPL and there, at the surface, metabolize the NAPL constituents. Bacteria that grow on aliphatic hydrocarbons not in aqueous solution often adhere to their substrates, and if these are droplets, the cells retained by the droplets may form agglomerates containing clumps of cells together with some of the hydrocarbon (Miura, 1978). Bacteria may also attach to such solid substrates as palmitic acid and sterol particles, and they then multiply over the surfaces to which they attach. Colonization of the surface may begin or become prominent only after the compound in solution is destroyed (Goswami *et al.*, 1983; Thomas and Alexander, 1987). Some microorganisms have a strong affinity for the NAPL, but others have a lesser degree of binding (Rosenberg and Rosenberg, 1985).

For certain microorganisms at least, this attachment of the cells is of great importance and may be a prerequisite for the degradation. For example, a strain of *Arthrobacter* has been described that biodegrades hexadecane dissolved in a NAPL (2,2,4,4,6,8,8-heptamethylnonane) without excreting products that increase the water solubility of hexadecane. In this instance, spontaneous partitioning of hexadecane into the water phase can be ruled out because no such partitioning is detectable. Instead, the bacterium becomes attached to the NAPL–water interface and there is able to obtain its substrate, hexadecane, from the NAPL (Efroymson and Alexander, 1991). This need for direct contact of the cells with the NAPL–water interface was further confirmed by showing that addition of Triton X-100, a surfactant that suppresses cell adherence but yet is not toxic to the bacterium at the concentration used, prevented mineralization of hexadecane dissolved in the heptamethylnonane (Fig. 9.6). A similar conclusion on the need for adherence comes from a study of *Acinetobacter calcoaceticus* in which comparisons were made of the parent culture with a mutant that did not adhere to a NAPL composed of a pure alkane. Cells of the adherent parent grow on hexane, but the mutant does not grow for a long period. When an emulsifying agent is added, however, the nonadhering mutant is able to utilize the insoluble C source (Rosenberg

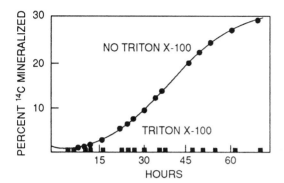

FIG. 9.6 Effect of Triton X-100 on the mineralization by *Arthrobacter* sp. of hexade-cane dissolved in heptamethylnonane. (From Efroymson and Alexander, 1991. Reprinted with permission from the American Society for Microbioogy.)

and Rosenberg, 1981). An identical conclusion was reached from a study of a strain of *Pseudomonas* that does not produce an extracellular surfactant or emulsifier but does adhere to the surface of hexadecane (Goswami and Singh, 1991). The substrate appears to be transferred from the NAPL directly to the bacterium. The cell surface of the organism may be critical in the adherence, possibly a lipophilic layer that facilitates binding of cell and substrate (Kaeppeli and Fiechter, 1976), or the process may involve an emulsifier that is present as a thin layer over the surface of the cell (Pines and Gutnick, 1986). In some species, hairlike structures or thin fimbriae appear to be formed by the microorganisms, and these thin appendages may permit the organism to transport the insoluble compound into the cell, so that the organism can make use of it and bring about its metabolism (Käppeli *et al.*, 1984; Rosenberg *et al.*, 1982).

Although most of the reactions involved in the degradation of constitents of NAPLs and the metabolism of sparingly soluble organic compounds probably take place within the cell, this may not always be the case. If an extracellular enzyme is involved in an initial step in the biodegradative sequence, that enzyme would probably function at the interface between the NAPL and the water (Mattson and Volpenhein, 1966).

ENHANCEMENT OF BIODEGRADATION

Several methods are known by which the biodegradation of chemicals in NAPLs may be enhanced. Three ways have been explored in some detail—the addition of surfactant preparations, inorganic nutrients, and bacteria. Most concern has been with oil. A commercial preparation known as Sugee 2, which disperses crude oil, promotes the degradation of those n-alkanes in crude oil with 17 to 28 C atoms in the molecule (Mulkins-Phillips and Stewart, 1974); a commercial dispersant known as Corexit together with a microbial enrichment enhances the degradation of lubricating oil in soil slurries (Rittmann and Johnson, 1989); nonionic surfactants stimulate the destruction by a strain of *Acinetobacter* of components of oil (Lupton and Marshall, 1979); and surfactants also stimulate the degradation of a defined mixture of aliphatic and aromatic hydrocarbons in a soil suspension (Oberbremer *et al.*, 1990).

The stimulation by inorganic nutrients is not unexpected. The NAPL is made up of one or a variety of organic compounds, and at the NAPL–water interface, the amount of C available to the potential degrading species is large. This results in a large microbial demand for inorganic nutrients, especially N and P. However, the concentration of these nutrients is frequently low, or the rate of diffusion from the surrounding water or soil to the vicinity of the NAPL, where the degradation is occurring, is too slow to meet the large demand. The concentration in the immediate vicinity of crude oil that provides a marked effect may range from 1 to 11 mg of N and about 0.07 mg of P per liter of seawater (Floodgate, 1984). An inadequate supply of inorganic nutrients in the immediate vicinity of the NAPL is common in subsoils and soils. For example, this is shown by a laboratory study in which phenanthrene was introduced into the subsoil either in a NAPL (DEHP or hexane) or without a NAPL. Addition of N and P markedly increases both the rate and extent of mineralization of phenanthrene dissolved in the first NAPL and had an effect, albeit less pronounced, on phenanthrene metabolism in the second NAPL. In contrast, addition of N and P has little or no influence on biodegradation of phenanthrene in subsoil having no NAPL (R. A. Efroymson and M. Alexander, unpublished data, 1994), apparently because the nutrient supply is adequate when the low concentration of phenanthrene is mixed throughout the subsoil. At the surface of marine or fresh waters polluted with oil, the supply of dissolved O_2 is usually sufficient and the diffusion of O_2 from the overlying air is usually rapid enough so that this requisite for aerobic bacteria does not limit the rate of biodegradation. The same is probably not true in soils, and an insufficient supply of O_2 probably often limits the rate of degradation of components of NAPLs. Because the Fe concentration in the ocean is low and much of the Fe is probably not readily assimilated at the pH values of marine water, this element may also sometimes limit the rate of degradation of oil or other NAPLs in seawater (Dibble and Bartha, 1976).

The possible enhancement by addition of inoculants is discussed in Chapter 14.

REFERENCES

Amin, P. M., Nigam, J. N., Lonsane, B. K., Baruah, B., Singh, H. D., Baruah, J. N., and Iyengar, M. S., *Folia Microbiol. (Prague)* **18**, 49–55 (1973).

Atlas, R. M., *Microbiol. Rev.* **45**, 180–209 (1981).

Bauer, J. E., and Capone, D. G., *Appl. Environ. Microbiol.* **50**, 81–90 (1985).

Becher, P., "Emulsions: Theory and Practice." Reinhold, New York, 1965.

Blanch, H. W., and Einsele, A., *Biotechnol. Bioeng.* **15**, 861–877 (1973).

Boehm, P. D., and Quinn, J. G., *Geochim. Cosmochim. Acta* **37**, 2459–2477 (1973).

Bossert, I. D., and Bartha, R., *Bull. Environ. Contam. Toxicol.* **37**, 490–495 (1986).

Cameotra, S. S., Singh, H. D., Hazarika, A. K., and Baruah, J. N., *Biotechnol. Bioeng.* **25**, 2945–2956 (1983).

Chakravarty, M., Singh, H. D., and Baruah, J. N., *Biotechnol. Bioeng.* **17**, 399–412 (1975).

Coates, M., Connell, D. W., and Barron, D. M., *Environ. Sci. Technol.* **19**, 628–632 (1985).

Dibble, J. T., and Bartha, R., *Appl. Environ. Microbiol.* **31**, 544–550 (1976).

Efroymson, R. A., and Alexander, M., *Appl. Environ. Microbiol.* **57**, 1441–1447 (1991).

Einsele, A., Schneider, H., and Fiechter, A., *J. Ferment. Technol.* **53**, 241–243 (1975).

Floodgate, G. D., in "Petroleum Microbiology" (R. M. Atlas, ed.), pp. 354–397. Macmillan, New York, 1984.

Fogel, S., Lancione, R., Sewall, A., and Boethling, R. S., *Chemosphere* **14**, 375–382 (1985).

Foght, J. M., Fedorak, P. M., and Westlake, D. W. S., *Can. J. Microbiol.* **36**, 169–175 (1990).

Gibbons, J. A., and Alexander, M., *Environ. Toxicol. Chem.* **8**, 283–291 (1989).

Goma, G., and Ribot, D., *Biotechnol. Bioeng.* **20**, 1723–1734 (1978).

Goswami, P. C., and Singh, H. D., *Biotechnol. Bioeng.* **37**, 1–11 (1991).

Goswami, P. C., Singh, H. D., Bhagat, S. D., and Baruah, J. N., *Biotechnol. Bioeng.* **25**, 2929–2943 (1983).

Haines, J. R., and Alexander, M., *Appl. Microbiol.* **28**, 1084–1085 (1974).

Inoue, A., and Horikoshi, K., *J. Ferment. Bioeng.* **71**, 194–196 (1991).

Kaeppeli, O., and Fiechter, A., *Biotechnol. Bioeng.* **18**, 967–974 (1976).

Käppeli, O., Walther, P., Mueller, M., and Fiechter, A., *Arch. Microbiol.* **138**, 279–282 (1984).

Kile, D. E., and Chiou, C. T., *Environ. Sci. Technol.* **23**, 832–838 (1989).

Laane, C., Boeren, S., Hilhorst, R., and Veeger, C., in "Biocatalysis in Organic Media" (C. Laane, J. Tramper, and M. D. Lilly, eds.), pp. 65–84. Elsevier, Amsterdam, 1987a.

Laane, C., Boeren, S., Vos, K., and Veeger, C., *Biotechnol. Bioeng.* **30**, 81–87 (1987b).

LePetit, J., and Barthelemy, M. H., *Ann. Inst. Pasteur, Paris* **114**, 149–158 (1968).

Lindley, N. D., and Heydeman, M. T., *Appl. Microbiol. Biotechnol.* **23**, 384–388 (1986).

Liu, D., *Water Res.* **14**, 1467–1475 (1980).

Lupton, F. S., and Marshall, K. C., *Geomicrobiol. J.* **1**, 235–247 (1979).

Mallee, F. M., and Blanch, H. W., *Biotechnol. Bioeng.* **19**, 1793–1816 (1977).

Mattson, F. H., and Volpenhein, R. A., *J. Am. Oil Chem. Soc.* **43**, 286–289 (1966).

McLee, A. G., and Davies, S. L., *Can. J. Microbiol.* **18**, 315–319 (1972).

Miura, Y., *Adv. Biochem. Eng.* **9**, 31–56 (1978).

Moo-Young, M., Shimuzu, T., and Whitworth, D. A., *Biotechnol. Bioeng.* **13**, 741–760 (1971).

Mulkins-Phillips, G. J., and Stewart, J. E., *Appl. Microbiol.* **28**, 547–552 (1974).

Nakahara, T., Hisatsuka, K., and Minoda, Y., *J. Ferment. Technol.* **59**, 415–418 (1981).

Oberbremer, A., and Müller-Hurtig, R., *Appl. Microbiol Biotechnol.* **31**, 582–586 (1989).

Oberbremer, A., Müller-Hurtig, R., and Wagner, F., *Appl. Microbiol Biotechnol.* **32**, 485–489 (1990).

Pines, O., and Gutnick, D., *Appl. Environ. Microbiol.* **51**, 661–663 (1986).

Rittmann, B. E., and Johnson, N. M., *Water Sci. Technol.* **21**(4/5), 209–219 (1989).

Robotham, P. W. J., and Gill, R. A., *in* "The Fate and Effects of Oil in Freshwater" (J. Green and M. W. Trett, eds.), pp. 41–79. Elsevier Applied Science, London, 1989.

Rosenberg, E., *CRC Crit. Rev. Biotechnol.* **3**, 109–132 (1986).

Rosenberg, E., Perry, A., Gibson, D. T., and Gutnick, D. L., *Appl. Environ. Microbiol.* **37**, 409–413 (1979).

Rosenberg, M., and Rosenberg, E., *J. Bacteriol.* **148**, 51–57 (1981).

Rosenberg, M., and Rosenberg, E., *Oil Petrochem. Pollut.* **2**, 155–162 (1985).

Rosenberg, M., Bayer, E. A., Delarea, J., and Rosenberg, E., *Appl. Environ. Microbiol.* **44**, 929–937 (1982).

Singer, M., and Finnerty, W. R., *in* "Petroleum Microbiology" (R. M. Atlas, ed.), pp. 1–59. Macmillan, New York, 1984.

Stucki, G., and Alexander, M., *Appl. Environ. Microbiol.* **53**, 292–297 (1987).

Sutton, C., and Calder, J. A., *Environ. Sci. Technol.* **8**, 654–657 (1974).

Takazawa, Y., Sato, S., and Takahashi, J., *Agric. Biol. Chem.* **48**, 2489–2495 (1984).

Thomas, J. M., and Alexander, M., *Microb. Ecol.* **14**, 75–80 (1987).

Thomas, J. M., Yordy, J. R., Amador, J. A., and Alexander, M., *Appl. Environ. Microbiol.* **52**, 290–296 (1986).

Wodzinski, R. S., and Bertolini, D., *Appl. Microbiol.* **23**, 1077–1081 (1972).

Wodzinski, R. S., and Coyle, J. E., *Appl. Microbiol.* **27**, 1081–1084 (1974).

Wodzinski, R. S., and Johnson, M. J., *Appl. Microbiol.* **16**, 1886–1891 (1968).

Wodzinski, R. S., and Larocca, D., *Appl. Environ. Microbiol.* **33**, 660–665 (1977).

Yalkowsky, S. H., Valvani, S. C., and Mackay, D., *Residue Rev.* **85**, 43–55 (1983).

10 BIOAVAILABILITY: SEQUESTERING AND COMPLEXING

The availability of many chemicals is affected by a series of ill-defined, often uncharacterized processes. In some of these processes, the compound is readily evident, and it can be easily removed from the soil, sediment, or aquifer by conventional extraction procedures; the evidence for reduced bioavailability of these compounds is the marked decline in the rate of biodegradation. In other processes, the compound is still present, but it can only be removed from the environmental sample by highly vigorous extraction techniques. The evidence for reduced bioavailability of such a compound is the marked decline in the rate of biodegradation with time or the almost complete resistance of the molecule to microbial destruction. This is evident by the data for an insecticide in Fig. 10.1. In the case of the illustration, the compound added to soil was aldrin, but the aldrin is converted by a simple epoxidation to a closely related molecule, dieldrin; the data shown are the percentage of aldrin plus dieldrin present in the soil.

The compound that remains is often termed an *aged* (or sometimes *weathered*) chemical or *aged residue*. Residue refers to the residual nature of the compound, that is, that it persists. A substrate that is only slowly metabolized, as shown in the figure, has considerable time to interact with physical or chemical components of soils, subsoils, or sediments and thus have its physical, chemical, or biological behavior altered. The term "aged" is unfortunate because it implies a change in the identity rather than the behavior of the molecule.

With other synthetic compounds, the molecule is converted to a form that cannot be removed by even the most vigorous extraction with nonpolar or polar organic solvents, yet the molecule remains in a form that is recognizable as the parent compound, or sometimes as a nonextractable metabolite derived from it. These highly persistent materials are commonly termed *bound residues*. Many insecticides, herbicides, fungicides, and undoubtedly other classes of chemicals undergo changes, especially in soil, that result in the formation of bound residues. In this instance, the

149

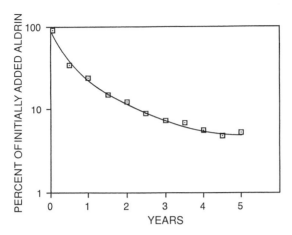

FIG. 10.1 Logarithmic plot of the disappearance of aldrin (plus the dieldrin formed from it) in 12 field sites in various countries. The data represent the mean of the percentages remaining at all sites in areas receiving 2.24 or 4.48 kg of aldrin per hectare. (From Elgar, 1975. Reprinted with permission from Georg Thieme Verlag, Stuttgart.)

word "bound" is also unfortunate and misleading because it implies a binding mechanism of rendering the compound far less available for bio-degradation. Many of the bound residues appear to result from complexing with humic materials in soils, and they may then be new molecular species and not the parent molecules in a strict sense.

At this stage in the development of knowledge of a subject that is still characterized by confusion, it might be prudent to envision four separate categories of molecules of reduced bioavailability. These are in addition to compounds that are poorly available because they are sorbed to solid surfaces or present within NAPLs *in the immediate vicinity of microbial cells* having the requisite catabolic enzymes.

(a) Nonsorbed compounds in micropores at some distance from cells having the requisite enzymes. Distance in the present context refers to micrometer distances, which are those that are important to cells in a porous or nonfluid environment.
(b) Compounds that are sorbed in micropores situated at some distance from the cells with biodegradative potential.
(c) Molecules that partition into the solids themselves, which may occur with nonpolar molecules which partition into humic substances that, in soils and sediments, exist in a particulate or solid form.
(d) Chemicals that complex with humic materials or other environmental constituents to form molecular species that, although containing

the parent molecules or metabolites generated from them, are in fact new molecular species.

One or more of these models may be inappropriate representations of what transpires in nature. Nevertheless, the confusing state of knowledge of these poorly available substrates, for the present at least, can best be understood by considering these four separate possibilities.

SEQUESTERED SUBSTRATES

A chemical may become less available or essentially wholly unavailable for biodegradation if it enters or is deposited in a micropore that is inaccessible to microorganisms. Soil, subsoils, and sediments are characterized by particles of various sizes, and between these particles are pores that obviously are also of various dimensions, both large and small. The various sizes of solids in a soil are depicted in Fig. 10.2. The pores may be filled entirely with water, as in sediments or below the water table in land areas, or they may be filled with air and water, as in soils above the water table. A chemical would move out of a micropore by diffusion to a site containing a bacterium with the capacity to bring about its destruction, but the path for diffusion of the molecule in an environment with small particles is far from straight. Instead, it is highly tortuous. This *tortuosity* may increase enormously the path the molecule must traverse before it reaches the appropriate cells (Fig. 10.3). The longer the path, the greater the influence on biodegradation. Models have been developed to describe the effect of

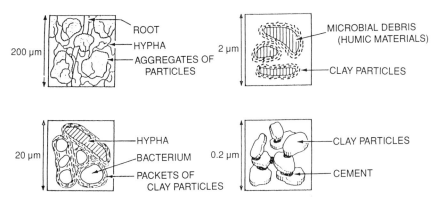

FIG. 10.2 Models of the various particle sizes in soil (From Tisdall and Oades, 1982. Reprinted with permission from Blackwell Scientific Publications.)

FIG. 10.3 Pathways of a molecule diffusing through an environment containing small particulates.

diffusion on biodegradation (Priesack, 1991; Scow and Alexander, 1992; Scow and Hutson, 1992).

In a habitat such as soil, most of the bacterial cells exist on the surfaces of particulate matter, and these cells are frequently present as small colonies or aggregates of cells. From the microscale vantage point of the bacterium, large portions of the clay and humus surfaces are free of living organisms. Most of the cells thus sorbed probably do not move freely, and a chemical occluded at a distant site, distant on a microscopical scale, is inaccessible to the cell fixed to the surface. Moreover, the pores between small particles have narrow necks that would impede movement of even mobile cells that otherwise might be transported to a point in a larger pore at the other side of the thin neck. An appreciation of the small dimensions of these pores can be gained from data indicating that pores with diameters less than 0.2 μm occupy 30% of the total volume of some soils (Hassink et al., 1993).

Some direct evidence exists that organic compounds may be physically sequestered and thereby protected from microbial attack. Thus, a thin layer of glass microbeads placed between a population of a *Pseudomonas* strain and chitin results in a long delay before the onset of mineralization of this polysaccharide by the organism. This delay in activity is not a result of sorption of the substrate but rather results from physical separation of bacterium from its C source (Ou and Alexander, 1974). Similar evidence suggesting the importance of accessibility of a substrate comes from a study in which starch deposited in micropores of artificially created soil aggregates was found to be protected from microbial attack (Adu and Oades, 1978). In both of these investigations, the substrates did not diffuse through the aqueous phase, but low-molecular-weight, nonsorbed compounds would diffuse to the physical sites occupied by bacteria so that a slow, diffusion-limited degradation would take place.

Bioavailability would be dramatically affected if a chemical is not only physically remote from potentially active microorganisms but is sorbed to solid surfaces associated with that remote micropore. The substrate would then need to be desorbed and diffuse through a tortuous pathway, all the while being subject to resorption. This combination of sorption–desorption, diffusion, and tortuosity imposes major limitations on the metabolism of an otherwise biodegradable molecule. An argument for entrapment in soil micropores of a sorbed compound was made to explain the persistence in soil of EDB for up to 19 years, although this fumigant, when freshly added to soil, is often quickly degraded (Fig. 10.4). The

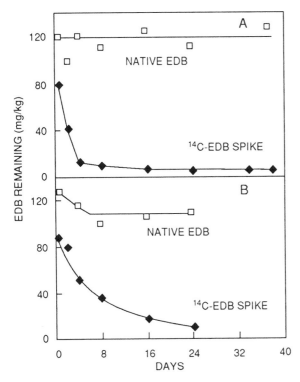

FIG. 10.4 Degradation of EDB in suspensions of field soil containing EDB applied earlier ("native EDB") or containing newly added ^{14}C-labeled EDB ("^{14}C-EDB spike"). The soils used in this particular study had received unlabeled EDB 0.9 (soil A) and 3 (soil B) years earlier, and they are not the soils in which EDB was found to persist for 19 years. (From Steinberg *et al.*, 1987. Reprinted with permission from the American Chemical Society.)

behavior of EDB illustrates eloquently the issue embodied in the term aged chemical—a change occurs associated with the persistence of the chemical such that it becomes highly resistant to microbial utilization. In this instance, Steinberg *et al.* (1987) explain the "aging" as a result of entrapment of the substrate in microscopic pores and the slow release of the sorbed and entrapped compound through a tortuous pathway to locales where EDB-metabolizing cells might be present. The low bioavailability of PAHs in soil has also been attributed to their presence in inaccessible sites in the soil matrix (Weissenfels *et al.*, 1992).

As discussed previously, sorption is a major reason for low availability of many compounds. It is relevant to add that the longer some compounds remain in soil, the more resistant they become to desorption. This is true of TCE (Pavlostathis and Mathavan, 1992), picloram (McCall and Agin, 1985), simazine (Scribner *et al.*, 1992), and EDB (Steinberg *et al.*, 1987). Moreover, the greater resistance to desorption of compounds present in soil for long periods of time is correlated with their greater resistance to degradation, at least for simazine and EDB (Scribner *et al.*, 1992; Steinberg *et al.*, 1987). The site of this sorption, however, remains an issue of contention. As discussed earlier, one view holds that the sorption is physical and the site of the retention—and presumably the site of the greater resistance to desorption and biodegradation associated with the aging of hydrophobic molecules—is within the micropores of the humic materials of soils. Another view maintains that the organic molecules partition into the solid portion of the organic matter itself. Diffusion of a molecule into a solid, such as humus, would be slow and could require years, which is the time period associated with the formation of aged residues. A molecule that is partitioned within the solid would be wholly unavailable for biodegradation and would only become available as it slowly diffuses within the solid to the solid–liquid interface, where the microflora would be available to function.

COMPLEXED SUBSTRATES

Many pesticides are reported to be converted to "bound residues" in soil. However, the very definition of bound residues serves to obscure chemical interpretation. Indeed, a common procedure for measuring the quantity of a pesticide converted to bound residue—namely, measuring the amount of ^{14}C remaining in soil amended with ^{14}C-labeled pesticide after incubation and then removing any solvent-extractable ^{14}C—shows the chemical and toxicological futility of the concept of bound residues, since the microbial biomass, polymers formed enzymatically from the parent molecule, and strongly sorbed products of microbial metabolism would all be considered as bound residues. The definition is merely a description of the results of an analytical procedure and is of no use in describing chemical behavior or toxicological significance.

Nevertheless, it is evident that a portion—whether large or small is often uncertain—of the fraction termed "bound residues" represents new molecular species formed from the parent compound, or from structurally similar products derived from it. These new molecules are often complexes of the parent or a metabolite with organic constituents of soils, and probably of sediments and subsoils as well. That the new molecules are derived from the parent compounds or sometimes structurally related metabolites is evident from the finding that hydrolysis of soil with strong alkali or acid yields the parent molecules or related metabolites (Fuchsbichler *et al.*, 1978; Hsu and Bartha, 1976; Singh and Agarwal, 1992).

Much of the complexing is with the organic or humus fraction of soils and sediments. This fraction is frequently subdivided into three complex, heterogeneous mixtures known as humin, fulvic acid, and humic acid. These represent portions that are (a) not soluble in the dilute alkali used for the initial extraction, (b) soluble in dilute base and weak acid, and (c) soluble in base but not acid, respectively. Because they can be thus brought into solution, more chemistry has been performed on the fulvic and humic acid fractions. Because of their poorly characterized structure, the heterogeneity of these three fractions, and their high molecular weights, the identities and structures of the complexes have been difficult to establish. Most of the complexes that have been studied have these characteristics: they are strongly associated with the humic materials, the stable linkages probably being a result of covalent bonding between the original chemical and the humic polymers; they are not readily extractable with organic solvents; part or sometimes all of the complexes may be cleaved with strong acid or alkali; and they are reasonably resistant to microbial degradation and persist for long periods.

The complexes probably are formed either by an attachment of the compound to reactive sites on the surfaces of the organic colloids or by incorporation of the compound into the structure of humic and fulvic acids that are being formed microbiologically (Stevenson, 1976). Much of the research has been done on amines (or on nitro compounds that presumably are reduced microbiologically to amines before the complex is formed), phenols, or quinones. The amines probably react with polyphenols or quinones in humus, and the phenols or catechols probably complex with amino groups of the humic materials (Bartha *et al.*, 1983; Calderbank, 1989). Such reactions do not account for the many complexes formed from molecules with neither amino nor hydroxyl groups. The complexes may have molecular weights ranging from 2100 to 100,000 (Meikle *et al.*, 1976).

Organic pollutants may also interact or complex with soluble humic substances or high-molecular-weight substances in aquatic environments. Such complexing has been reported for PCBs and PAHs (Eadie *et al.*, 1990; Jota and Hassett, 1991) as well as other compounds. The complexes formed between test chemicals and such water-soluble humic substances may have molecular weights of 700–5000 (Madhun *et al.*, 1986). It is also likely that these complexes are resistant to biodegradation, although this has not been investigated.

In some and possibly many instances, complex formation involves microbial metabolism to a greater or lesser degree. For example, because the presence of amino or hydroxyl groups is necessary for the formation of some complexes, microbial reduction of the nitro moiety of a molecule to an amino group or the hydroxylation of an aromatic ring may be a necessary first step. A comparison of the formation of complexes generated from parathion in sterile and nonsterile soil provides more direct evidence for a key role of microorganisms, since sterilization markedly reduces the magnitude of complex formation (Katan *et al.*, 1976; Katan and Lichtenstein, 1977). Although the conversion of most organic pollutants to simple products is probably chiefly an intracellular process, it is possible that the biosynthesis of complexes results in part from extracellular enzymes. This possibility is suggested by the observation that phenol oxidase of the fungus *Rhizoctonia solani* catalyzes reactions that lead to the formation of a series of oligomers from mixtures containing chloroanilines or 2,4-dichlorophenol and hydroxylated aromatic acids (Bollag *et al.*, 1980, 1983).

The compounds contained in the complexes are far less readily degraded than the free molecules. Under laboratory or field conditions, their prolonged persistence is immediately and strikingly evident. The bonds associated with creating the complexes therefore are obviously not readily cleaved enzymatically, since such cleavage would yield the parent molecules, which are usually metabolized at reasonable rates. The resistance is evident from data showing that only 1% of the ^{14}C from labeled dinitroaniline complexes is converted to $^{14}CO_2$ in soil in 21 weeks (Helling, 1976) and that only 1.2–10.0% of the added ^{14}C of a series of chlorophenol complexes is mineralized in 13 weeks (Dec and Bollag, 1988). An atrazine complex is even found in soil 9 years after this herbicide was applied in the field (Capriel *et al.*, 1985). Nevertheless, the complexes are slowly destroyed, and mineralization tests with individual bacteria or fungi show a slow release of $^{14}CO_2$ from ^{14}C-labeled substrates converted to complexes in soil or in liquid culture.

TOXICOLOGICAL SIGNIFICANCE

Because many of the parent molecules are acknowledged pollutants, some are pesticides, and others probably affect animals, plants, or micro-organisms, considerable attention has been devoted to assessing whether the complexes are cleaved in nature to give detectable levels of the original compound and whether the complexes are assimilated by animals and plants. Are they problems of present or future toxicological importance? From a practical viewpoint, no convincing answer can be given, although governmental agencies typically do not regulate them, viewing the complexes as nontoxic and as yielding the parent molecules in concentrations too low to represent a hazard. Whether one can be confident of the absence of an environmental concern is questionable because at least some complexes can be assimilated and some converted from a form not extractable with organic solvents to a solvent-extractable state. For example, plants assimilate and translocate butralin complexes formed in soil (Helling and Krivonak, 1978), and both earthworms and oat plants not only assimilate complexes of parathion formed in soil but modify them to yield products that are soluble in organic solvents and even in water and thus have simpler structures (Racke and Lichtenstein, 1985). To obtain more convincing evidence of the hazards, or lack thereof, will require further investigation.

REFERENCES

Adu, J. K., and Oades, J. M., *Soil Biol. Biochem* **10**, 109–115 (1978).
Bartha, R., You, I.-S., and Saxena, A., *in* "Pesticide Chemistry: Human Welfare and the Environment" (S. Matsunaka, D. H. Hutson, and S. D. Murphy, eds.), Vol. 3, pp. 345–350. Pergamon, Oxford, 1983.
Bollag, J.-M., Liu, S.-Y., and Minard, R. D., *Soil Sci. Soc. Am. J.* **44**, 52–56 (1980).
Bollag, J.-M., Minard, R. D., and Liu, S.-Y., *Environ. Sci. Technol.* **17**, 72–80 (1983).
Calderbank, A., *Rev. Environ. Contam. Toxicol.* **108**, 71–103 (1989).
Capriel, P., Haisch, A., and Khan, S. U., *J. Agric. Food Chem.* **33**, 567–569 (1985).
Dec, J., and Bollag, J.-M., *Soil Sci. Soc. Am. J.* **52**, 1366–1371 (1988).
Eadie, B. J., Morehead, N. R., and Landrum, P. F., *Chemosphere* **20**, 161–178 (1990).
Elgar, K. E., *Environ. Qual. Saf.* **3** Suppl., 250–257 (1975).
Fuchsbichler, G., Süss, A., and Wallnöfer P., *Z. Pflanzenkr. Pflanzenschutz* **85**, 724–734 (1978).
Hassink, J., Bouwman, L. A., Zwart, K. B., and Brussard, L., *Soil Biol. Biochem.* **25**, 47–55 (1993).
Helling, C. S., *in* "Bound and Conjugated Pesticide Residues" (D. D. Kaufman, G. G. Still, G. D. Paulson, and S. K. Bandal, eds.), pp. 366–367. American Chemical Society, Washington, DC, 1976.
Helling, C. S., and Krivonak, A. E., *J. Agric. Food Chem.* **26**, 1164–1172 (1978).
Hsu, T.-S., and Bartha, R., *J. Agric. Food Chem.* **24**, 118–122 (1976).

Jota, M. A. T., and Hassett, J. P., *Environ Toxicol. Chem.* **10,** 483–491 (1991).

Katan, J., and Lichtenstein, E. P., *J. Agric. Food Chem.* **25,** 1404–1408 (1977).

Katan, J., Fuhremann, T. W., and Lichtenstein, E. P., *Science* **193,** 891–894 (1976).

Madhun, Y. A., Young, J. L., and Freed, V. H., *J. Environ. Qual.* **15,** 64–68 (1986).

McCall, P. J., and Agin, G. L., *Environ. Toxicol. Chem.* **4,** 37–44 (1985).

Meikle, R. W., Regoli, A. J., Kurihara, N. H., and Laskowski, D. A., *in* "Bound and Conjugated Pesticide Residues" (D. D. Kaufman, G. G. Still, G. D. Paulson, and S. K. Bandal, eds.), pp. 272–284. American Chemical Society, Washington, DC, 1976.

Ou, L.-T., and Alexander, M., *Soil Sci.* **118,** 164–167 (1974).

Pavlostathis, S. G., and Mathavan, G. N., *Environ. Sci. Technol.* **26,** 532–538 (1992).

Priesack, E., *Soil Sci. Soc. Am. J.* **55,** 1227–1230 (1991).

Racke, K. D., and Lichtenstein, E. P., *J. Agric. Food. Chem.* **33,** 938–943 (1985).

Scow, K. M., and Alexander, M., *Soil Sci. Soc. Am. J.* **56,** 128–134 (1992).

Scow, K. M., and Hutson, J., *Soil Sci. Soc. Am. J.* **56,** 119–127 (1992).

Scribner, S. L., Benzing, T. R., Sun, S., and Boyd, S. A., *J. Environ. Qual.* **21,** 115–120 (1992).

Singh, D. K., and Agarwal, H. C., *J. Agric. Food Chem.* **40,** 1713–1716 (1992).

Steinberg, S. M., Pignatello, J. J., and Sawhney, B. L., *Environ. Sci. Technol.* **21,** 1201–1208 (1987).

Stevenson, F. J., *in* "Bound and Conjugated Pesticide Residues" (D. D. Kaufman, G. G. Still, G. D. Paulson, and S. K. Bandal, eds.), pp. 180–207. American Chemical Society, Washington, DC, 1976.

Tisdall, J. M., and Oades, J. M., *J. Soil Sci.* **33,** 141–163 (1982).

Weissenfels, W. D., Klewer, H. J., and Langhoff, J., *Appl. Microbiol. Biotechnol.* **36,** 689–696 (1992).

11 EFFECT OF CHEMICAL STRUCTURE ON BIODEGRADATION

Shortly following the beginning of the widespread use of pesticides and detergents, it was recognized that members of individual classes of organic compounds had markedly different periods of persistence in soils and waters. In many instances, a slight modification in the structure of the molecule was found to make it considerably more or less susceptible to destruction in these environments. Because it was soon evident that the compounds with short lives were destroyed microbiologically, the conclusion was reached that these modest alterations in chemical structure changed the suitability of the molecules as substrates for growth or metabolism by the resident community of microorganisms. An apparently modest alteration in the molecule, for example, the substitution of one atom or substituent for another, rendered the molecule appreciably more or less susceptible to microbial metabolism. Because of the enormous economic importance of these pesticides and surfactants and the growing concern with their contribution to environmental deterioration if they persisted, considerable effort was directed to establishing the structural features that governed the suitability of these chemicals for microbial degradation. Since that time, studies have been conducted on a wide variety of other chemical classes, including molecules that have different uses, and these have provided a large literature on the relationship between structure and biodegradation.

The choices of compounds for these investigations have rarely been made to establish generalizations or scientific principles relative to the underlying mechanisms, but rather have fit in with the needs of the particular industry involved. Thus, a large base of information exists on certain pesticides and surfactants that are major components of widely used detergents, but only a limited number of compounds of other classes have been investigated. As a result, the types of compounds that might be selected to establish more widely useful generalizations have not been tested, and the generalizations that are possible are still only a few in number.

In the two major industries (those concerned with pesticides and surfactants) initially concerned with the relationship between chemical structure and biodegradation, the compelling reason to conduct research was to replace the more persistent chemicals with new but often structurally similar molecules that were more readily metabolized. The public outcry against surfactants that persisted for long periods in water and the concern with the ecological or health consequences of the highly persistent pesticides motivated industry and, in many instances, regulatory agencies of government to seek replacements. The continued use of these classes of chemicals was threatened because of their longevity in waters and soils, and this longevity was specifically associated with characteristics of the molecules that made them less suitable for microbial metabolism and growth. Their replacement with new compounds is witness to the success in the search for more readily decomposable compounds.

The approach to finding biodegradable replacements for persistent but efficacious compounds largely remains one of trial and error. Such trial-and-error approaches are characteristic of the industrial research focused on pesticides, surfactants, the detergent builders that accompany the surfactants, polymers, and other classes of materials. Such approaches are expensive and time-consuming, and they often fail. A far more suitable approach would rely on basic principles that explain the relationships between structure and biodegradability. Unfortunately, the information base for establishing meaningful relationships is still quite small. The underlying issue is the specificity of microorganisms for their substrates, a specificity that is linked, in large part, with the specificity of enzymes to catalyze only certain types of chemical reactions. Each enzyme is largely restricted to carrying out only a single type of reaction on a narrow and often unpredictable range of substrates of very similar structures.

GENERALIZATIONS

At the present time, only a few generalizations are possible on the influence of structure on biodegradation, and exceptions to these generalizations or to other generalizations that might be made are many. Several reasons may be advanced for the few generalizations and the many exceptions. (a) Different microorganisms are present in dissimilar environments, and the development of one population may result in one group of related chemicals being degraded in the first environment, but because of the proliferation of other organisms in a different habitat, another set of compounds may be destroyed in the second environment. (b) Structural features of organic substrates often alter their availability to microorganisms (e.g., by sorption or by partitioning into a NAPL), and thus a molecule of one structure may be readily degraded in environments in which it is freely available but persist where its bioavailability is low. (c) In those instances in which a lengthy acclimation period occurs before biodegradation is detected, it is likely that a somewhat unique population appears for the degradation; those populations arising as a result of acclimation probably will not be the same in dissimilar environments. (d) The physical or chemical characteristics of two environments are quite different, for example, because one is aerobic and another anaerobic or one is at low pH and another is at neutral pH, and the populations that assume dominance in the transformation will not only often be different but will rely on different enzymes. These dissimilar enzymes will likely act on different members of groups of closely related chemical structures.

A few of these difficulties are well illustrated in studies of cultures of individual bacterial species. Investigations of several organisms that are able to utilize the same organic substrate as a C source for growth show that they proliferate by using or cometabolizing a somewhat different range of chemicals of an individual class of substrates. Some of the molecules are metabolized by one species but not a second, and the second organism will use a few but not others that support growth or metabolism of the first species. Such investigations establish the catabolic potential of an isolate and provide more definitive answers on the effect of chemical structure on microbial utilization than would be obtained in studies of natural environments that contain a multitude of species with dissimilar degradative potentials. Nevertheless, generalizations derived from studies of individual microorganisms suffer from the fact that they may not apply to an environment where the tested species is not present and where a population with an entirely different range of substrates assumes dominance in a particular biodegradation. Individual organisms have their physiological and catabolic idiosyncrasies, and the idiosyncracies may not be related to the intrinsic resistance of chemicals to biodegradation. Because microbial strains, species, and genera have enzymes with dissimilar substrate specificities and probably different cell permeabilities, it is more difficult to establish generalizations than in chemistry, in which the role of structure on chemical reactivity is also addressed. For the present, it is prudent to assume that susceptibility to biodegradation is an attribute of a chemical class in a particular ecosystem with particularly important environmental variables, including O_2 status, and that the biochemical potential of the entire community, and not just individual species, needs to be assessed.

The difficulty in making generalizations is particularly evident among chemicals that are quite persistent and then, after a long acclimation period, suddenly disappear. In these instances, probably no organism was initially able to grow rapidly on the compound. However, after some time, a rare organism becomes prominent, or a genetic change occurs in one of the indigenous species such that it is now able to metabolize the chemical. That newly emerging population then acts on a range of chemicals that otherwise would have appeared to be persistent in tests of the original environment.

REASONS FOR PERSISTENCE

Life appeared on earth hundreds of millions of years ago, and the biochemistry of living organisms has evolved in a limited number of ways. Comparisons of ancient fossils with modern organisms disclose that the chemistry of cells has not changed drastically, suggesting that only some of the countless reactions possible of organic molecules have been exploited in the processes necessary for metabolism. The resulting anabolic, and presumably catabolic, processes are thus few in number. Many possible chemical reactions are thus foreign to macro- and microorganisms. All the catabolic reactions that characterize living cells are consistent with chemical principles, but because of the few reactions that can be catalyzed by the enzymes that have appeared during the course of biochemical evolution as well as the very complexity of enzymes as catalysts and microorganisms as integrated assemblages of catalysts, the reactions that microorganisms effect may not be those that would be expected based on the chemical principles established for the same organic molecules whose change is brought about by nonbiological agencies.

Given the relatively few catabolic pathways that characterize microbial cells, it is not surprising that an organic chemical that is not a product of biosynthesis, sometimes termed a *xenobiotic,* will be degraded to an *appreciable extent* only if an enzyme or an enzyme system exists that is able to catalyze its conversion to a product that is an intermediate or a substrate of one of those pathways. (In contrast, a chemical may be modified to a *slight extent,* i.e., cometabolized, but the one or few enzymes involved convert the substrate to a product having many of the features of the parent molecule.) The greater the difference in structure of the xenobiotic from the constituents of living organisms or the less common the substituent in living matter, the less the likelihood of extensive biodegradation or the slower the transformation.

Furthermore, if nearly all microorganisms able to metabolize a chemical do so by one or a few similar metabolic pathways, then modification of the molecule to render it somewhat different from the intermediates or substrates of those pathways probably will prevent, slow, or delay the initiation of the biodegradation. If the altered molecule can be modified enzymatically to yield a natural intermediate, then the degradation will proceed—although the process may be slow if the organism that converts the xenobiotic to a natural intermediate grows slowly when that molecule is the C source. On the other hand, biodegradation will be evident only after a long period if (a) the microorganism that can convert the xenobiotic is present initially at low cell densities and must proliferate before a significant loss of the parent compound is detected, (b) a mutant must appear, or (c) the organism that destroys the chemical does so by ignoring the change that makes the natural product into a xenobiotic, that is, acts on a portion of the molecule not affected by the change.

Organic compounds that are mineralizable may be rendered partially or possibly wholly resistant to mineralization by the addition of a single substituent. These substituents may be termed *xenophores,* that is, substituents that are physiologically uncommon or that are entirely nonphysiological. Thus, because addition of a single Cl, NO_2, SO_3H, Br, CN, or CF_3 to simple aromatic molecules, fatty acids, or other readily utilizable substrates greatly increases their resistance, they are xenophores. These substituents are alien to most organisms, and hence their removal is not effected by many species, or their presence hinders the functioning of otherwise common pathways. Sometimes CH_3, NH_2, OH, and OCH_3 act as xenophores. Often the degradation of organic compounds is stimulated by the presence of OH, COOH, or an amide, ester, or anhydride functional group.

Stated in other terms, the identity of the substituent added to a molecule affects its biodegradation, or at least its mineralization. This is evident by the findings in Table 11.1 for a group of compounds. The resistance associated with the presence of halogens and NO_2 is widespread, but only a few compounds containing CN or CF_3 have been tested. The impact of the substituent will vary, however, with the structure of the rest of the molecule, witness that some compounds with NO_2 (e.g., 4-nitrophenol), SO_3H (e.g., some surfactants), and CN groups are quickly destroyed.

The presence of several potential xenophores on a molecule makes it less likely for a xenobiotic to be transformed to intermediates in normal metabolic pathways or results in slower rates of transformation; hence, they are more resistant to biodegradation. Typically, the addition of two xenophores, whether they be identical or different, to a biodegradable molecule makes it even less likely for rapid degradation to occur or results in a still longer acclimation period before a population of sufficient size develops and causes rapid degradation. Addition of a third xenophore, the same or a different one, renders the molecule even more resistant or results in the need for a still longer acclimation period for aerobic

TABLE 11.1

Substituents Whose Addition Slows Extensive Aerobic Biodegradation of
Organic Compounds

Test chemical	Substituent slowing degradation[a]	Test environment	Reference
Aniline	3-Br, 3-CH$_3$, 3-NO$_2$, 3-CN, 3-OCH$_3$	Water	Paris and Wolfe (1987)
Aniline	4-Cl	Soil	Süss et al. (1978)
Benzene	NO$_2$, SO$_3$H[b]	Soil suspension	Alexander and Lustigman (1966)
Benzoic acid	2-, 3-, or 4-NO$_2$ 2-, 3-, or 4-Cl	Wastewater	Haller (1978)
Cyanuric acid	NH$_2$	Soil	Hauck and Stephenson (1964)
Diphenylmethane	3-CF$_3$	Water	Saeger and Thompson (1980)
IPC	3-Cl	Soil	Clark and Wright (1970)
Phenylacetic acid	4-NO$_2$, 4-Cl	Soil suspension	Subba-Rao and Alexander (1977)
Pyridine	2-, 3-, or 4-NH$_2$	Soil suspension	Sims and Sommers (1986)
Pyridine	2-Cl; 2-, 3-, or 4-OH	Soil suspension	Naik et al. (1972)
Valeric acid	3-CH$_3$	Soil suspension	Hammond and Alexander (1972)

[a]The number designates the position of the substituent.
[b]Compared to phenol.

biodegradation. Some typical examples in which the added substituents
are Cl, NH$_2$, NO$_2$, OH, or CH$_3$ are shown in Table 11.2. The entries in
the third column indicate whether the molecule is mono-, di-, or trisubsti-
tuted with the xenophore or the position of the substituent on the otherwise
readily degradable substrate. Such retarding effects by increasing number
of substituents may not occur under anaerobic conditions, at least not for
chlorophenols (Mikesell and Boyd, 1985) and chlorophenoxyacetic acids
(Suflita et al., 1984).

The position of the xenophore on the molecule has a pronounced influence on the degradation. At some positions, it may have little impact; in
others, it may drastically reduce the rate of microbial utilization. Because
different environments contain dissimilar populations, the effect on biodegradation of the position of the substituent may not be the same in all
localities. This is shown by the data in Table 11.3, from which it is evident
that a substituent in one position may enhance degradation in one environ-

TABLE 11.2
Effect of Several Xenophores on Extensive Aerobic Biodegradation of Organic Compounds

Test chemical	Substituent	Degradation rate	Test environment	Reference
Acetic acid	Cl	Mono > di > tri	Soil	Kaufman (1966)
Aniline	NH_2	4- > 3,4-	Soil	Süss et al. (1978)
Benzoic acid	Cl	Mono > 2,4-	Sewage	DiGeronimo et al. (1979)
Benzoic acid	NO_2	3-, 4- > 3,5-	Sewage	Hallas and Alexander (1983)
Cyanuric acid	NH_2[a]	Mono > di > tri	Soil	Hauck and Stephenson (1964)
Diphenylmethane	OH	4- > 4,4'-	Soil suspension	Subba-Rao and Alexander (1977)
Fatty acid	Cl	Mono > di	Sewage	Dias and Alexander (1971)
Fatty acid	CH_3	Mono > di	Soil suspension	Hammond and Alexander (1972)
IPC	Cl[b]	4- > 2,4- > 2,4,5-	Soil	Kaufman (1966)
Phenoxyacetic acid	Cl	2,4- > 2,4,5-	Soil	Burger et al. (1962)
Propionic acid	Cl	2- > 2,2- > 2,3,3-	Soil	Kaufman (1966)
Pyridine	OH	Mono > di	Soil suspension	Sims and Sommers (1986)

[a]NH_2 replaces the OH of cyanuric acid.
[b]Cl on the ring.

TABLE 11.3
Effect of Position of Substituent on Biodegradation

Compound	Added substituent	Effect of substituent position on biodegradation		Environment	Reference
		Rapid	Slow		
Phenol	Cl	Not meta	Meta	Soil suspension	Alexander and Aleem (1961)
	Cl	Not meta	Meta	Soil	Baker and Mayfield (1980)
	Cl	2, 3-	4-	Wastewater	Haller (1978)
	Cl	2-	4-	Soil	Boyd et al. (1983)
	Cl	2-	4-	Sludge	Mikesell and Boyd (1985)
Benzoic acid	Cl	3-	2-, 4-	Wastewater	Haller (1978)
	Cl	3,4-	2,4-	Sewage	DiGeronimo et al. (1979)
Aliphatic acid	Halogen	ω-	α-, β-	Sewage	Dias and Alexander (1971)
Phenol	CH$_3$	2-	4-	Soil	Boyd et al. (1983)
	CH$_3$	4-	3-, 2-	Soil	Smolenski and Suflita (1987)
Phenoxyalkanoic acid	Cl	4-	3-	Soil	Burger et al. (1962)
Aliphatic acid	Cl	4-	2-	Soil	Audus (1960)
	Phenoxy	ω-	α-	Soil	Burger et al. (1962)
Benzamide	Cl	3,6-	2,6-	Soil	Fournier (1974)
Diphenylmethane	Cl	2,4-	2,5-	Sludge	Saeger and Thompson (1980)

ment and depress it in another. Moreover, one xenophore in a given position on the molecule may have a different effect than another in the same location. Thus, generalizations on the effect of position of substituents do not appear to be applicable to all environments. That the effect is dependent on the idiosyncracies of the particular populations that appear is suggested by the finding that some bacteria readily degrade di-, tri-, tetra-, and pentachlorophenols (Apajalahti and Salkinoja-Salonen, 1986), yet the populations in most environments do not act readily on polychlorinated phenols. Thus, if conditions favor development of a population that behaves atypically, as may occur because of prior additions of specific compounds, the populations in that environment will degrade substituted compounds at different relative rates than in some other environment in which that type of population is not favored.

An early, highly dramatic, and widely recognized effect of structure on biodegradation was evident when synthetic detergents first became widely used. The lakes and rivers into which the washwaters were introduced showed an obvious failing of the microflora: they were covered with froth and foam. The reason: those early surfactants in the detergent preparations were not quickly destroyed. The ensuing public outcry compelled industry to seek the cause for the persistence, and industrial researchers quickly found that extensive methyl branching on the alkyl moiety of the surfactant created obstacles for the mineralizing populations.

These early detergents contained alkylbenzene sulfonates (ABSs) as the surfactant constituent (Fig. 11.1). The sulfonate was at different positions on the benzene ring, but the alkyl portion invariably contained many methyl branches. This large number of methyl groups on the alkyl moiety was responsible for the longevity of the surfactants in surface waters and in sewage treatment facilities. A single methyl group usually had little or no noticeable influence among the many tested chemicals that had alkyl chains of various lengths, but many methyl branches on the alkyl portion rendered the molecule less readily biodegradable. Moreover, if these ABSs had two methyl groups on the penultimate C, making it into a quaternary

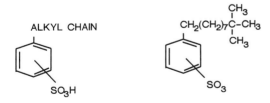

FIG. 11.1 ABS and ABS with a quaternary C.

C (Fig. 11.1), the compound showed considerable resistance. In the face of widespread public pressure and the threat of government regulation, the soap and detergent industry quickly not only found the structures that were refractory but also provided replacements. These readily degraded alternatives did not have alkyl moieties with many methyl branches but rather were linear [$RCH_2CH_2(CH_2)_nCH_2CH_3$] and did not contain the quaternary C atoms that were the culprits (Huddleston and Allred, 1967; Swisher, 1987).

Methyl branching is also associated with persistence of aliphatic hydrocarbons, aliphatic acids, alcohols, and other chemicals. Among the alkanes that are major components of oil, the nonbranched molecules entering soils and waters usually are more readily destroyed than are those alkanes having several or many methyl branches on the chain. Among the more resistant of these hydrocarbons are the highly branched phytane and pristane (Bossert and Bartha, 1984; Jobson *et al.*, 1974; Kator, 1973):

$$\underset{CH_3}{\overset{CH_3}{|}}\quad \underset{CH_3}{\overset{CH_3}{|}}\quad \underset{CH_3}{\overset{CH_3}{|}}\quad \underset{CH_3}{\overset{CH_3}{|}}$$
$$CH_3CH(CH_2)_3CH(CH_2)_3CH(CH_2)_3CHCH_3 \qquad \text{Pristane}$$

$$\underset{CH_3}{\overset{CH_3}{|}}\quad \underset{CH_3}{\overset{CH_3}{|}}\quad \underset{CH_3}{\overset{CH_3}{|}}\quad \underset{CH_3}{\overset{CH_3}{|}}$$
$$CH_3CH(CH_2)_3CH(CH_2)_3CH(CH_2)_3CHCH_2CH_3 \quad \text{Phytane}$$

Similarly, mono- and dicarboxylic acids with no methyl branches are quickly degraded by soil microorganisms. These molecules have the structures

$$CH_3(CH_2)_nCOOH$$

$$HOOC(CH_2)_nCOOH$$

However, if the two hydrogens on one of the methylene (CH_2) carbons are replaced with two methyl groups, thereby giving a quaternary C, mineralization is markedly retarded (Hammond and Alexander, 1972). In like fashion, linear aliphatic alcohols with the structure

$$CH_3(CH_2)_nCH_2OH$$

are quickly destroyed microbiologically, but alcohols with quaternary carbons tend to be attacked only slowly (Dias and Alexander, 1971; Fukuda and Brannon, 1971; McKinney and Jeris, 1955).

The reason for this influence of structure on biodegradation has been established. Alkanes or compounds, like ABSs, with alkyl side chains are usually metabolized to give the corresponding carboxylic acids:

$$R(CH_2)_nCH_2CH_2CH_3 \rightarrow R(CH_2)_nCH_2CH_2CH_2OH \rightarrow$$

$$R(CH_2)_nCH_2CH_2CHO \rightarrow R(CH_2)_nCH_2CH_2COOH$$

Alcohols are intermediates in this sequence, and primary linear alcohols [$R(CH_2)_nCH_2CH_2CH_2OH$] in nature appear to be metabolized as depicted in this sequence. The carboxylic acid [$R(CH_2)_nCH_2CH_2COOH$] is then destroyed by a sequence known as β-oxidation. The sequence was so named because the β-carbon is oxidized as shown in Fig. 11.2. The new carboxylic acid thus generated has two carbons fewer than the first one, and it, in turn, is metabolized in the same fashion to give CH_3COOH and a third carboxylic acid, which has two carbons fewer than its predecessor. The process is repeated until the molecule is converted to simple carboxylic acids, which are readily oxidized. The reaction inside the cell, in fact, involves coenzyme A, but the coenzyme-complexed acids are not found outside the cell.

Consider the α and β carbons in this sequence. In the pathway of conversion, they bear H, OH, or O but no methyl branches. The enzymatic transformation will not proceed if a methyl is present. However, individual methyl groups may be removed enzymatically. If two methyl groups are present on either of the C atoms, the likelihood is small of their both being removed readily. Hence, such a hindrance to β-oxidation operates if the C containing two methyl groups (the quaternary C) is near the end of the chain—either where the COOH is present or where a COOH will be formed from the CH_3 or CH_2OH. The quaternary C apparently does not initially pose a hindrance if it is some distance from the end of the chain. Under these conditions, the chain is shortened, but only until the shortening brings the COOH close to the quaternary C, at which point the further degradation is difficult to achieve. This shorter chain bearing the quaternary C then persists (Catelani *et al.*, 1977; Hammond and Alexander, 1972) (Fig. 11.3). With time, however, these compounds probably are destroyed, possibly because of the growth of microorganisms that possess a different mechanism of degrading such compounds, as by the oxidation of the methyl groups attached to the quaternary C.

Another influence of structure on biodegradation is evident among molecules containing both an aromatic and an alkyl or aliphatic acid moiety.

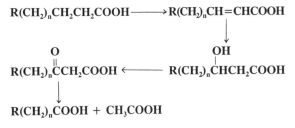

FIG. 11.2 Products formed in β-oxidation.

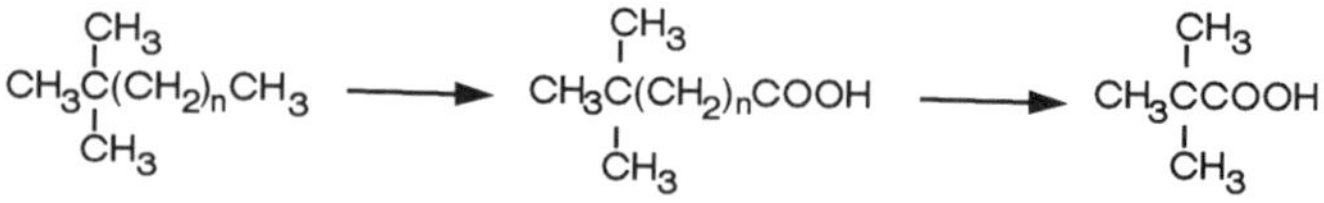

$$\underset{\underset{CH_3}{|}}{\overset{\overset{CH_3}{|}}{CH_3C(CH_2)_nCH_3}} \longrightarrow \underset{\underset{CH_3}{|}}{\overset{\overset{CH_3}{|}}{CH_3C(CH_2)_nCOOH}} \longrightarrow \underset{\underset{CH_3}{|}}{\overset{\overset{CH_3}{|}}{CH_3CCOOH}}$$

FIG. 11.3 Conversion of dimethylalkanes and dimethylalkanoic acids to trimethylacetic acid (synonym, pivalic acid).

This effect is related to the place at which the alkyl or aliphatic acid portion is linked to the benzene ring. Thus, among the phenoxy herbicides in which biodegradation is not delayed because of a *meta* chlorine on the ring, detoxication and loss of the parent molecule are rapid if the phenoxy portion is linked to the last (the ω-position) and not the α C of the aliphatic acid (Burger *et al.*, 1962) (Fig. 11.4). Similarly, 4-phenylbutyric acid is readily destroyed by sewage microorganisms, but 2-phenylbutryic acid is resistant (Dias and Alexander, 1971) (Fig. 11.4). An analogous relationship is evident among some of the ABSs that do not have methyl branches; that is, the isomers with the phenyl linked near the end of the alkyl moiety are destroyed more quickly than those in which the linkage is with the central portion of the chain (Swisher, 1987). These observations are consis-

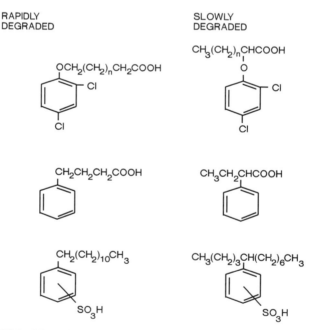

FIG. 11.4 Effect of the point of linkage of the aromatic ring to the alkane or aliphatic acid on biodegradation.

tent with the view that the phenyl or phenoxy moiety hinders β-oxidation, the obstruction more close to the carboxyl or the end of the chain being more likely to hinder this pathway. On the other hand, the point of attachment of the alkyl moiety to the aromatic ring sometimes may have no effect (Larson, 1990).

A marked effect on biodegradability is evident among the PAHs in soil. Anthracene, phenanthrene, and acenaphthylene contain three rings (as well as pyrene, a tetracyclic molecule) and are destroyed at reasonable rates when O_2 is present (Fig. 11.5). Other compounds with four rings as well as the pentacyclic hydrocarbons, in contrast, are highly persistent.

Generalizations about structure–biodegradability relationships, which are sometimes termed SBRs, in aerobic environments do not seem to be applicable to anaerobic environments. For example, the findings that 3-chlorobenzoate is degraded but not 2- and 4-chlorobenzoates and that 2,4,5-trichlorobenzoate is metabolized but not mono- and dichlorobenzoates (DeWeerd et al., 1986) are totally dissimilar from the patterns in aerobic habitats. Similar differences in structure–biodegradability relationships have been noted during the metabolism of phenols, benzoates, and phenoxyalkanoates in anaerobic sludge from municipal sewage (Buisson et al., 1986) and in anaerobic sediments (Genthner et al., 1989).

Other structural characteristics of chemicals are associated with slow mineralization. As a rule, however, the observations with a single class of chemicals are few in number so that generalizations are still risky. On the other hand, many studies have been performed with pure cultures, but it is not clear which of the data represent valid generalizations for microbial communities and which merely reflect the idiosyncracies of the particular species being tested. Consider, for example, a group of homologous chemicals (designated A, B, C, D, E, and F) that differ in some slight way from one another, and assume that four species exist in nature with somewhat different abilities to use the group of homologues as sources of C and energy for growth. Based on their dissimilar capacities to utilize these potential nutrients, environments containing one or more of these species would show markedly different patterns of degradation in an evaluation of the effect of chemical structure on the fate of the chemical. Some of these possibilities are illustrated in Table 11.4. In the examples given in Table 11.4, the first species metabolizes only A for growth, the second uses A, B, or C, the third can grow on A or D, and the fourth can use C, D, or E as nutrients for growth. The outcomes of these activities in nature are given below Table 11.4.

Generalizations on the effect of structure on biodegradation often do not apply following growth of a single population able to destroy a less readily degradable isomer or member of a homologous series of com-

MOST RAPIDLY DEGRADED

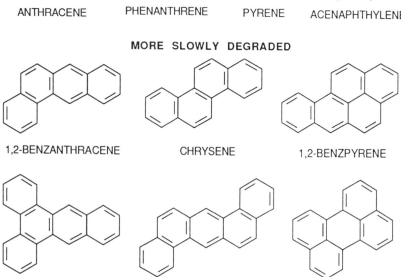

ANTHRACENE   PHENANTHRENE   PYRENE   ACENAPHTHYLENE

MORE SLOWLY DEGRADED

1,2-BENZANTHRACENE   CHRYSENE   1,2-BENZPYRENE

1,2,3,4-DIBENZANTHRACENE   1,2,5,6-DIBENZANTHRACENE   PERYLENE

FIG. 11.5 Relative rates of biodegradation of polyaromatic hydrocarbons in soil (Reprinted with permission from Bossert and Bartha, 1986).

TABLE 11.4
Biodegradability of a Homologous Series of Compounds That Support Growth of Different Species

Species	Growth rate of species	Occurrence of species	Homologous compound supporting growth					
			A	B	C	D	E	F
1	Rapid	Widespread	+	−	−	−	−	−
2	Rapid	Rare	+	+	+	−	−	−
3	Slow	Widespread	+	−	−	+	−	−
4	Slow	Rare	−	−	+	+	+	−

Outcomes: A: always degraded rapidly; B: degraded rapidly but only in some environments; C: degraded rapidly, slowly, or not at all; D: always degraded slowly; E: degraded slowly but only in some environments; F: not degraded.

pounds. The activity then represents the metabolic potential of the previously rare organism, and the spectrum of substrates acted on by this population may be quite different from that of the original microbial community. This is evident in the finding that once samples of sewage become acclimated to pentachlorophenol, the populations that proliferate during the acclimation period are able to metabolize tri- and tetrachlorophenols (Etzel and Kirsch, 1975) that, in an environment with no such acclimation, would be highly persistent.

PREDICTING BIODEGRADABILITY

Given the broad array of organic chemicals that are of potential ecological and public health concern, attempts have been made to develop an overview of why structure affects degradability. This overview is needed for the development of a predictive capacity for previously untested chemicals. Such a predictive ability is important (a) in industry so that chemicals may be synthesized that will not only be useful for the particular purpose but will also have relatively short lives in nature and (b) for regulatory agencies of government that must decide whether a new chemical will be persistent or not before it is introduced commercially. The need by regulatory agencies for a predictive capacity is obvious in light of the fact that probably less than 1% of the more than 2000 new compounds submitted to the U.S. Environmental Protection Agency each year for regulatory review contain data on biodegradability (Boethling and Sabljić, 1989). For those new commercial products that will be released into natural environments, either deliberately or inadvertently, that information is important to assess potential exposure.

One approach to predicting biodegradability from the properties of a molecule invokes the similarity of the test chemical to substrates and intermediates in known metabolic pathways. The chemicals that do not differ to a marked extent or have only a single xenophore are presumably decomposed reasonably quickly. Those that differ to a significant extent are more resistant, presumably because they are converted to these intermediates slowly owing to the need for several enzymes (found only in rare species) to convert the synthetic molecules to natural products. This approach is biochemical because it relies on comparisons with biochemical precedents.

A variety of other approaches to predicting biodegradability rely on physical or chemical properties of the compound rather than on biochemical principles. Among the properties suggested to predict biodegradability are water solubility, melting point, boiling point, molecular weight, molar refractivity, density, and log K_{ow} or some other descriptor of hydrophobicity. Such approaches are designed to predict degradability, qualitatively at least, based on chemical properties that can be determined either experimentally or from considerations of the structure of the compound. For example, the biodegradation rates of some compounds are correlated with the rate constants for their alkaline hydrolysis, but the biodegradability of N-methyl arylcarbamates is not correlated with their susceptibility to abiotic hydrolysis, even though the initial metabolic step is a hydrolysis of the carbamate (Pussemier et al., 1989). The rates of anaerobic biodegradation of halogenated aromatic compounds in sediments is correlated with the strength of the C–halogen bond that is cleaved, a not unsurprising finding because this is the bond broken in the rate-determining step of the microbial conversion (Peijnenburg et al., 1992). Use also is made of the group contribution approach, which entails dividing the molecule into functional groups or their fragments (Dessai et al., 1990), and of electronic and steric parameters of the molecule (Degner et al., 1991).

Predictions of biodegradability also are based on molecular topology, which deals with such structural features of molecules as their shapes and sizes, the presence of branching, or the types of atom-to-atom connections. Of particular interest in molecular topology is molecular connectivity. Molecular connectivities can be determined from the structural formula of the chemical, and their use appears to have promise for some classes of compounds (Boethling, 1986; Boethling and Sabljić, 1989; Dearden and Nicholson, 1986; Sabljić and Piver, 1992). The rates of decomposition also appear to be correlated with the van der Waals radius (the van der Waals radius being a property of the substituents of the compound) for phenols with several substituents in the para position (Paris et al., 1982, 1983) and anilines with several substituents in the meta position (Paris and Wolfe, 1987). The rate of decomposition of a few chemicals is also correlated with the Hammett substituent constant (Reineke and Knackmuss, 1978; Pitter, 1985).

In face of the enormous number and variety of organic molecules that have yet to be tested, it must be admitted that predicting biodegradability among classes for which there are few precedents remains difficult. Nevertheless, models have been devised that have good predictive abilities for increasingly larger numbers of compounds, and these correctly classify the relative biodegradability of many compounds under aerobic conditions (Howard et al., 1991; 1992). As the data base grows and more generalizations and models appear, it should be possible to better select structures that not only have efficacy for the purposes they were made but also degrade sufficiently rapidly to prevent untoward consequences.

REFERENCES

Alexander, M., and Aleem, M. I. H., *J. Agric. Food Chem.* **9,** 44–47 (1961).

Alexander, M., and Lustigman, B. K., *J. Agric. Food Chem.* **14,** 410–413 (1966).

Apajalahti, J. H. A., and Salkinoja-Salonen, M. S., *Appl Microbiol. Biotechnol.* **25,** 62–67 (1986).

Audus, L. J., *in* "Herbicides and the Soil" (E. K. Woodford and G. R. Sagar, eds.), pp. 1–19. Blackwell, Oxford, 1960.

Baker, M. D., and Mayfield, C. I., *Water, Air, Soil Pollut.* **13,** 411–424 (1980).

Boethling, R. S., *Environ. Toxicol. Chem.* **5,** 797–806 (1986).

Boethling, R. S., and Sabjlić, A., *Environ. Sci. Technol.* **23,** 672–679 (1989).

Bossert, I., and Bartha, R., *in* "Petroleum Microbiology" (R. M. Atlas, ed.), pp. 435–473. Macmillan, New York, 1984.

Bossert, I. D., and Bartha, R., *Bull. Environ. Contam. Toxicol.* **37,** 490–495 (1986).

Boyd, S. A., Shelton, D. R., Berry, D., and Tiedje, J. M., *Appl. Environ. Microbiol.* **46,** 50–54 (1983).

Buisson, R. S. K., Kirk, P. W. W., Lester, J. N., and Campbell, J. A., *Water Pollut. Control* **85,** 387–394 (1986).

Burger, K., MacRae, I. C., and Alexander, M., *Soil Sci. Soc. Am. Proc.* **26,** 243–246 (1962).

Catelani, D., Colombi, A., Sorlini, C., and Treccani, V., *Appl. Environ. Microbiol.* **34,** 351–354 (1977).

Clark, C. G., and Wright, S. J. L., *Soil Biol. Biochem.* **2,** 19–26 (1970).

Dearden, J. C., and Nicholson, R. M., *Pestic. Sci.* **17,** 305–310 (1986).

Degner, P., Nendza, M., and Klein, W., *Sci. Total Environ.* **109/110,** 253–259 (1991).

Dessai, S. M., Govind, R., and Tabak, H. H., *Environ. Toxicol. Chem.* **9,** 473–477 (1990).

DeWeerd, K. A., Suflita, J. M., Linkfield, T., Tiedje, J. M., and Pritchard, P. H., *FEMS Microbiol. Ecol.* **38,** 331–339 (1986).

Dias, F. F., and Alexander, M., *Appl. Microbiol.* **22,** 1114–1118 (1971).

DiGeronimo, M. J., Nikaido, M., and Alexander, M., *Appl. Environ. Microbiol.* **37,** 619–625 (1979).

Etzel, J. E., and Kirsch, E. J., *Dev. Ind. Microbiol.* **16,** 287–295 (1975).

Fournier, J.-C., *Chemosphere* **3,** 77–82 (1974).

Fukuda, D. S., and Brannon, D. R., *Appl. Microbiol.* **21,** 550–551 (1971).

Genthner, B. R. S., Price, W. A., II, and Pritchard, P. H., *Appl. Environ. Microbiol.* **55,** 1466–1471 (1989).

Hallas, L. E., and Alexander, M., *Appl. Environ. Microbiol.* **45,** 1234–1241 (1983).

Haller, H. D., *J. Water Pollut. Control Fed.* **50,** 2771–2777 (1978).

Hammond, M. W., and Alexander, M., *Environ. Sci. Technol.* **6,** 732–735 (1972).

Hauck, R. D., and Stephenson, H. F., *J. Agric. Food Chem.* **12,** 147–151 (1964).

Howard, P. H., Boethling, R. S., Stiteler, W., Meylan, W., and Beauman, J., *Sci. Total Environ.* **109/110,** 635–641 (1991).

Howard, P. H., Boethling, R. S., Stiteler, W. M., Meylan, W. M., Hueber, A. E., Beauman, J. A., and Larosche, M. E., *Environ. Toxicol. Chem.* **11,** 593–603 (1992).

Huddleston, R. L., and Allred, R. C., *in* "Soil Biochemistry" (A. D. McLaren and G. H. Peterson, eds.), pp. 343–370. Dekker, New York, 1967.

Jobson, A., McLaughlin, M., Cook, F. D., and Westlake, D. W. S., *Appl. Microbiol.* **27,** 166–171 (1974).

Kator, H., *in* "The Microbial Degradation of Oil Pollutants" (D. G. Ahearn and S. P. Meyers, eds.), pp. 47–65. Louisiana State University, Center for Wetlands Resources, Baton Rouge, 1973.

Kaufman, D. D., *in* "Pesticides and Their Effects on Soils and Water" (M. E. Bloodworth, ed.), pp. 85–94. Soil Science Society of America, Madison, WI, 1966.

Larson, R. J., *Environ. Sci. Technol.* **24,** 1241–1246 (1990).

McKinney, R. E., and Jeris, J. S., *Sewage Ind. Wastes* **27,** 728–735 (1955).

Mikesell, M. D., and Boyd, S. A., *J. Environ. Qual.* **14,** 337–340 (1985).

Naik, M. N., Jackson, R. B., Stokes, J., and Swaby, R. J., *Soil Biol. Biochem.* **4,** 313–323 (1972).

Paris, D. F., and Wolfe, N. L., *Appl. Environ. Microbiol.* **53,** 911–916 (1987).

Paris, D. F., Wolfe, N. L., and Steen, W. C., *Appl. Environ. Microbiol.* **44,** 153–158 (1982).

Paris, D. F., Wolfe, N. L., Steen, W. C., and Baughman, G. L., *Appl. Environ. Microbiol.* **45,** 1153–1155 (1983).

Peijnenburg, W. J. M., Hart, M. J. 'T., den Hollander, H. A., van de Meent, D., Verboom, H. H., and Wolfe, N. L., *Environ. Toxicol. Chem.* **11,** 301–314 (1992).

Pitter, P., *Acta Hydrochim. Hydrobiol.* **13,** 453–460 (1985).

Pussemier, L., DeBorger, R., Cloos, P., and van Bladel, R., *J. Environ. Sci. Health Part B,* **B24,** 117–129 (1989).

Reineke, W., and Knackmuss, H.-J., *Biochim. Biophys. Acta* **542,** 412–423 (1978).

Sabjlić, A., and Piver, W. T., *Environ. Toxicol. Chem.* **11,** 961–972 (1992).

Saeger, V. W., and Thompson, Q. E., *Environ. Sci. Technol.* **14,** 705–709 (1980).

Sims, G. K., and Sommers, L. E., *Environ. Toxicol. Chem.* **5,** 503–509 (1986).

Smolenski, W. J., and Suflita, J. M., *Appl. Environ. Microbiol.* **53,** 710–716 (1987).

Subba-Rao, R. V., and Alexander, M., *J. Agric. Food Chem.* **25,** 327–329 (1977).

Suflita, J. M., Stout, J., and Tiedje, J. M., *J. Agric. Food Chem.* **32,** 218–221 (1984).

Süss, A., Fuchsbichler, G., and Eben, C., *Z. Pflanzenernaehr. Bodenkd.* **141,** 57–66 (1978).

Swisher, R. D., "Surfactant Biodegradation." Dekker, New York, 1987.

12 COMETABOLISM

Microorganisms have long been known to have the ability to transform organic molecules to yield organic products that accumulate in culture media. Such conversions have achieved prominence in industrial microbiology because of the importance of the products, especially pharmaceutical agents, thus generated. The first evidence of analogous transformations with environmentally important chemicals came from studies of chlorinated aliphatic acids. In this early investigation, it was noted that a strain of *Pseudomonas* that grew on monochloroacetate was able to dehalogenate trichloroacetate but not use the later compound as a C source for growth (Jensen, 1963). This transformation of an organic compound by a microorganism that is unable to use the substrate as a source of energy or of one of its constituent elements is termed *cometabolism* (Alexander, 1967).

The active populations thus derive no nutritional benefit from the substrates they cometabolize. Energy sufficient to fully sustain growth is not acquired even if the conversion is an oxidation and releases energy, and the C, N, S, or P that may be in the molecule is not used as a source, or at least a significant source, of these elements for biosynthetic purposes. Because of the preface *co*, which often is appended to a word to indicate that something is done jointly or together (as in copilot or cooperate), there have been some semantic disagreements. Specifically, some authorities propose that the term cometabolism should be applied only to circumstances in which a substrate that is not used for growth is metabolized in the presence of a second substrate that is used to support multiplication. According to this view, the transformation of a substance that is not used as a nutrient or energy source but which occurs in the absence of a chemical supporting growth should be designated by another term, for example, *fortuitous metabolism* (Dalton and Sterling, 1982). However, the prefix *co* also has another meaning, namely, the same or similar (as in coconscious). The latter usage implies that the cometabolic transformation is similar to some other metabolic reaction, which is consistent with one explanation for the phenomenon (see the following). Fortuitous metabolism is, indeed, a more attractive term because it suggests an explanation for cometabolism, but the term will be used here as in the original definition, if for no other reason than it has gained wide acceptance. Thus, the term will be used to describe the metabolism of an organic substrate by a microorganism that is unable to use that compound as a source of energy or an essential nutrient element. Covered by the term will be cases in which the organism is simultaneously growing on a second compound and instances in which multiplication is not occurring at the time the chemical of interest is being metabolized (Horvath, 1972).

The term *cooxidation* is sometimes used in studies of pure cultures of bacteria, this term referring specifically to oxidations of substrates that do not support growth in the presence of a second compound that does support multiplication (Perry, 1979). Cooxidation has historical precedence in the semantic debate (Foster, 1962), but since it is restricted to oxidation, the word does not have sufficient breadth to include many reactions that are not oxidations.

Nevertheless, it should be pointed out that two types of reactions of these sorts take place in pure cultures of bacteria. In one, the cometabolized compound is transformed only in the presence of a second organic substrate, which indeed may be the compound that supports growth (Malashenko et al., 1976; You and Bartha, 1982; Schukat et al., 1983). In the second, the compound is metabolized even in the absence of a second substrate (Horvath and Alexander, 1970a).

Particularly cogent reasons for using the more general definition, and even for maintaining cometabolism as a term apart from bioconversion or biotransformation, are the environmental consequences of cometabolism. Cometabolic reactions have impacts in nature that are different from growth-linked biodegradations, and when the transformations take place, it is usually totally unclear whether the microorganisms do or do not have a second substrate available on which they are growing.

SUBSTRATES AND REACTIONS

A large number of chemicals are subject to cometabolism in culture. Among the compounds thus acted on are cyclohexane (Beam and Perry, 1974), PCBs (Brunner et al., 1985), 3-trifluoromethylbenzoate (Knackmuss, 1981), several chlorophenols (Liu et al., 1991), 3,4-dichloroaniline (You and Bartha, 1982), 1,3,5-trinitrobenzene (Mitchell et al., 1982), such pesticides as propachlor (Novick and Alexander, 1985), alachlor (Smith and Phillips, 1975), ordram (Golovleva et al., 1978), 2,4-D (Bauer et al., 1979), and dicamba (Ferrer et al., 1985), as well as the compounds listed in Table 12.1. The organisms that carry out these reactions in laboratory media includes species of *Pseudomonas, Acinetobacter, Nocardia, Bacillus, Mycococcus, Achromobacter, Methylosinus,* and *Arthrobacter* among the bacteria and *Penicillium* and *Rhizoctonia* among the fungi. Among cometabolic conversions that appear to involve a single enzyme, the reactions may be hydroxylations, oxidations, denitrations, deaminations, hydrolyses, acylations, or cleavages of ether linkages, but many of the conversions are complex and involve several enzymes. Even a substrate that will, in nature, support growth of microorganisms may be metabolized by some bacteria in culture with no incorporation of the C into their cells (Schmitt *et al.,* 1992).

TABLE 12.1
Cometabolism of Various Substrates in Pure Culture

Substrate	Product	Reference
Tetrachloroethylene	Trichloroethylene	Fathepure and Boyd (1988)
Benzothiophene	Benzothiophene-2,3-dione	Fedorak and Grbić-Galić (1991)
3-Hydroxybenzoate	2,3-Dihydroxybenzoate	Daumy *et al.* (1980)
Cyclohexane	Cyclohexanol	deKlerk and van der Linden (1974)
3-Chlorophenol	4-Chlorocatechol	Engelhardt *et al.* (1979)
Chlorobenzene	3-Chlorocatechol	Klečka and Gibson (1981)
Bis(tributyltin) oxide	Dibutyl tin	Barug (1981)
3-Nitrophenol	Nitrohydroquinone	Raymond and Alexander (1971)
Trinitroglycerine	1- and 2-Nitroglycerine	Cornell and Kaplan (1977)
Parathion	4-Nitrophenol	Daughton and Hsieh (1977)
4-Chloroaniline	4-Chloroacetanilide	Engelhardt *et al.* (1977)
Metamitron	Desaminometamitron	Engelhardt and Wallnöfer (1978)
Propane	Propionate, acetone	Leadbetter and Foster (1959)
2-Butanol	2-Butanone	Patel *et al.* (1979)
Phenol	*cis,cis*-Muconate	Knackmuss and Hellwig (1978)
DDT	DDD,DDE,DBP	Pfaender and Alexander (1973)
o-Xylene	*o*-Toluic acid	Raymond *et al.* (1967)
2,4,5-T	2,4,5-Trichlorophenol	Rosenberg and Alexander (1980)
4-Fluorobenzoate	4-Fluorocatechol	Clarke *et al.* (1979)
4,4'-Dichlorodiphenyl-methane	4-Chlorophenylacetic acid	Focht and Alexander (1971)
2,3,6-Trichlorobenzoate	3,5-Dichlorocatechol	Horvath and Alexander (1970a)
3-Chlorobenzoate	4-Chlorocatechol	Horvath and Alexander (1970b)
m-Chlorotoluene	Benzyl alcohol	Higgins *et al.* (1979)
Kepone	Monohydrokepone	Orndorff and Colwell (1980)
4-Trifluoromethyl-benzoate	4-Trifluoromethyl-2,3-dihydroxybenzoate	Engesser *et al.* (1988)

Some of the cometabolic reactions brought about by bacteria and fungi in culture are given in Table 12.1. Even this incomplete list illustrates the wide range of conversions, reaction types, and products associated with cometabolism. The kinds of transformation come as no surprise in view of the vast array of biological transformations that heterotrophic bacteria and fungi bring about in culture (Kieslich, 1976). The methane monooxygenase of methylotrophic bacteria is able to oxidize alkanes, alkenes, secondary alcohols, di- or trichloromethane, dialkyl ethers, cycloalkane, and aromatic compounds (Haber *et al.*, 1983), and a single strain of *Nocardia corallina* can cometabolize tri- and tetramethylbenzenes, diethylbenzenes, biphenyl, tetralin, and dimethylnaphthalenes to yield a diversity of products (Jamison *et al.*, 1971).

Cometabolism yields organic products, but the C in the substrate is not converted to typical cell constituents. This is evident in studies of pure cultures and in samples from natural environments. For example, during the metabolism of ^{14}C-labeled 2,5,2'-trichlorobiphenyl, strains of *Alcaligenes* and *Acinetobacter* do not incorporate ^{14}C into cell constituents nor do they generate ^{14}CO$_2$ (Furukawa *et al.*, 1978). Similarly, none of the C is assimilated by bacteria that cometabolize propachlor (Novick and Alexander, 1985). The same lack of use of the C is evident as the natural microflora of sewage cometabolizes the herbicides trifluralin, profluralin, fluchloralin, and nitrofen, the C in the substrate that had been transformed being converted to low-molecular-weight products instead (Jacobson *et al.*, 1980).

Several lines of evidence suggest that many compounds are cometabolized in soils, waters, and sewage. Only one or a few of the following items of evidence have been obtained for any one chemical, however. (a) The chemical is converted to organic products in nonsterile but not in sterile samples of the environment (or is more readily transformed in nonsterile samples), but a microorganism able to use that substrate as a source of energy, C, or another element essential for growth cannot be isolated from that environment. For example, propachlor is converted in sewage and lake water to organic products but not CO$_2$, and a microorganism able to use it as a sole source of C and energy has not been isolated (Novick and Alexander, 1985). (b) Microorganisms that use other organic molecules as C sources for growth metabolize the chemical in culture to yield products identical to those found in nature (Beam and Perry, 1974). (c) Carbon from the chemical is not incorporated into cell components. The almost quantitative transformation of specific compounds to organic products and the lack of incorporation of ^{14}C from the radioactive substrate into microbial cells are strong lines of evidence for cometabolism (Jacobson *et al.*, 1980). Similar evidence for cometabolism was obtained in a study of the metabolism of ^{14}C-labeled carbon monoxide in soil. Because the ^{14}C is not converted to organic matter in soil (this fraction containing microbial cells), cometabolism is indicated. The populations oxidizing CO to CO_2 in soil apparently do not grow using CO as a C or energy source because prior exposure of the soil to this air pollutant does not result in an enhanced oxidation of later increments of CO; had they grown, the rate presumably should have increased (Bartholomew and Alexander, 1982). A similar argument can be made for the transformation of EPTC in soil since little of the ^{14}C from ^{14}C-labeled EPTC is incorporated into biomass (Moorman *et al.,* 1992). (d) Often but not always, the products known to be generated by cometabolism in culture media also accumulate and persist in nature.

Caution needs to be exercised in concluding that cometabolism is occurring merely because an organism cannot be isolated from an environment in which a chemical is undergoing a biological reaction. The isolation of bacteria acting on specific substrates is usually performed by enriching for the organism in a medium whose only C source is the test chemical, and the agar medium used to plate the enrichments contains that single organic supplement. Yet, many bacteria that are able to grow at the expense of that substrate will not develop in such simple media because they require amino acids, B vitamins, or other growth factors. These essential growth factors are not routinely included in such liquid media, and hence bacteria and fungi needing them fail to proliferate. If the only organisms in the environment able to metabolize a test chemical need these growth factors, no isolate will be obtained, and the conclusion will be reached that the compound is cometabolized; that conclusion may thus be erroneous. If a chemical supports the growth of many species, some will undoubtedly require no growth factors (these organisms are called *prototrophs*), and they will be enriched and ultimately can be isolated. If the compound is acted on by only one species, in contrast, it is likely that the responsible organism will need amino acids, B vitamins, and other growth factors; these species are termed *auxotrophs*. Hence, the failure to isolate a bacterium or fungus capable of using the molecule as the sole C source for growth is not sufficient evidence for cometabolism.

EXPLANATIONS

Several reasons have been advanced to explain cometabolism, that is, why an organic chemical that is a substrate does not support growth but is converted to products that accumulate. Three have experimental support: (a) the initial enzyme or enzymes convert the substrate to an organic product that is not further transformed by other enzymes in the microorganism to yield the metabolic intermediates that ultimately are used for biosynthesis and energy production; (b) the initial substrate is transformed to products that inhibit the activity of late enzymes in mineralization or that suppress growth of the organisms; and (c) the organism needs a second substrate to bring about some particular reaction. It is likely that the first explanation is the most common, especially at concentrations of organic chemicals that are not likely to be metabolized to yield products that have antimicrobial effects. The basis of this explanation is the fact that many enzymes act on several structurally related substrates; thus, an enzyme naturally present in the cell—because it functions in processes characterizing normal growth of the organism on other than synthetic molecules—will catalyze reactions that alter chemicals that are not typical cellular intermediates. These enzymes are not absolutely specific for their substrates. Consider a normal metabolic sequence involving the conversion of A to B by enzyme a, B to C by enzyme b, and C to D by enzyme c in a sequence that ultimately yields CO_2, energy for biosynthetic reactions, and intermediates that are converted to cell constituents.

$$A \xrightarrow{a} B \xrightarrow{b} C \xrightarrow{c} D \rightarrow \rightarrow \rightarrow CO_2 + energy + cell\text{-}C.$$

The first enzyme (a) may have a low substrate specifity and act on a molecule structurally similar to A, namely, A'. The product (B') would differ from B in the same way that A differs from A'. However, if enzyme b is unable to act on B' (because the structural features controlling which substrates it modifies differ from those controlling the substrate specificity of enzyme a), B' will accumulate:

$$A' \xrightarrow{a} B' \xrightarrow{\quad} \not\rightarrow.$$

In addition, CO_2 and energy will not be generated, and because cell-C is not formed, the organisms do not multiply. The formation of B' is thus entirely fortuitous (Raymond and Alexander, 1971; Alexander, 1979). The initial evidence for this explanation came from studies of the metabolism of 2,4-D. This herbicide is usually converted first to 2,4-dichlorophenol, but the enzyme further metabolizing 2,4-dichlorophenol acts on some but not all the phenols generated by the initial enzyme acting on other phenoxyacetic acids (Bollag et al., 1968; Loos et al., 1967) (Fig. 12.1). When this occurs, the product of cometabolism accumulates in almost quantitative yield, at least in pure culture. A typical case is the bacterial conversion of 3-chlorobenzoate to 4-chlorocatechol, the yield of the catechol being 98% of the substrate that is transformed (Fig. 12.2).

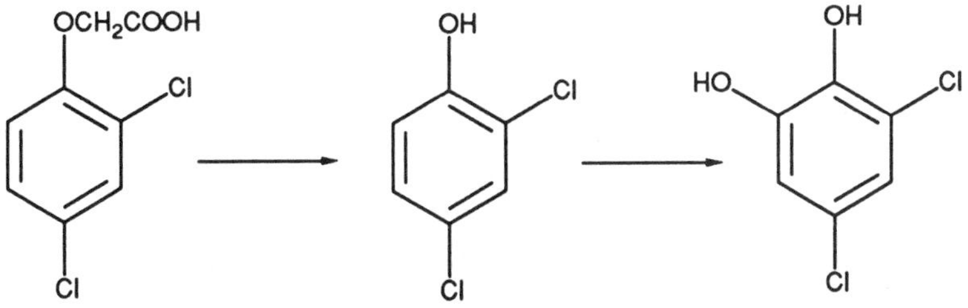

FIG. 12.1 Conversion of 2,4-D to 2,4-dichlorophenol and 3,5-dichlorocatechol.

In instances in which the chemical concentration is high, cometabolism may result from the conversion of the parent compound to toxic products. In the sequence just depicted, if the rate of reaction catalyzed by enzyme *a* is faster than the process catalyzed by enzyme *b*, B will accumulate because it is not destroyed as readily as it is generated. For example, a strain of *Pseudomonas* that grows on benzoate but not 2-fluorobenzoate converts the latter to fluorinated products that are toxic (Taylor *et al.*, 1979). The inhibitor that accumulates may affect a single enzyme that is important for the further metabolism of the toxin. For example, *Pseudomonas putida* cometabolizes chlorobenzene to 3-chlorocatechol, but the latter is not degraded because it suppresses the enzymes involved in further degradation (Klečka and Gibson, 1981). *Pseudomonas putida* also converts 4-ethylbenzoate to 4-ethylcatechol, and the latter inactivates enzymes necessary for subsequent metabolic steps (Ramos *et al.*, 1987). As a result, growth of this bacterium does not occur on chlorobenzene and 4-ethylbenzoate.

In some instances, in pure culture at least, an organism may not be able to metabolize an organic compound because it needs a second substrate to bring about a particular reaction. The second substrate may provide something that is present in insufficient supply in the cells for the reaction to proceed, for example, as an electron donor for the transformation (Lütjens and Gottschalk, 1980; Schukat *et al.*, 1983).

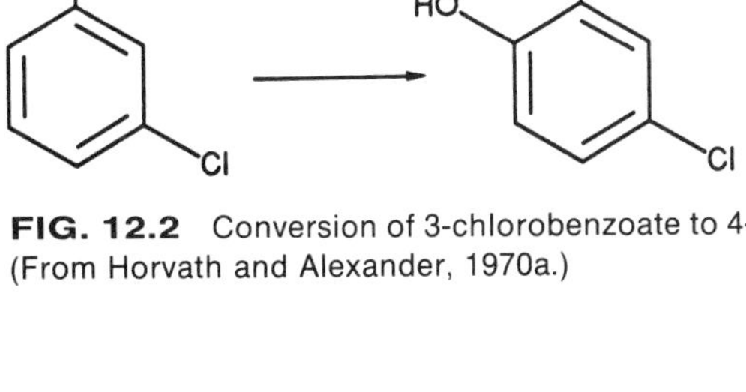

FIG. 12.2 Conversion of 3-chlorobenzoate to 4-chlorocatechol by *Arthrobacter* sp. (From Horvath and Alexander, 1970a.)

ENZYMES WITH MANY SUBSTRATES

The first explanation is linked to the existence of enzymes acting on more than a single substrate. Many enzymes are not absolutely specific for a single substrate. As a rule, they act on a series of closely related molecules, but some carry out a single type of reaction on a variety of somewhat dissimilar molecules. The following are examples of single enzymes acting on a range of substrates.

(a) Methane monooxygenase of methylotrophic bacteria. When grown on methane, methanol, or formate, these aerobic bacteria are able to cometabolize a large array of organic molecules, including several major pollutants. Some of the reactions carried out by these bacteria are shown in Fig. 12.3. In each instance, it is methane monooxygenase that is the responsible catalyst. Other chlorinated aliphatic hydrocarbons transformed by one such methylotroph, *Methylosinus trichosporium*, are *cis*- and *trans*-1,2-dichloroethylene, 1,1-dichloroethylene, 1,2-dichloropropane, and 1,3-dichloropropylene (Oldenhuis *et al.*, 1989). Apparently the same enzyme in other bacteria, after growth on methane, will catalyze the oxidation of *n*-alkanes with two to eight C atoms, *n*-alkenes with two to six C atoms, and mono- and dichloroalkanes with five or six C atoms (Imai *et al.*, 1986), as well as dialkyl ethers and cycloalkanes (Haber *et al.*, 1983).

(b) Toluene dioxygenase of a number of aerobic bacteria. This enzyme incorporates both atoms of oxygen from O_2 (hence, it is a dioxygenase) into toluene as it catalyzes the first step in the degradation of toluene by bacteria grown on that aromatic hydrocarbon (Fig. 12.4). However, that enzyme has very low specificity and also is able to bring about the degradation of TCE (Nelson *et al.*, 1988; Li and Wackett, 1992), to convert 2- and 3-nitrotoluene to the corresponding alcohols, and to hydroxylate the ring of 4-nitrotoluene (Robertson *et al.*, 1992).

$CHCl_3 \longrightarrow CO_2$
$CH_2Cl_2 \longrightarrow CO$
$CH_3Cl \longrightarrow HCHO$
$CH_3CH_3 \longrightarrow CH_3CH_2OH + CH_3CHO$
$CH_3CH_2CH_3 \longrightarrow CH_3CH_2CH_2OH$
$\qquad\qquad\qquad + CH_3CH_2OHCH_3$
$\qquad\qquad\qquad + CH_3COCH_3$
$ClCH{=}CCl_2 \longrightarrow HOClC{=}CCl_2$

FIG. 12.3 Reactions catalyzed by the methane monooxygenase of methylotrophic bacteria. (Reprinted with permission from Haber *et al.*, 1983; Oldenhuis *et al.*, 1989.)

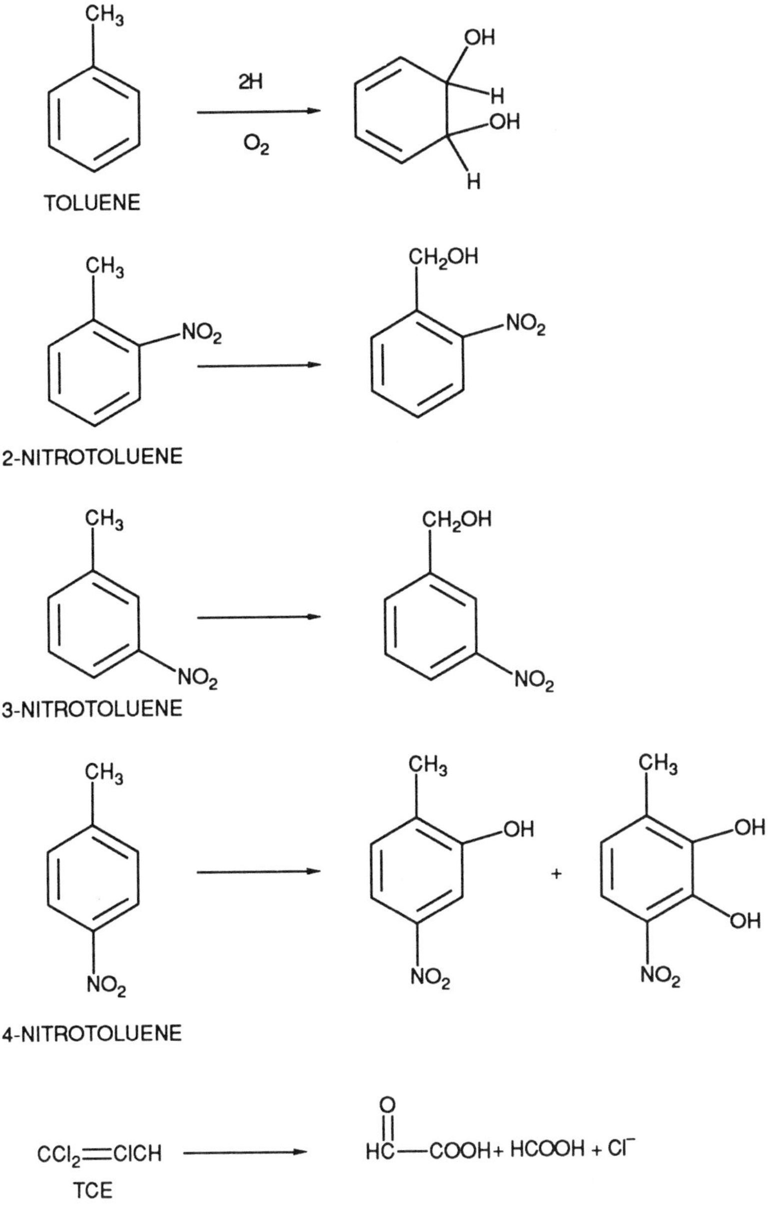

FIG. 12.4 The reactions catalyzed by toluene dioxygenase.

(c) Toluene monooxygenase of several aerobic bacteria. Differing from the dioxygenase, this enzyme incorporates only one atom of oxygen from O_2 into toluene to give o-cresol (Fig. 12.5). However, because of this enzyme, bacteria can cometabolize TCE, convert 3- and 4-nitrotoluenes to the corresponding benzyl alcohols and benzaldehydes (Delgado *et al.*, 1992), and add hydroxyl groups to other aromatic compounds (Shields *et al.*, 1991).

(d) Oxygenase of propane-utilizing bacteria. Aerobes using propane as C and energy source for growth also have an oxygenase of broad specificity. This enzyme cometabolizes TCE, vinyl chloride, and 1,1-di- and *trans*- and *cis*-1,2-dichloroethylene (Wackett *et al.*, 1989).

(e) Ammonia monooxygenase of *Nitrosomonas europaea*. This bacterium, which is a chemoautotroph whose energy source in nature is NH_3 and whose C source is CO_2, cometabolizes TCE, 1,1-dichloroethylene, di- and tetrachloroethanes, chloroform, other monochloroethanes, and even fluoro-, bromo-, and iodoethane (Rasche *et al.*, 1990, 1991).

(f) Halidohydrolase acting on simple halogenated fatty acids. Depending on the specific organism, this enzyme may cleave halogens from fluoro-, chloro-, and iodoacetate (Goldman, 1965), dichloroacetate, 2-chloropropionate, and 2-chlorobutyrate (Goldman *et al.*, 1968), and all monohaloacetates except fluoroacetate (Klages *et al.*, 1983).

(g) Halidohydrolase that removes the halogen from 1-iodomethane, 1-iodoethane, 1-chlorobutane, 1-bromobutane, and 1-chlorohexane to yield the corresponding *n*-alcohols (Scholtz *et al.*, 1987).

(h) A dehalogenase that removes the halogens from CH_2Cl_2, CH_2BrCl, CH_2Br_2, and CH_2I_2 (Kohler-Staub and Leisinger, 1985).

(i) A dehalogenase that acts on 4-chloro-, 4-bromo-, and 4-iodo- but not 4-fluorobenzoate (Thiele *et al.*, 1987).

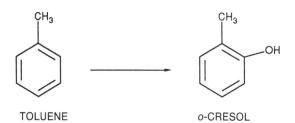

TOLUENE   o-CRESOL

FIG. 12.5   The reaction catalyzed by toluene monooxygenase.

(j) A catechol dioxygenase that oxidizes catechol, 3- and 4-methylcat-echol, and 3-fluoro- but not 3-chlorocatechol (Klečka and Gibson, 1981).

(k) A benzoate hydroxylase that metabolizes benzoate and 4-amino-, 4-nitro-, 4-chloro-, and 4-methylbenzoates (Reddy and Vaidyana-than, 1976).

(l) An enzyme that cleaves the nitrile from a number of aromatic nitriles to yield ammonia (Harper, 1977).

(m) A phosphatase that hydrolyzes parathion, paraoxon, diazinon, durs-ban, and fenitrothion, but not several related insecticides (Mun-necke, 1976).

(n) An alcohol dehydrogenase that oxidizes normal aliphatic alcohols containing 1 to 11 carbon atoms (Sperl et al., 1974).

(o) A deaminase cleaving the amine moiety of a number of purines (Sakai and Jun, 1978).

The organism containing these enzymes may be able to use one or several of the enzyme's substrates for growth. However, many of the substrates are transformed but do not support growth. The product of the reaction then accumulates.

ENVIRONMENTAL SIGNIFICANCE

In a sense, cometabolism is merely a special type of microbial transfor-mation. As such, it might seem to be of mere academic interest; however, such transformations have considerable importance in nature. These eco-logical consequences have been the basis of the great attention given to cometabolism, and it is because of these environmental issues that cometabolism is considered to be a special type of biological transforma-tion. These environmental consequences are readily evident from the characteristics of the process, specifically the inability of the organisms to grow at the expense of the organic compound and the conversion of the substrate to an organic product that often accumulates. Two effects are immediately evident. First, because the size of the population or the biomass of organisms acting on most synthetic chemicals is small in surface and subsurface soils and waters, a chemical subject to cometabolism by these organisms is transformed slowly, and the rate of conversion does not increase with time. This contrasts with chemicals used as C and energy sources because the rate of metabolism of such substrates increases as the responsible organisms multiply (Fig. 12.6). Second, many organic products accumulate as a result of cometabolism, and these products tend

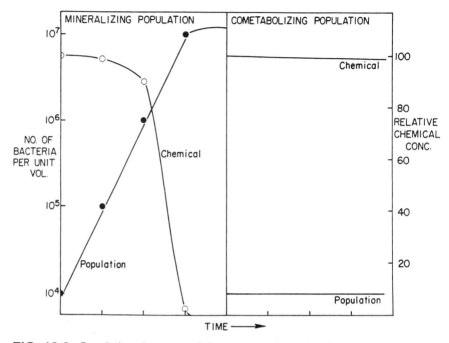

FIG. 12.6 Population changes and disappearance of a chemical acted on by (left) bacteria growing logarithmically and using the compound as a C source or (right) bacteria cometabolizing the chemical. (From Alexander, 1981. Reprinted with permission from the American Association for The Advancement of Science.)

to persist. This accumulation is an outcome of cometabolism by a single species because it cannot further metabolize the product. Moreover, since the outcome of cometabolism frequently is only a small alteration in the structure of the molecule, a toxic parent compound is often converted to a harmful product (Alexander, 1979).

Few estimates have been made of the numbers or biomass of microorganisms able to cometabolize individual substrates in nature. However, the number of cells in one soil able to cometabolize 2,4-D ranges from 0.3 to 0.8 million per gram (Fournier *et al.*, 1981). In contrast, 20 to 75% of the bacteria isolated from sewage have the ability to cometabolize DDT, and 90 million cells per milliliter of sewage can cometabolize the insecticide (Pfaender and Alexander, 1973).

Although the products of cometabolism accumulate in culture, the same is not necessarily true in nature. Those products may be acted on by a second species and may thereby either be cometabolized or mineralized. Under such circumstances, the initial products may not persist. An exam-

ple is shown in Fig. 12.7 of a compound, sucralose, that is initially cometabolized to yield organic products, but the latter are subsequently mineralized, presumably by species not acting on the original trichlorinated
disaccharide. Indeed, if the second population grows by using the cometabolic products of the first population, those products may not be detected
at all because the population will grow to the size permitted by the yield
of its C source, that is, the cometabolic product.

Several cases have been described in which a second species destroys
the metabolites excreted by the first in culture (Fig. 12.8). Six examples
will be cited. (a) Parathion is cometabolized by *Pseudomonas stutzeri* to
yield 4-nitrophenol and diethyl phosphate, and *Pseudomonas aeruginosa*
uses the phenol as a source of C and energy (Daughton and Hsieh, 1977).
(b) Cyclohexane is cometabolized to cyclohexanol by one pseudomonad,
and cyclohexanol is mineralized by a different species of *Pseudomonas*
(deKlerk and van der Linden, 1974). (c) 4,4'-Dichlorobiphenyl is cometabolized to yield 4-chlorobenzoate, and the latter is a C and energy source
for an *Acinetobacter* strain (Adriaens *et al.*, 1989). (d) DDT is converted
by cometabolism to 4-chlorophenylacetic acid by a strain of *Pseudomonas,*
and the product is then used for growth by *Arthrobacter* sp. (Pfaender
and Alexander, 1972). (e) 2,4,5-T is cometabolized by *Pseudomonas fluo-
rescens* to 2,4,5-trichlorophenol, which is then further metabolized by
other microorganisms (Rosenberg and Alexander, 1980). (f) 4-Chloro-3,5-
dinitrobenzoic acid is cometabolized to yield 2-hydroxymuconic semialde-
hyde, which is then mineralized by *Streptomyces* sp. (Jacobson *et al.*,
1980). In such instances, a single organism able to mineralize the initial

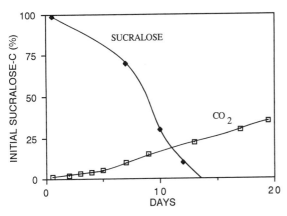

FIG. 12.7 Transformation of sucralose in samples of soil. (From M. P. Labare and
M. Alexander, unpublished data.)

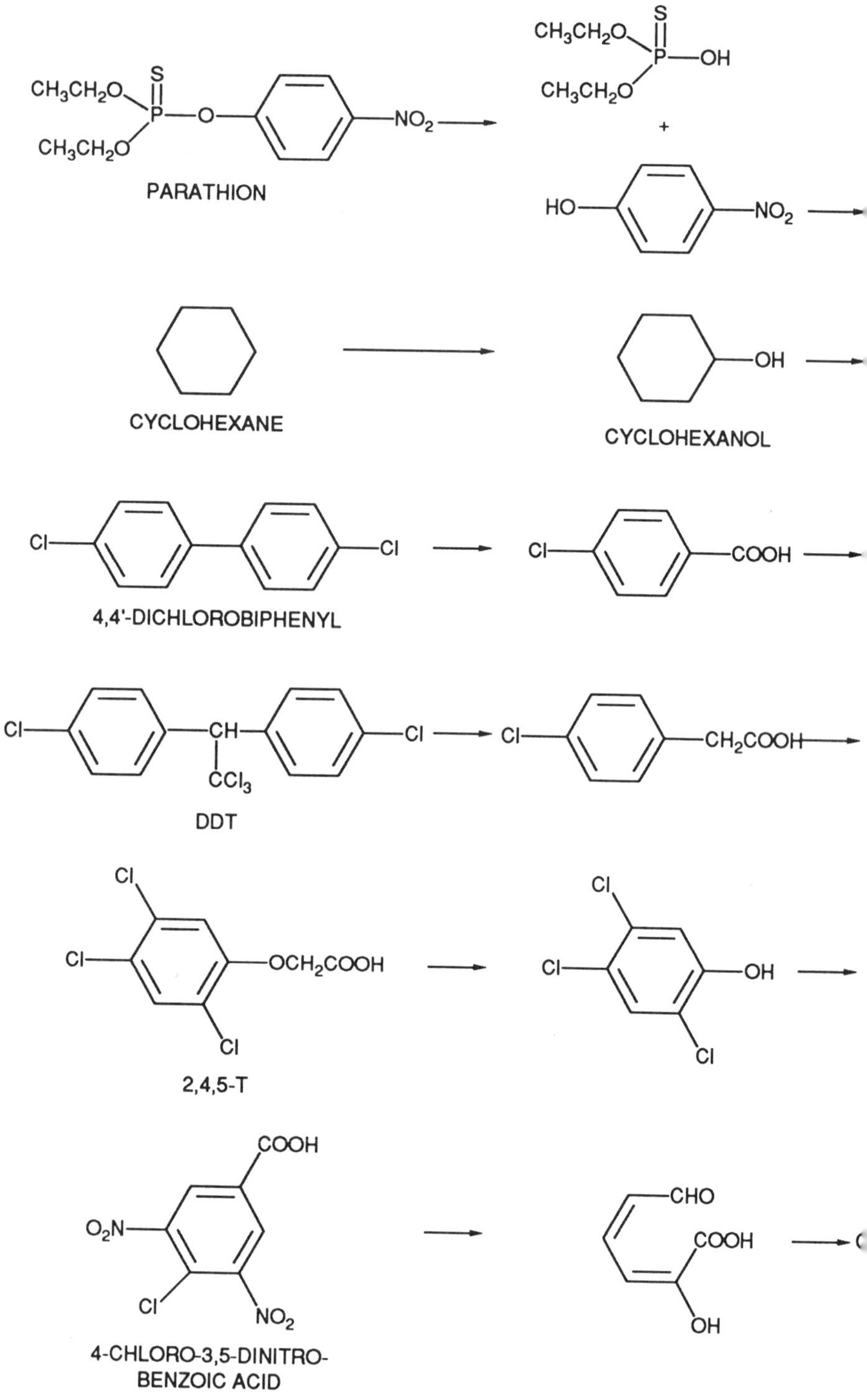

CH₃CH₂O, S, P, O, NO₂ → reaction

PARATHION

CH₃CH₂O—, S, P, OH, CH₃CH₂O—

+

HO———NO₂ →

CYCLOHEXANE → **CYCLOHEXANOL** —OH →

Cl———Cl → Cl———COOH →

4,4'-DICHLOROBIPHENYL

Cl———CH———Cl → Cl———CH₂COOH →
CCl₃

DDT

Cl, Cl, —OCH₂COOH, Cl → Cl, Cl, —OH, Cl →

2,4,5-T

COOH, O₂N, Cl, NO₂ → CHO, COOH, OH →

4-CHLORO-3,5-DINITRO-BENZOIC ACID

substrate would not be obtained in pure culture, yet the product of the first population and the substrate for the second might not be detected in nature. The two phases of the transformation, in effect, are *complementary catabolic pathways,* complementary in the sense that they complete an entire degradative sequence. These complementary catabolic pathways may be exploitable by genetic engineering to construct organisms that are able to mineralize the parent molecule. This might be done by transferring genes into one species so that its cells contain the enzymes bringing about the initial cometabolic sequence but also those enzymes that allow the organism to mineralize and grow on the products of the initial cometabolic sequence (Ramos *et al.,* 1987; Rubio *et al.,* 1986).

A chemical that is cometabolized at one concentration may be mineralized in the same environment at another concentration, or it may be cometabolized in one environment and mineralized in another. This suggests that the organic products of cometabolism may accumulate at only certain concentrations or only in some environments. For example, IPC is cometabolized at 1.0 mg/liter but mineralized at 0.4 μg/liter in lake water, and monuron is cometabolized apparently to 4-chloroaniline at 10 mg/liter but mineralized at 10 μg/liter in sewage. Chlorobenzilate is cometabolized in samples from the water column of lakes but is mineralized in the presence of the microflora of freshwater sediments (Wang *et al.,* 1984, 1985). Thus, caution must be exercised in predicting that cometabolism will take place at concentrations or in environments other than those specifically tested.

The kinetics of cometabolism have received scant attention. If the microbial populations are neither growing nor declining and the concentration of substrate for cometabolism is below the K_m of the active organisms, it is likely that the conversion would be first order, as discussed in Chapter 6 on kinetics. The transformation of propachlor may be first or zero order in lake water or sewage (Novick and Alexander, 1985). In a biofilm bioreactor inoculated with methane-oxidizing bacteria, the cometabolism of TCE, 1,1,1-trichloroethane, and *cis-* and *trans*-1,2-dichloroethylene is first order at concentrations up to 1 mg/liter (Arvin, 1991). However, in environments in which the transformations are slow, the C source for growth probably is being depleted, so the kinetic patterns may change with time. Other models have been developed for cometabolism by nongrowing or growing populations (Alvarez-Cohen and McCarty, 1991; Criddle, 1993).

FIG. 12.8 Conversions involving cometabolism by one species followed by mineralization of the cometabolic products by a second species.

Because cometabolism generally leads to a slow destruction of the substrate, attention has been given to enhancing its rate. The stimulation of such activities is especially important if the substrate is toxic to humans, agricultural crops, or species in natural ecosystems. The addition of a number of organic compounds to soil or sewage promotes the rate of cometabolism of DDT and a number of chlorinated aromatic compounds and chlorinated fatty acids (Jacobson and Alexander, 1981; Pfaender and Alexander, 1973), but the responses to such additions are not predictable. No relation is known to exist between the metabolic pathways involved in destruction of the added mineralizable substrate and the compound that is cometabolized in these studies. The added molecules are randomly chosen in these trials, and sometimes they do and sometimes they do not stimulate cometabolism. In instances in which stimulation occurs, the benefit probably results from an unpredicted increase in the biomass of organisms, some of which fortuitously cometabolize the compound of interest.

An alternative approach is to add mineralizable compounds that are structurally analogous to the compound whose cometabolism one wishes to promote. Presumably, the microflora that grows on the mineralizable compound contains enzymes transforming the analogous molecule, that is, the one that is cometabolized. This larger biomass thus has more of the degradative enzyme than present in the unsupplemented water or soil. This method of *analogue enrichment* has been used to enhance the cometabolism of PCBs by additions of biphenyl. The unchlorinated biphenyl was selected for addition to soil since it is mineralizable, nontoxic, and serves as a C source for microorganisms that are able to cometabolize PCBs (Brunner *et al.*, 1985). A similar approach has been used to enhance the cometabolism of trifluoromethylbenzoates by addition of alkyl-substituted benzoates (Knackmuss, 1981) and the metabolism of 2,4-dichloroaniline in soil by additions of aniline (You and Bartha, 1982).

Analogue enrichment is a procedure that is similar to the usual means of isolating bacteria that can cometabolize a compound. The enrichment culture contains a C source that supports growth, and the pure cultures thus obtained also cometabolize structurally related compounds that would not support growth. For example, bacteria isolated on diphenylmethane and containing enzymes to degrade it also cometabolize chlorinated diphenylmethanes. Many of the latter do not sustain growth (Focht and Alexander, 1970a,b).

Various approaches have been devised recently to enhance cometabolism as part of actual bioremediation efforts. These will be presented in Chapter 15.

REFERENCES

Adriaens, P., Kohler, H.-P. E., Kohler-Staub, D., and Focht, D. D., *Appl. Environ. Microbiol.* **55**, 887–892 (1989).

Alexander, M., *in* "Agriculture and the Quality of Our Environment" (N. C. Brady, ed.), pp. 331–342. American Association for the Advancement of Science, Washington, DC, 1967.

Alexander, M., *in* "Microbial Degradation of Pollutants in Marine Environments" (A. W. Bourquin and P. H. Pritchard, eds.), pp. 67–75. U.S. Environmental Protection Agency, Gulf Breeze, FL, 1979.

Alexander, M., *Science* **211**, 132–138 (1981).

Alvarez-Cohen, L., and McCarty, P. L., *Environ. Sci. Technol.* **25**, 1381–1387 (1991).

Arvin, E., *Water Res.* **25**, 873–881 (1991).

Bartholomew, G. W., and Alexander, M., *Environ. Sci. Technol.* **16**, 301–302 (1982).

Barug, D., *Chemosphere* **10**, 1145–1154 (1981).

Bauer, S. R., Wood, E. M., and Traxler, R. W., *Int. Biodeterior. Bull.* **15**, 53–56 (1979).

Beam, H. W., and Perry, J. J., *J. Gen. Microbiol.* **82**, 163–169 (1974).

Bollag, J.-M., Helling, C. S., and Alexander, M., *J. Agric. Food Chem.* **16**, 826–828 (1968).

Brunner, W., Sutherland, F. H., and Focht, D. D., *J. Environ. Qual.* **14**, 324–328 (1985).

Clarke, K. F., Callely, A. G., Livingstone, A., and Fewson, C. A., *Biochim. Biophys. Acta* **404**, 169–179 (1979).

Cornell, J. H., and Kaplan, A. M., *Abstr. Annu. Meet., Am. Soc. Microbiol.*, 276 (1977).

Criddle, C. S., *Biotechnol. Bioeng.* **41**, 1048–1056 (1993).

Dalton, H., and Sterling, D. I., *Philos. Trans. R. Soc. Lond., Ser. B* **297**, 481–495 (1982).

Daughton, C. G., and Hsieh, D. P. H., *Appl. Environ. Microbiol.* **34**, 175–184 (1977).

Daumy, G. O., McColl, A. S., and Andrews, G. C., *J. Bacteriol.* **141**, 293–296 (1980).

deKlerk, H., and van der Linden, A. C., *Antonie van Leeuwenhoek* **40**, 7–15 (1974).

Delgado, A., Wubbolts, M. G., Abril, M. A., and Ramos, J. L., *Appl. Environ. Microbiol.* **58**, 415–417 (1992).

Engelhardt, G., and Wallnöfer, P. R., *Chemosphere* **7**, 463–466 (1978).

Engelhardt, G., Wallnöfer, P., Fuchsbichler, G., and Baumeister, W., *Chemosphere* **6**, 85–92 (1977).

Engelhardt, G., Rast, H. G., and Wallnöfer, P. R., *FEMS Microbiol. Lett.* **5**, 377–383 (1979).

Engesser, K. H., Rubio, M. A., and Ribbons, D. W., *Arch. Microbiol.* **149**, 198–206 (1988).

Fathepure, B. Z., and Boyd, S. A., *Appl. Environ. Microbiol.* **54**, 2976–2980 (1988).

Fedorak, P. M., and Grbić-Galić, D., *Appl. Environ. Microbiol.* **57**, 932–940 (1991).

Ferrer, M. R., del Moral, A., Ruiz-Berraquero, F., and Ramos-Cormenzana, A., *Chemosphere* **14**, 1645–1648 (1985).

Focht, D. D., and Alexander, M., *Science* **170**, 91–92 (1970a).

Focht, D. D., and Alexander, M., *Appl. Microbiol.* **20**, 608–611 (1970b).

Focht, D. D., and Alexander, M., *J. Agric. Food Chem.* **19**, 20–22 (1971).

Foster, J. W., *Antonie van Leeuwenhoek* **28**, 241–274 (1962).

Fournier, J. C., Coddaccioni, P., and Soulas, G., *Chemosphere* **10**, 977–984 (1981).

Furukawa, K., Matsumura, F., and Tonomura, K., *Agric. Biol. Chem.* **42**, 543–548 (1978).

Goldman, P., *J. Biol. Chem.* **240**, 3434–3438 (1965).

Goldman, P., Milne, G. W. A., and Keister, D. B., *J. Biol. Chem.* **243**, 428–434 (1968).

Golovleva, L. A., Golovlev, E. L., Zyakun, A. M., Shurukhin, Y. V., and Finkelshtein, Z. I., *Izv. Akad. Nauk SSSR, Ser. Biol.* **1**, 44–51 (1978).

Haber, C. L., Allen, L. N., Zhao, S., and Hanson, R. S., *Science* **221**, 1147–1153 (1983).

Harper, D. B., *Biochem. J.* **167**, 685–692 (1977).

Higgins, I. J., Sariaslani, F. S., Best, D. J., Tryhom, S. F., and Davies, M. M., *Soc. Gen. Microbiol. Q.* **6**, 71 (1979).

Horvath, R. S., *Bacteriol. Rev.* **36**, 146–155 (1972).

Horvath, R. S., and Alexander, M., *Can. J. Microbiol.* **16**, 1131–1132 (1970a).

Horvath, R. S., and Alexander, M., *Appl. Microbiol.* **20**, 254–258 (1970b).

Imai, T., Takigawa, H., Nakagawa, S., Shen, G.-J., Kodama, T., and Minoda, Y., *Appl. Environ. Microbiol.* **52**, 1403–1406 (1986).

Jacobson, S. N., and Alexander, M., *Appl. Environ. Microbiol.* **42**, 1062–1066 (1981).

Jacobson, S. N., O'Mara, N. L., and Alexander, M., *Appl. Environ. Microbiol.* **40**, 917–921 (1980).

Jamison, V. W., Raymond, R. L., and Hudson, J. O., *Dev. Ind. Microbiol.* **12**, 99–105 (1971).

Jensen, H. L., *Acta Agric. Scand.* **13**, 404–412 (1963).

Kieslich, K., "Microbial Transformations of Non-steroid Cyclic Compounds." Thieme, Stuttgart, 1976.

Klages, U., Krauss, S., and Lingens, F., *Hoppe-Seyler's Z. Physiol. Chem.* **364**, 529–535 (1983).

Klečka, G. M., and Gibson, D. T., *Appl. Environ. Microbiol.* **41**, 1159–1165 (1981).

Knackmuss, H.-J., in "Microbial Degradation of Xenobiotics and Recalcitrant Compounds," (T. Leisinger, A. M. Cook, R. Hütter, and J. Nüesch, eds.), pp. 189–212. Academic Press, New York, 1981.

Knackmuss, H.-J., and Hellwig, M., *Arch. Microbiol.* **117**, 1–7 (1978).

Kohler-Staub, D., and Leisinger, T., *J. Bacteriol.* **162**, 676–681 (1985).

Leadbetter, E. R., and Foster, J. W., *Arch. Biochem. Biophys.* **82**, 491–492 (1959).

Li, S., and Wackett, L. P., *Biochem. Biophys. Res. Commun.* **185**, 443–451 (1992).

Liu, D., Maguire, R. J., Pacepavicius, G., and Dutka, B. J., *Environ. Toxicol. Water Qual.* **6**, 85–95 (1991).

Loos, M. A., Roberts, R. N., and Alexander, M., *Can. J. Microbiol.* **13**, 679–690 (1967).

Lütjens, M., and Gottschalk, G., *J. Gen. Microbiol.* **119**, 63–70 (1980).

Malashenko, Y. R., Romanovskaya, V. A., Sokolov, I. G., and Kryshtab, T. P., *Mikrobiologiya* **45**, 1105–1107 (1976).

Mitchell, W. R., Dennis, W. H., and Burrows, E. P., "Microbial Interactions with Several Munitions Compounds: 1,3-Dinitrobenzene, 1,3,5-Trinitrobenzene, and 3,5-Dinitroaniline." Publ. TR-820.1. U.S. Army Bioengineering Research and Development Laboratory, Ft. Detrick, MD, 1982.

Moorman, T. B., Broder, M. W., and Koskinen, W. C., *Soil Biol. Biochem.* **24**, 121–127 (1992).

Munnecke, D. M., *Appl. Environ. Microbiol.* **32**, 7–13 (1976).

Nelson, M. J. K., Montgomery, S. O., and Pritchard, P. H., *Appl. Environ. Microbiol.* **54**, 604–606 (1988).

Novick, N. J., and Alexander, M., *Appl. Environ. Microbiol.* **49**, 737–743 (1985).

Oldenhuis, R., Vink, R. L. J. M., Janssen, D. B., and Witholt, B., *Appl. Environ. Microbiol.* **55**, 2819–2826 (1989).

Orndorff, S. A., and Colwell, R. R., *Appl. Environ. Microbiol.* **39**, 398–406 (1980).

Patel, R. N., Hou, C. T., Laskin, A. I., Derelanko, P., and Felix, A., *Appl. Environ. Microbiol.* **38**, 219–223 (1979).

Perry, J. J., *Microbiol. Rev.* **43**, 59–72 (1979).

Pfaender, F. K., and Alexander, M., *J. Agric. Food Chem.* **20**, 842–846 (1972).

Pfaender, F. K., and Alexander, M., *J. Agric. Food Chem.* **21**, 397–399 (1973).

Rasche, M. E., Hicks, R. E., Hyman, M. R., and Arp, D. J., *J. Bacteriol.* **172**, 5368–5373 (1990).

Rasche, M. E., Hyman, M. R., and Arp, D. J., *Appl. Environ. Microbiol.* **57**, 2986–2994 (1991).

Ramos, J. L., Wasserfallen, A., Rose, K., and Timmis, K. N., *Science* **235**, 593–596 (1987).

Raymond, D. G. M., and Alexander, M., *Pestic. Biochem. Physiol.* **1**, 123–130 (1971).

Raymond, R. L., Jamison, V. W., and Hudson, J. O., *Appl. Microbiol.* **15**, 857–865 (1967).

Reddy, C. C., and Vaidyanathan, C. S., *Arch. Biochem. Biophys.* **177**, 488–498 (1976).

Robertson, J. B., Spain, J. C., Haddock, J. D., and Gibson, D. T., *Appl. Environ. Microbiol.* **58**, 2643–2648 (1992).

Rosenberg, A., and Alexander, M., *J. Agric. Food Chem.* **28**, 297–302 (1980).

Rubio, M. A., Engesser, K.-H., and Knackmuss, H.-J., *Arch. Microbiol.* **145**, 116–122 (1986).

Sakai, T., and Jun, H.-K., *J. Ferment. Technol.* **56**, 257–265 (1978).

Schmitt, P., Diviès, C., and Cardona, R., *Appl. Microbiol. Biotechnol.* **36**, 679–683 (1992).

Scholtz, R., Schmuckle, A., Cook, A. M., and Leisinger, T. M., *J. Gen. Microbiol.* **133**, 267–274 (1987).

Schukat, B., Janke, D., Krebs, D., and Fritsche, W., *Curr. Microbiol.* **9**, 81–86 (1983).

Shields, M. S., Montgomery, S. O., Cuskey, S. M., Chapman, P. J., and Pritchard, P. H., *Appl. Environ. Microbiol.* **57**, 1935–1941 (1991).

Smith, A. E., and Phillips, D. V., *Agron. J.* **67**, 347–349 (1975).

Sperl, G. T., Forrest, H. S., and Gibson, D. T., *J. Bacteriol.* **118**, 541–550 (1974).

Taylor, B. F., Hearn, W. L., and Pincus, S., *Arch. Microbiol.* **122**, 301–306 (1979).

Thiele, J., Müller, R., and Lingens, F., *FEMS Microbiol. Lett.* **41**, 115–119 (1987).

Wackett, L. P., Brusseau, G. A., Householder, S. R., and Hanson, R. S., *Appl. Environ. Microbiol.* **55**, 2960–2964 (1989).

Wang, Y.-S., Subba-Rao, R. V., and Alexander, M., *Appl. Environ. Microbiol.* **47**, 1195–1200 (1984).

Wang, Y.-S., Madsen, E. L., and Alexander, M., *J. Agric. Food Chem.* **33**, 495–499 (1985).

You, I.-S., and Bartha, R., *Appl. Environ. Microbiol.* **44**, 678–681 (1982).

13 ENVIRONMENTAL EFFECTS

The microbial populations destroying synthetic chemicals are subject to a variety of physical, chemical, and biological factors that influence their growth, their activity, and their very existence. The environments in which these species function vary enormously, and these differences in environmental properties and characteristics have a profound impact on the resident populations, the rate of biochemical transformations, and the identities and persistence of products of biodegradation.

The great impact of site factors is evident from studies showing that a specific compound is biodegraded in samples from one but not another environment. For example, TCE was found to be metabolized by the indigenous microorganisms in only 1 of 43 samples of water and soil (Nelson *et al.,* 1986), 2,4-D was mineralized in samples from a eutrophic (rich in inorganic nutrients) but not an oligotrophic (nutrient-poor) lake (Rubin *et al.,* 1982), methyl parathion was transformed in sediments but not in samples from the water column of an estuary (Pritchard *et al.,* 1987), IPC was mineralized in samples from only some lakes (Hoover *et al.,* 1986), and the reductive dehalogenation of aromatic compounds occurred in only some sewage sludges, pond sediments, and aquifer solids under anaerobic conditions (Gibson and Suflita, 1986). Sometimes a compound may be mineralized in one environment but only cometabolized at a different site (Wang *et al.,* 1985) or, even ignoring temperature effects, transformed at one but not another time of year (Rubin *et al.,* 1982). More often than not, the reasons for the sporadic or nonuniversal occurrence of a biodegradative sequence are unknown. In some instances, the random occurrence of biodegradation may be a result of the presence of organisms at only some sites, the existence of fastidious populations whose growth factor requirements are met in not all environments, the presence of toxins, the availability of O_2, or the impact of other environmental characteristics that promote, restrict, or prevent biodegradation. It is inappropriate to assume that a compound biodegraded in one environment is *ipso facto* going to be transformed in another.

A vast amount of information exists on the biochemical activities of bacteria and fungi grown in pure culture at high substrate concentrations in laboratory media. This research has created a foundation for the understanding of the nutrition, genetics, and catabolic potential of microorganisms. Yet, in nature, bacteria and fungi are exposed to enormously different conditions. They may have an insufficient supply of inorganic nutrients, a paucity of essential growth factors, temperatures and pH values at their extremes of tolerance, and toxins that retard their growth or result in loss of viability. They may benefit from the activities of other microorganisms or be consumed by species residing in the same habitat. As a consequence, extrapolations from tests of laboratory-grown pure cultures to nature are fraught with peril. Not only must there be information on the characteristics of the biodegrading species *in vitro,* but there also must be an understanding of those factors in nature that determine the occurrence, rate, and products of biodegradation.

ABIOTIC FACTORS

Every strain of microorganisms has a range of tolerances to ecologically important factors (e.g., temperature, pH, salinity, etc.) affecting its growth and activity. That range is bounded by the maximum level tolerated and, for some species, a minimum tolerance level. If a particular environment contains several species able to bring about a particular transformation, the tolerance range often is broader than that of a single species, encompassing the tolerances of all the indigenous populations. Outside of the tolerance ranges of all the inhabitants able to perform the degradation of concern, no activity will occur.

Apart from the supply of nutrients and factors that control the bioavailability of organic compounds, the chief abiotic factors influencing microbial transformations are temperature, pH, moisture level (in the case of soil), salinity in some environments, toxins, and hydrostatic pressure if the compounds are in deep marine sediments or at sites deep below the soil surface. An organic pollutant that is quickly destroyed in one environment will persist at another site if these factors preclude or retard microbial activity.

The prevailing temperature is of paramount importance. If the compound of interest exists near the surfaces of soils or water, the low temperatures of winter and even the time immediately preceding and following the winter season are typically associated with little or no biodegradation of many organic substrates. In the frozen soils of the northern parts of North America, Europe, and Asia, organic molecules will persist for long periods. As the temperature rises with the change of seasons, microbial activity will increase in response to the more favorable circumstances. The magnitude of response to a particular increase or decrease in temperature varies with the compound and the environment, which reflects the physiology of the individual populations at the site. To a great degree, the changes in rate of degradation associated with seasons of year are a consequence of the concomitant changes in temperature. On occasion, however, the anticipated increase or decrease in activity with rise or fall in temperature is not evident. This lack of response to the warmer conditions may sometimes be attributable to some other factor becoming limiting during the warmer period; for example, nutrient deficiencies may severely limit degradation of some substances, such as oil, in lakes. At times, the activity in the winter months is not diminished because a factor other than temperature comes into play. This is evident in a study of the degradation of an ester of 2,4-D in a stream in which the transformation rate increased in the winter. The anomalously greater rates in the colder months were a consequence of leaf fall and deposition of leaves in the stream. The surface of the leaves provided abundant sites for microbial colonization, and the enhanced biomass compensated for the otherwise detrimental effects of the cooler water (Lewis *et al.*, 1986).

At extremes of acidity or alkalinity, activity declines. At more moderate pH values, biodegradation tends to be fastest. If a compound in a particular environment can be metabolized by a diverse group of organisms, the range of pH values at which degradation occurs frequently is broader than if only one species can bring about the transformation. Apart from pesticides, however, the effect of pH on biodegradation of polluting chemicals has received scant attention, although it is common practice to add lime to bioremediate acid soils or subsoil materials containing harmful organic compounds.

The microorganisms carrying out a metabolic transformation require adequate moisture for their growth and activity. Moisture obviously is not a limiting factor in the oceans, in fresh waters, or in subterranean aquifers. However, an inadequate supply of water can severely restrict biodegradation in surface soils, in which drying to suboptimal water levels is common. In one study, for example, the optimum moisture level for the biodegradation of oily sludges was found to be at 30 to 90% of the soil's water-holding capacity (Dibble and Bartha, 1979). The optimum moisture level will depend on the properties of the soil, the compound in question, and whether the transformation is aerobic or anaerobic. The last factor is of particular significance since excess water displaces air from the pores in soil, and a waterlogged soil soon becomes anaerobic and unfavorable for aerobic processes. Decreasing the moisture content of soil diminishes rates of degradation (Walker, 1976), a result of an inadequate supply of water to sustain proliferation, metabolism, or both.

Salinity sometimes is sufficiently high to become harmful. Soils and inland waters in certain areas of the world are rich in salts, and microbial processes in such environments are inhibited. It is possible that the salinity in estuaries and in oceans may also be detrimental to some species involved in the biodegradation of organic pollutants, but no strong argument can presently be advanced for the salinity in such waters being major deterrents to biodegradation.

Components of oil and other pollutants that have specific gravities greater than that of marine waters will move downward and sink to the deep benthic zone. At these depths, the hydrostatic pressure is notably high. It is likely that a combination of high pressure and low temperatures in the deep ocean will result in low microbial activity, slow biodegradation, and consequently prolonged persistence of substances that reach deep benthic zones (Atlas, 1981).

It is not economically feasible to modify or control some of these factors. Yet, it is important to understand them, qualitatively at least, in order to predict the likely persistence of organic molecules in environments differing in the intensity of these factors. Understanding the quantitative impacts of these abiotic factors would be far more useful for predictive purposes, but knowledge for attaining such objectives remains scant.

Conversely, some practical technologies to promote biodegradation do entail manipulation of several of these abiotic factors. This is evident at sites in the field encompassing limited areas or in bioreactors. Temperatures may be made more favorable, moisture levels can be improved, and undesirable acidities can be corrected. Optimization is also evident in the controlled microbiological treatment of industrial wastes, a practice that a few prudent chemical companies have followed for many years.

NUTRIENT SUPPLY

To grow, heterotrophic bacteria and fungi require—in addition to an organic compound that serves as a source of C and energy—a group of other nutrient elements and an electron acceptor. That electron acceptor is O_2 for aerobes, but it may be nitrate, sulfate, CO_2, ferric iron, or organic compounds for specific bacteria able to utilize these substances to accept the electrons released in the oxidation of the energy source. Many bacteria and fungi also require low concentrations of one or more amino acids, B vitamins, fat-soluble vitamins, or other organic molecules; these trace organic nutrients are termed growth factors. The absence from a particular environment of any of these essential nutrients will prevent growth of organisms requiring that substance or prevent any microbial replication if the requisite, such as an inorganic nutrient, is needed by all species.

Soils, sediments, and marine and fresh waters contain low concentrations of readily metabolizable organic matter. This may not seem to be true for soils or sediments, which may contain 1% or more organic matter, but that organic C exists in complex forms that bacteria and fungi either cannot use or utilize only slowly. Typically, the supply of all other nutrient elements exceeds the need of the resident microbial communities given the little readily available C, and hence the limiting nutrient element for heterotrophs in soils, sediments, and natural waters is commonly C.

However, the situation changes markedly if a pollutant that is potentially readily utilizable is introduced into the environment, provided that its concentration is sufficiently high to make one or more previously nonlimiting nutrients into a limiting factor. At very low pollutant concentrations, such a change may not occur. However, even at what might appear to be a low concentration, a pollutant that is in a NAPL or otherwise does not mix throughout the site is, in fact, at high concentrations in the microenvironment in which it is deposited. Thus, at the interface between crude oil, gasoline, or an organic solvent and the surrounding environment, the C concentration is high. Under these circumstances, the supply of one or several nutrients, which previously may have been nonlimiting, may be in concentrations too low to meet the now higher demand. Usually, the nutrients now in short supply are N, P, or both, and a frequent concomitant of the greater growth on the pollutant C is a greater demand for an electron acceptor. For hydrocarbons and many other C compounds, that electron acceptor is O_2. Nearly always, the supply of K, S, Mg, Ca, Fe, and micronutrient elements is greater than the demand.

Oceanic waters, lakes, rivers, soils, and aquifers containing oil, gasoline, or organic solvents from leaking underground storage tanks typically have too low concentrations of inorganic nutrients, O_2, or both at the interface between the water-insoluble pollutant and the aqueous phase to support the activity that is otherwise possible.

The release into marine and estuarine waters of crude oil from leaking tankers and corrosion and the subsequent leakage of petroleum or oil products from underground storage tanks have prompted studies designed to establish means to bioremediate the surface or ground waters. These investigations show that crude oil degradation in seawater is slow unless both N and P are added. Individually, N or P alone fails to cause appreciable stimulation (Atlas and Bartha, 1972). Similarly, additions of N and P to samples of groundwater contaminated with gasoline stimulate the growth of bacteria (Jamison et al., 1975). The nutrient level for optimal activity varies with the type of oil and the particular water body, but stimulations in seawater have been reported over a range of concentrations (Atlas and Bartha, 1972; Floodgate, 1984; LePetit and N'Guyen, 1976). The concentration of water-soluble salts of N and P introduced into surface waters at or very near the oil–water interface rapidly declines because of turbulence of many waters. Therefore, a number of "oleophilic" fertilizers were developed, and these hydrophobic preparations, after addition, remain associated with the oil and stimulate hydrocarbon-degrading bacteria. The N- and P-containing compounds in these early fertilizer materials include octyl phosphate, decyl phosphate, paraffinized urea, and dodecyl urea (Atlas and Bartha, 1972; Olivieri et al., 1978).

The addition of N and P to soil also stimulates the biodegradation of oil and individual hydrocarbons and increases bacterial abundance. The effect of inorganic N and P on the mineralization in subsoil of phenanthrene present in two NAPLs or in soil with no NAPL is shown in Fig. 13.1. The stimulation is sometimes apparent immediately, but it may require some time for a benefit to be evident (Bossert and Bartha, 1984; Jobson et al., 1974). On the other hand, fertilizer additions sometimes may be without benefit, possibly because of high N and P levels in the soil, the presence of N and P in the organic pollutants, or the low concentration of chemical whose biodegradation is being determined.

It is widely believed that only one nutrient element is limiting at any one time, and that only when that one deficiency is overcome does another nutrient element become limiting. This view frequently may be incorrect since microbial growth may be simultaneously limited by two nutrients (Egli, 1991). It is not uncommon to find that additions of combinations of inorganic nutrients have a greater effect on biodegradation than single nutrients (Swindoll et al., 1988), although often such responses may result from the second nutrient element becoming limiting as the deficiency of the first is overcome by the supplement.

Even in the absence of added N and P, biodegradation continues in waters, soils, and sediments, albeit at a slow rate. This probably is a consequence of nutrient regeneration, that is, a recycling of the elements as they are first assimilated into microbial cells and then are converted back to the inorganic forms as the cells lyse or are consumed by predators or parasites, both of which release some of the N and P contained in their prey or hosts. Under such circumstances, the rate of biodegradation will be governed by the rate at which the limiting nutrient is recycled. Protozoa probably are especially important for nutrient regeneration in oceans and lakes, and possibly also in soil.

The variation in rate of biodegradation with time is often the result of diurnal or seasonal changes in temperature; however, it may have other

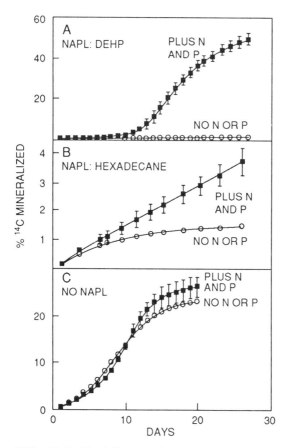

FIG. 13.1 The influence of added N and P on mineralization of phenanthrene in a subsoil. The compound was added in di(2-ethylhexyl) phthalate (A) or hexadecane (B) as NAPLs or was added with no NAPL (C). (Reprinted from *Environ. Toxicol. Chem.*, in press. Efroymson and Alexander, 1994, with kind permission from Pergamon Press Ltd., Headington Hill Hall, Oxford OX3 OBW, U.K.)

causes. For example, the concentration of N and P in lake and river water varies with rainfall as drainage in the watershed carries soil materials into streams, rivers, and lakes, thereby making the water more fertile. In turn, the rate of a N- or P-limited transformation in the water may be enhanced, as has been observed for 4-nitrophenol mineralization (Zaidi *et al.,* 1988).

The concentration of N and P needed for the biodegradation of oil or other materials present at high concentrations, either throughout the environment or within the NAPL that is the oil itself, is usually assumed to reflect the amount of those elements that must be incorporated into the biomass that would be formed as the microorganisms use the organic materials as C sources for growth. For example, consider the mineralization of 1000 g of organic C. If the active organisms assimilate 30% of the substrate-C to make 300 g of biomass-C and elemental analysis of those cells shows that they contain a C : N ratio of 10 : 1 and a C : P ratio of 50 : 1, then the amount of N and P needed to be incorporated into the biomass is 30 g of N and 6 g of P. This is a convenient assumption and is probably appropriate for predicting the N and P needed to totally destroy the C source, but it is probably not valid for predicting the concentration of N and P to support the maximum rate of degradation. It is important to distinguish between the optimum nutrient level for extent of degradation and that needed for the highest rate. This argument of rate versus extent is particularly important in considering explanations for the occasional need for high N or P concentrations to degrade organic molecules whose level is too low to give large biomasses.

It is not yet clear why phosphate or inorganic N stimulates biodegradation of chemicals present at concentrations appreciably below 1.0 mg/liter, for example, of 4-nitrophenol at concentrations of 2–200 μg/liter in lake water. With no added P, the degradation was observed to be slow or failed to proceed (Jones and Alexander, 1988a,b). The slow biodegradation of comparably low levels of chlorophenols by marine plankton communities (Kuiper and Hanstveit, 1984), of phenols in lake water (Rubin and Alexander, 1983), and of IPC at 400 ng/liter and 2,4-D at 200 ng/liter (Wang et al., 1984) was also limited by the supply of inorganic nutrients, and the rates were enhanced by providing those nutrients. The requirement for high concentrations of P and N in waters may be related not to the amount of these elements needed to be incorporated into the biomass but rather to the K_s value for P or N. As with C compounds, microbial growth at P or N concentrations below the K_s (of the P or N source) is slower than at higher concentrations. If the K_s value for the rate of P or N utilization is high, the maximum rate of degradation would require a high P or N concentration. The K_s values for P for different microorganisms may range from as low as 0.4 to as high as 500 μg/liter (Owens and Legan, 1987). Alternatively, the need for high P levels may result from nonbiological reactions that reduce phosphate availability. These reactions could be the precipitation of phosphate as insoluble salts of Ca, Fe, or Mg. However, phosphate in solution may not be represented solely by $H_2PO_4^-$ and HPO_4^{2-} because Ca, Fe, and other metallic salts are also present in soluble form, and the dependency of microbial nutrition on the solution chemistry of inorganic P is unexplored.

Calcium and Mg are abundant in many inland waters, and reactive Ca, Fe, and Mg exist in soils and sediments. These cations alter the availability of P. Moreover, pH affects the identities of the Ca, Fe, and other salts of P in the aqueous phase and also alters the relative abundance of $H_2PO_4^-$ and HPO_4^{2-}. These changes in solution chemistry of P may explain why a strain of *Pseudomonas* requires high P concentrations for phenol mineralization at pH 8.0 but only low concentrations at pH 5.2 (Robertson and Alexander, 1992).

Rarely do the additions of elements other than P, N, and O_2 stimulate biodegradation in natural or polluted environments. However, Fe may sometimes limit the rate of microbial destruction of oil in seawater, in which the available forms of that element are often present at very low concentrations.

Nevertheless, the disappearance of many pesticides in soil and probably of many other chemicals at low concentrations in soils and waters is not known to be stimulated by supplementary N and P. The reason is either presence of an adequate supply of these nutrients to support growth of the active species or the existence of some other limiting factor, for example, sorption.

Little attention has been given to the possible role of growth factors in controlling microbial activity. In an environment containing several species able to degrade a particular compound, it is likely that both auxotrophs and prototrophs will coexist, and the absence of growth factors will not affect the transformation because the prototrophs will flourish. However, in environments containing only one or two species active on the compound of concern, it is likely that the supply or rate of excretion of growth factors will limit the rate of degradation since one or both species are quite possibly auxotrophs. The abundance of auxotrophs is evident from findings that about 90% of the bacteria in marine waters, 75–80% of the bacteria in marine sediments (Skerman, 1963), a high percentage of those in lakes (Fondén, 1969), and more than 90% of the bacteria in soil (Rouatt and Lochhead, 1955) need one or more B vitamins, amino acids, or other growth factors to multiply. In these environments, the growth factors would be excreted by bacteria, fungi, or algae, or they would be generated as these organisms are grazed by protozoa or higher animals or are parasitized. The rates of such excretions are unknown, but their importance to auxotrophic populations must be great.

Growth factors may also affect the threshold concentration of a C source for growth and biodegradation. Thus, the concentration of glucose below which a bacterium would not multiply was lowered by a mixture of amino acids (Law and Button, 1977), and the threshold for phenol mineralization by lake water bacteria was reduced by a single amino acid (Rubin and Alexander, 1983).

Because bacteria and fungi that cometabolize organic compounds need a substrate for growth, it is not surprising that additions of organic materials or individual chemicals to natural environments often stimulate degradation. Some examples are presented in Table 13.1. However, the mechanism by which such stimulations occur is rarely known. In the case in which the addition of biphenyl to soil promotes PCB transformation, the effect may be the result of the larger population of biphenyl degraders, which grow using biphenyl as C source but can cometabolize PCBs, since biphenyl is an analogue of the chlorinated biphenyls. However, most of the stimulatory organic amendments are not analogues of the compounds being cometabolized (such as DDT, heptachlor, endrin, and BHC) so any benefit must be nonspecific, for example, by increasing the biomass of organisms that only coincidentally carry out a cometabolic reaction. In some instances, the effect may result from the added material causing a depletion of O_2, at least when the transformation is favored by anaerobiosis. Moreover, some of the substrates listed in Table 13.1 presumably are acted on by growth-linked and not cometabolic processes (e.g., MCPA, *m*-cresol, and 4-chlorophenol) so that the explanation may not be one associated with cometabolism. On the other hand, some organic amendments may reduce the rate of degradation (Subba-Rao *et al.*, 1982).

In many environments, the supply of the electron acceptor is not sufficient to meet the need, especially if the microflora places a large demand on it because of an abundance of organic substrates. For hydrocarbons

TABLE 13.1
Stimulation of Biodegradation of Test Substrates by Additions of Individual Compounds or Complex Organic Materials

Substrate	Environment	Amendment	Reference
BHC	Soil suspension	Peptone	Ohisa and Yamaguchi (1978)
m-Cresol, 4-Chlorophenol	Lake water	Amino acids	Shimp and Pfaender (1985)
DDT	Sewage	Glucose	Pfaender and Alexander (1973)
DDT, heptachlor	Flooded soil	Plant residues	Guenzi *et al.* (1971)
2,6-Dichlorobenzamide	Soil	Benzamide	Fournier (1975)
Malathion	Soil	Heptadecane	Merkel and Perry (1977)
MCPA	Soil	Straw	Duah-Yentumi and Kuwatsuka (1980)
PCB	Soil	Biphenyl	Focht and Brunner (1985)

and several other chemical classes, the only or preferred electron acceptor is O_2, and the transformations are only aerobic or the most rapid conversions are carried out by obligate aerobes. This need for O_2 is especially evident in the degradation of crude oil and individual hydrocarbons, particularly where O_2 diffusion from the atmosphere to replenish the supply is restricted or physically prevented. In groundwaters contaminated with gasoline or oil, the O_2 initially in the aqueous phase is rapidly consumed, and the subsequent degradation either is extremely slow or does not occur. As a result, remediation strategies typically involve the introduction of O_2 from forced air, pure O_2, or H_2O_2. The biodegradation of fractions of crude oil that sink to the sediments of marine and freshwater environments often is limited because the small amount of O_2 dissolved in the pore water or immediately overlying water column is consumed by the aerobic bacteria that act on the hydrocarbons. In soils, the O_2 dissolved in the liquid phase and in the gas-filled pores is also quickly consumed if much hydrocarbon or oil is present, and the rate of O_2 entry from the overlying air is too slow to sustain appreciable transformation. Although some bacteria metabolize hydrocarbons anaerobically, such reactions in most natural ecosystems proceed very slowly, if at all (Atlas, 1981). In contrast, the supply of O_2 rarely limits biodegradation of hydrocarbons at the surfaces of marine and fresh waters because O_2 from the atmosphere is accessible and diffuses readily into the top of the water column, especially if there is turbulence from wave action.

The biodegradation of many organic compounds is independent of the O_2 supply, and anaerobic conversions are common. Indeed, the transformation of some substrates is more rapid or only occurs in anoxic environments. In many instances, the electron acceptor is an organic molecule, but sometimes it is nitrate, sulfate, or CO_2. However, the supply of nitrate or sulfate may be totally consumed, so that further conversions may stop or be governed by the reentry of additional electron acceptors.

MULTIPLE SUBSTRATES

Laboratory studies are typically conducted with individual organic substrates, but natural and polluted environments characteristically contain a multiplicity of organic compounds that can be used by one or more of the indigenous bacteria or fungi. These substrates may be synthetic compounds, discrete natural products, the complex materials associated with humic fractions of soil or sediments, or the dissolved organic C (DOC) of natural waters. Their concentrations may be either quite high or extremely low, and the levels may be sufficiently high to be toxic or so low that they will not support growth. Because of the number of coexisting species and compounds, biodegradation of individual substrates will differ from the transformations brought about by a single species acting on one chemical.

Several organic substrates may be used at the same time. Such simultaneous metabolism has been reported for mixtures containing widely different classes of compounds. For example, marine bacteria degrade linear alkanes with 16 to 30 carbons simultaneously in oil-contaminated marine sediments and waters (Kator, 1973), the rate of metabolism of glucose in activated sludge is unaffected by the ongoing degradation of acetate (Painter *et al.*, 1968), mixed cultures simultaneously degrade 2,4-D and mecoprop (Hallberg *et al.*, 1991), *Pseudomonas fragi* utilizes several C sources simultaneously at concentrations from 5 to 80 mM (Molin, 1985), and *Pseudomonas putida* metabolizes phenol and *p*-cresol at the same time (Hutchinson and Robinson, 1988). In the latter study, the rate of biodegradation of each substrate depends on its concentration relative to the second compound.

Frequently one substrate enhances the rate of degradation of a second. Such stimulations occur in bioreactors, mixtures containing two organisms, or pure culture. Thus, the addition of glucose to a sludge bioreactor enhances the anaerobic transformation of pentachlorophenol (Hendriksen *et al.*, 1992), the addition of glucose promotes the biodegradation of 2,4-dinitrophenol by an actinomycete and a strain of *Janthinobacterium* (Hess *et al.*, 1990), and toluene stimulates the degradation of benzene and *p*-xylene by a pseudomonad (Alvarez and Vogel, 1991).

The converse may also occur, namely, one substrate may slow the degradation of a second. This is evident in the reduced rates of pentachlorophenol utilization by an enrichment culture in solutions containing phenol or 2,4,5-trichlorophenol, both of which could be utilized (Klečka and Maier, 1988), and of the degradation of low concentrations of acetate by a *Pseudomonas* strain also provided with methylene chloride as a substrate (LaPat-Polasko *et al.*, 1984). In some instances, the suppression by one compound of the metabolism of a second is manifested in a sequential use of the substrates, one disappearing only after the second is largely or wholly destroyed, as in the sequential destruction of linear alkanes by *Cladosporium resinae* (Lindley and Heydeman, 1986). Such sequential utilization is similar to the common finding that some components of oil or prepared mixtures of hydrocarbons are degraded by microorganisms in natural waters, soils, or enrichments before others (Mechalas *et al.*, 1973; Raymond *et al.*, 1976), although the differences in rates of disappearance in these instances probably are frequently, but not always, a result of the differences in intrinsic resistance of the various hydrocarbons to biodegradation.

Sequential utilization of substrates by pure cultures is frequently the result of diauxie, and it is evident when one substrate delays utilization of a second. However, in diauxie, the first substrate is being utilized during the period of apparent inhibition of the second conversion. Diauxie, which has been considered previously as a cause of acclimation, occurs in cultures of a number of bacterial genera in media containing high concentrations of organic substrates, and it sometimes has been noted in enrichment cultures that probably are dominated by a single bacterial species (Gaudy et al., 1964; Stumm-Zollinger, 1966). The substrate that supports the faster growth is generally the one that is used first when two growth-supporting compounds are included in culture media. For molecules for which individual species exhibit diauxie at high substrate concentrations, the two compounds are used simultaneously when their concentrations are low (Harder and Dijkhuizen, 1982). Similarly, one P source may be used in preference to another in a diauxie-type relationship, as in the preferential use of inorganic phosphate over methylphosphonate by *Pseudomonas testosteroni* (Daughton et al., 1979).

The explanations for the effect of one substrate on the biodegradation of a second in natural and polluted environments are largely unknown and have scarcely been explored. The reasons for the absence of an influence of one compound on the metabolism of a second are likewise uncertain. The absence of an effect in nature probably can be attributed frequently to the action of two different species functioning independently on the substrates of concern. The two may act independently unless they are limited by some common factor (e.g., grazing by protozoa or a deficiency of O_2 or an inorganic nutrient). Alternatively, if a single species is degrading the two compounds, their concentrations may be too low for diauxie to come into play. Because diauxie involves repression of synthesis of the enzymes catalyzing degradation of the second substrate as the first is being metabolized (Harder et al., 1984), diauxie may not be important if the catabolic pathways for the two C sources or the enzyme-regulatory mechanisms in the organisms are not subject to control by the physiological processes associated with diauxie.

A number of hypotheses have been advanced to account for the stimulation of biodegradation of one compound by a second, but few have experimental support. A likely cause in many instances is the greater population size or biomass arising because of the additional C source; if the resulting organisms can act readily on both substrates, growth on the second C source would enhance destruction of the first. If the second compound is only degraded cometabolically, then the first would obviously be beneficial inasmuch as a large mass of cometabolizing cells is produced. In some instances, if the organisms acting on one substrate are auxotrophs, the stimulation may result from the excretion of growth factors by the population acting on the second. Alternatively, the second compound may be beneficial because it induces enzymes necessary for catabolizing the other molecule. Should one of the two chemicals be at concentrations below the threshold for growth of the requisite bacteria or fungi, the second may serve as an energy source and thus facilitate destruction of the trace contaminant (Bouwer and McCarty, 1984).

More attention has been given to explanations of how second compounds inhibit biodegradation. (a) Undoubtedly, the suppression in many highly polluted sites results from toxicity of the second compound—a toxicity that slows or prevents growth or that diminishes activity of the microorganisms. If both compounds individually are at levels just below those that are toxic, a combination of the two could then exceed the tolerance of the active microorganisms (Smith *et al.,* 1991). (b) One substrate could be converted to products that are detrimental to the population acting on the second, as with the products of 4-nitrophenol metabolism by a pseudomonad, which inhibits phenol oxidation by a different bacterium (Murakami and Alexander, 1989). (c) Studies of the biodegradation of two substrates by two bacterial species, each of which can metabolize only one of the molecules, show that competition between the organisms for limiting concentrations of P may be reflected in a reduction in the rate of biodegradation of one or both of the compounds as compared to media with only one substrate. The two bacteria are competing for an inadequate supply of a limiting factor, and this competition is manifested in an effect on the transformation (W. S. Steffensen and M. Alexander, unpublished data, 1994). (d) Similarly, competition for O_2 or another electron acceptor, if present in amounts insufficient for the microbial demand, may be the reason why the organisms degrading one substance apparently have an effect on the utilization of a second. (e) The number of bacterial cells will be greater if two rather than a single C source is present, and this larger population would result in more intense grazing in environments in which protozoa are active; a likely outcome would be a lower rate or extent (or both) of biodegradation of a compound when a second substrate is being degraded by a different bacterial species. This reason for an inhibition is suggested by an investigation of degradation of a mixture of two substrates by two bacteria in the presence of the ciliate *Tetrahymena thermophila* (W. S. Steffensen and M. Alexander, unpublished data, 1994). (f) Should a single species be responsible for the biodegradation of both organic molecules, the inhibition may result from a repression of further synthesis of enzymes needed for the catabolism of one substrate by an intermediate formed in the catabolism of the second (catabolite repression), the inhibition by an intermediate of the activity of already existing enzymes, or by an interference by one substrate in the uptake by the cell of the second substrate (Harder and Dijkhuizen, 1982).

The effects of one compound on the biodegradation of a second are frequent and occur in many environments. However, given the number of compounds, the undefined populations causing their destruction, and the multitude of chemical mixtures, generalizations on whether there will or will not be an effect, whether the effect will be stimulatory or inhibitory, and the reasons for the enhancement or suppression are premature.

SYNERGISM

Many biodegradations require the cooperation of more than a single species. These interactions may be necessary for the initial step in the conversion, a later phase of the transformation, or the mineralization of the compound. These various interactions represent several types of *synergism*, in which two or more species carry out a transformation that one alone cannot perform or in which the process carried out by the multispecies mixture is more rapid than the sums of the rates of reactions effected by each of the separate species. Thus, some reactions take place in mixtures of species but not in pure culture or take place more readily in multispecies associations.

Several examples of synergism will serve as adequate illustrations of the phenomenon. Isolates of *Arthrobacter* and *Streptomyces* together are able to mineralize diazinon, but neither bacterium alone produces CO_2 from this insecticide (Gunner and Zuckerman, 1968). A mixture of *Pseudomonas* and *Arthrobacter* together degrades the herbicide silvex, although neither alone has this capacity (Ou and Sikka, 1977). Synergism is also shown by the more rapid degradation of dodecyl-1-decaethoxylate, a surfactant, by a mixture of two bacteria than by either organism alone (Watson and Jones, 1979).

A number of mechanisms for synergistic relationships have been described, but undoubtedly other mechanisms have yet to be discovered. (a) One or more species provide B vitamins, amino acids, or other growth factors to one or more of the other organisms. (b) One species grows on the test compound and carries out an incomplete degradation to yield one or several organic products, and the second species mineralizes the products that otherwise would accumulate. The second species commonly grows on the intermediate in the sequence. (c) The initial species cometabolizes the test compound to yield a product that it can no longer metabolize, and the second species destroys that product. This mechanism differs from the second in the type of activity of the initial population, that is, whether it uses the parent compound as a C source for growth or only cometabolizes it. (d) The first species converts the substrate to a toxic metabolite that then slows the transformation, but the reaction proceeds rapidly if the second member of the association destroys the inhibitor. This detoxication may sometimes be a consequence of the use of the inhibitor as a C source for growth, so that it is somewhat analogous to mechanism (b). On the other hand, it may involve interspecies hydrogen transfer, in which H_2 or reducing equivalents generated by one population are used by another.

Many bacteria and fungi that are capable of degrading toxicants require one or more growth factors. These auxotrophs will not grow in liquid media with the test substrate as sole C source because they need a single or several B vitamins, amino acids, purines, pyrimidines, or more complex growth factors. This inability of an organism to grow in simple media does not necessarily mean that the organism is unimportant in nature because a high percentage of the heterotrophic microorganisms in soils, sediments, and waters excrete growth factors (Burkholder, 1963; Lochhead, 1958), thereby permitting proliferation of the auxotrophs responsible for biodegradation. In such environments, there exist both a continuous formation and a continuous utilization of growth factors, resulting in a constant turnover or flux of trace nutrients essential for the auxotrophic inhabitants. Indeed, two fastidious species may coexist because each excretes and provides its associate with the required growth factor, a mutual feeding that is known as *syntrophy*.

Many growth factors are responsible for synergistic interactions, but B vitamins and amino acids are most often implicated. Excretion of vitamin B_{12}, for example, is required for a bacterium to grow on and dechlorinate trichloroacetic acid (Jensen, 1957), and growth factors excreted by a strain of *Pseudomonas* appear to be needed by an isolate of *Xanthomonas* to grow and degrade dodecyltrimethylammonium bromide (Dean-Raymond and Alexander, 1977).

A large number of synthetic molecules are converted to organic products, and little or no mineralization is evident. Even aerobic bacteria and fungi in pure culture often convert little or sometimes none of their C and energy source to CO_2. However, such species in nature often coexist with others that use the products of their associates as sources of C and energy for growth. Several examples are presented in Fig. 13.2. The first species

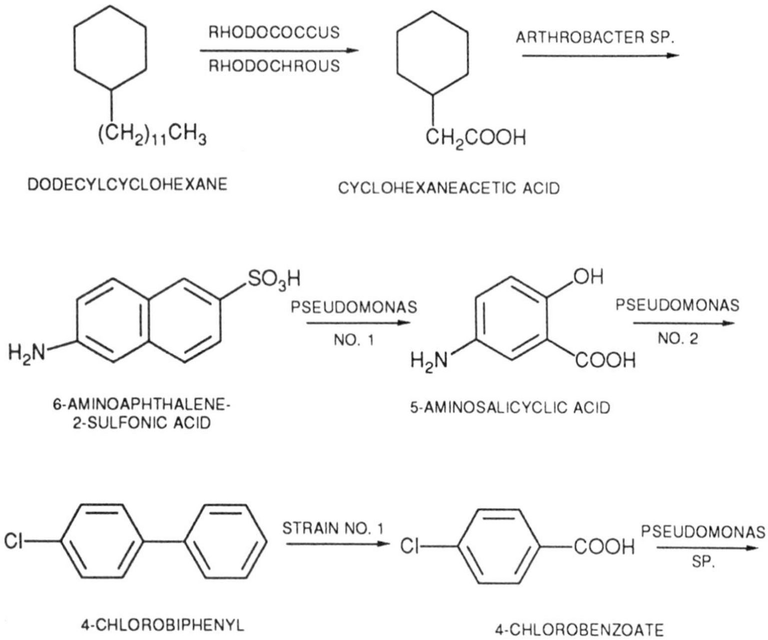

FIG. 13.2 Synergistic associations leading to the complete destruction of dodecyl-cyclohexane (Feinberg *et al.*, 1980), 6-aminonaphthalene-2-sulfonic acid (Nörtemann *et al.*, 1986), and 4-chlorobiphenyl (Sylvestre *et al.*, 1985).

in each instance converts the initial compound to a product that accumulates in a one-membered culture, but that compound is destroyed if the second species is present.

Cometabolizing microorganisms do not mineralize their organic substrates, and hence they are responsible for the accumulation of organic products. However, many of those products are C sources for other organisms and support their growth; hence the accumulation is transitory. Several examples are depicted in Fig. 13.3. Each of these illustrations comes from studies of defined cultures. However, the same types of relationships undoubtedly occur in nature, so that the intermediate in the conversion would be at low levels or undetectable in natural habitats, and the parent compound is converted to CO_2. Thus, although cyclohexane is apparently acted on by cometabolism to yield cyclohexanone, the latter can be used as a C source by aerobic bacteria; as a consequence, cyclohexane is mineralized in samples of soil and marine sediments (Beam and Perry, 1973, 1974). Nevertheless, the second species in some associations, as

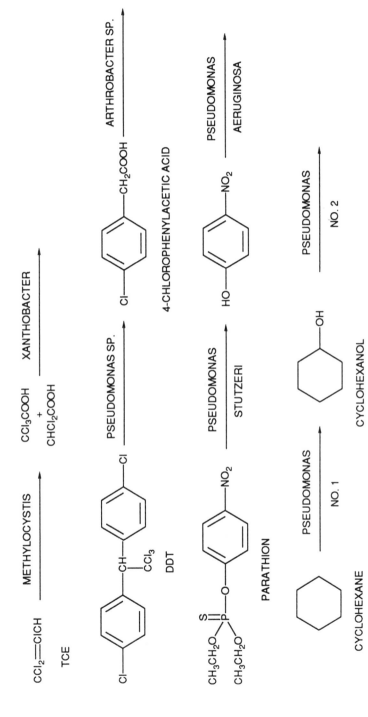

FIG. 13.3 Two-member associations in which the second species grows on the product of cometabolism of TCE (Uchiyama et al., 1992), DDT (Pfaender and Alexander, 1972), parathion (Daughton and Hsieh, 1977), and cyclohexane (deKlerk and van der Linden, 1974).

well as the first, may act only by cometabolism, so products of the second accumulate. This topic is also considered in Chapter 12.

The basis for some synergisms is the destruction of a toxin that is formed by one species. In some biodegradative reactions, the population acting on the parent chemical forms a metabolite that inhibits either the initial species or a species that is involved in a later step in the metabolic sequence. Yet, if the toxin producer has nearby an organism that destroys the inhibitor, the sensitive species continues to function without hindrance. In most of such synergisms that have been investigated, the first species generates a molecule that is harmful to itself, and the second species uses that compound as a source of C, energy, or both. However, the second organism may act in a somewhat different manner, as in the conversion of N-(3,4-dichlorophenyl)propionamide (the herbicide propanil) by *Penicillium piscarium* to 3,4-dichloroaniline, a compound that inhibits further propanil biodegradation, but the toxicity is relieved by *Geotrichum candidum*, a fungus that dimerizes the aniline derivative to yield 3,3',4,4'-tetrachloroazobenzene (Bordeleau and Bartha, 1971). Species that metabolize nitro compounds commonly produce nitrite, which is toxic to many micro- as well as macroorganisms, but many bacteria and fungi are able to destroy nitrite—by converting it to ammonium, N oxides, N_2, or sometimes nitrate.

Interspecies hydrogen transfer represents a unique type of synergism. In part, such an association also relies on the destruction by the second species of a toxin produced by the first. In this instance, the inhibitor is H_2, which metabolically represents reducing power not needed by the initial population. The reactions carried out by each of the two anaerobic bacteria are

$$2CH_3CH_2OH + 2H_2O \rightarrow 2CH_3COOH + 4H_2$$

$$4H_2 + CO_2 \rightarrow CH_4 + 2H_2O$$

The net effect of these two organisms is to make methane and acetate from ethanol, and the H_2 toxicity to the first organism is relieved by the second (Reddy *et al.*, 1972):

$$2CH_3CH_2OH + CO_2 \rightarrow 2CH_3COOH + CH_4$$

Other substrates may be acted on synergistically by mixtures of anaerobes in sequences that also involve interspecies hydrogen transfer (Laube and Martin, 1981; McCarty and Smith, 1986). A three-membered association has also been described in which 3-chlorobenzoate is converted anaerobically to methane. The different bacteria convert (a) 3-chlorobenzoate to benzoate; (b) benzoate to acetate, H_2, and CO_2; and (c) $H_2 + CO_2$ to methane. The reducing power for the reductive dehalogenation of 3-chlorobenzoate comes from the organism responsible for the second step (Dolfing and Tiedje, 1986).

PREDATION

An environment with a high density of bacteria or a large fungal biomass usually will also contain microorganisms that act as predators or parasites and some that will cause lysis. These predatory, parasitic, or lytic inhabitants may affect the biodegradation carried out by the bacteria and fungi. The impact is often deleterious, but it may be beneficial.

Among the predators and parasites found in soils, sediments, and surface and groundwaters are protozoa, bacteriophages, viruses affecting fungi, *Bdellovibrio*, mycobacteria, Acrasiales, and organisms that excrete enzymes that destroy cell walls of fungi and bacteria and thereby cause their lysis. Of these several groups, only the protozoa are known to affect biodegradation. This does not mean that the other groups are not important, only that evidence for their role has not been obtained.

Protozoa typically multiply by feeding on bacteria. In environments in which these microscopic animals are abundant, their grazing may markedly reduce the number of bacteria since 10^3 to 10^4 bacteria may be consumed to permit the division of a single protozoan. However, not only may protozoa affect bacterial activity by grazing but they may facilitate the cycling of limiting inorganic nutrients (especially P and N) and excrete essential growth factors. In some environments, protozoa are sparse and not particularly active, so that their role is highly dependent on prevailing conditions.

Active grazing requires a prey density greater than 10^6 to 10^7 bacterial cells per milliliter or, for nonaqueous environments, per cubic centimeter. Below this threshold density of bacteria, protozoan feeding is inconsequential. Similarly, when a substrate is provided in high concentrations in enrichment cultures containing protozoa, the predators will feed on the bacteria to bring their density down to about 10^6 cells per milliliter. Thus, if one assumes that approximately 1 pg of a C source will support 1 bacterial cell, a population size of 10^8 bacteria per milliliter would be expected in an enrichment containing 100 μg of an organic substrate per milliliter (100 mg/liter); instead, the cell density is sometimes only about 10^6 cells/ml (DiGeronimo *et al.*, 1979). However, the 10^6 cell/ml threshold is not for a single bacterial species but rather for all prey, so the abundance of a single species may be reduced below the threshold when other species are at high cell densities. For example, if species A is present at 10^8 cells/ml, species B is at 10^6/ml, and species C is at 10^3/ml, nonselective grazing would reduce the three populations to a final density of about 10^6/ml, but the mixture would contain 10^6 of A, 10^4 of B, and 10 of C per milliliter (Mallory *et al.*, 1983).

In environments in which they are active and abundant, the impact of protozoa depends on their grazing rate and the rate of biodegradation or, for transformations that proceed parallel to growth, bacterial multiplication. If grazing is slow and bacterial multiplication is rapid, protozoa will have little or no effect. If the predation rate is rapid (as occurs when the community of all bacteria is large) and the growth of the particular bacterial species causing the degradation is slow, protozoa may have a large impact. Such slow multiplication characterizes bacteria growing on a low concentration of an organic molecule, typically at levels below the K_s. At these low concentrations, the density of the biodegradative species may even decline despite the presence of a substrate that it alone, of the organisms present, can utilize. This is illustrated by the finding that indigenous protozoa have no effect on 4-nitrophenol mineralization at 50, 75, or 100 mg/liter in lake water samples inoculated with a nitrophenol-utilizing *Corynebacterium*, but they markedly suppress the transformation and prevent growth of the bacterium when the compound is added at 26 mg/liter (Zaidi *et al.*, 1989). However, a sufficient number of cells may survive the attack by protozoa so that, when the period of active predation is over, bacteria with the requisite metabolic capacity can grow and destroy the chemical (Ramadan *et al.*, 1990). Active grazing typically ends when the protozoa have reduced the density of all susceptible bacteria to approximately 10^6 cells/ml.

As discussed in Chapter 3, in sewage or other wastewaters containing many protozoa and a large number of indigenous bacteria to serve as prey, the acclimation period prior to active biodegradation may be the result of feeding by these unicellular animals. The acclimation phase lasts as long as the grazing is intense. Once the predation rate declines, usually because of the decrease in total number of bacteria to serve as prey, the remaining cells of bacteria able to degrade the chemical of interest begin to multiply and the active period of biodegradation commences (Wiggins and Alexander, 1988; Wiggins *et al.*, 1987).

Conversely, protozoa may sometimes stimulate microbial activity. This is evident, for example, in the enhanced decomposition of crude oil by a bacterial mixture in the presence of the ciliate *Colpidium colpoda* (Rogerson and Berger, 1983). Similar stimulatory effects on the rate of degradation of plant constituents or particulate matter have been observed with a number of flagellates and ciliates. A likely reason for this enhancement is the regeneration of P, N, or both. In environments in which the concentrations of inorganic P or N are so low that they limit microbial growth, the P or N is assimilated by bacteria, algae, and fungi, and little is available for the species important in a specific biodegradation; hence, the rate of microbial transformation is low. However, the grazers consume part of that microbial biomass and excrete some of the P and N in the material they consume. That P and N are then available for use by bacteria and fungi active in biodegradation. Such a regeneration of P and N, which represents P and N mineralization, is believed to be important in soil and fresh and marine waters (Anderson *et al.*, 1986; Cole *et al.*, 1978; Johannes, 1968). Protozoa also excrete growth factors as they ingest and digest bacteria, and they may thereby enhance biodegradation by auxotrophs (Huang *et al.*, 1981), which rely on other species for the vitamins, amino acids, and other growth factors they need.

GROWING PLANTS

Soil immediately surrounding the roots of growing plants, a site that is known as the rhizosphere, is a zone of intense microbial activity. This activity is a consequence of the large number of bacteria that utilize the simple organic compounds continuously excreted by the roots of plants during their active stages of development. In view of the large and metabolically active bacterial community of the rhizosphere, it is not unexpected that the rates of biodegradation, of some compounds at least, are more rapid in rhizosphere than adjacent nonrhizosphere soil or in soil under vegetation than in comparable fallow soil. For example, the mineralization of several surfactants is 1.1- to 1.9-fold faster in the rhizosphere of several plants than in nonrhizosphere soil (Knaebel and Vestal, 1992), and more benz(a)anthracene, chrysene, benzo(a)pyrene, and dibenz(a,h)anthracene disappear from soils supporting deep-rooted prairie grasses than from fallow soil (Aprill and Sims, 1990). Moreover, because TCE is more readily destroyed in rhizosphere than nonrhizosphere soil, the use of growing plants to promote bioremediation of TCE-contaminated soils has been proposed (Walton and Anderson, 1990).

ANAEROBIC BIODEGRADATION

Biodegradation under aerobic conditions has been the subject of intense inquiry, but it was only in recent years that anaerobic transformations have begun to receive their long overdue attention. These more recent studies have demonstrated that bacteria that function under anaerobiosis are frequently highly versatile, and they can destroy a variety of compounds. Many of these substrates can also be metabolized in the presence of O_2, often more rapidly but sometimes more slowly than under anaerobiosis. With certain molecules, however, the only known transformations occur when O_2 is not present, and such compounds persist in aerobic sites but disappear, albeit often slowly, under anaerobic conditions.

Some environments characteristically are devoid of O_2. In others, the O_2 is depleted by the aerobes that initially act on the organic materials. Aerobic bacteria and fungi are then displaced, and species able to function with other electron acceptors become prominent. These new populations may use organic compounds as electron acceptors, but many sites contain bacteria that use nitrate, sulfate, or CO_2 as electron acceptors. As a result of the use of the inorganic acceptors, N_2 and N_2O, sulfide, and methane are generated. When the level of pollution is high, provided that the pollutants are not highly toxic to microorganisms, the processes of concern would thus be anaerobic. Not only are such conversions important in biodegradation, but several bioremediation technologies are specifically designed to exploit anaerobic activities, especially for reactions that only occur, or only take place rapidly, when the system is O_2 free.

Individual species of anaerobes rarely bring about an extensive conversion of most compounds to CO_2. As a rule, a single species carries out only part of the sequence of steps necessary to mineralize organic molecules, but the species responsible for the initial transformation frequently coexists with other anaerobes that carry out the later steps. In some cases, three different bacteria may be involved, as in the destruction of 3-chlorobenzoate cited earlier as an example of synergism. Several of the compounds metabolized anaerobically are listed in Table 13.2. Some of the chemicals, however, are degraded anaerobically in one environment but not in another, at least in the time periods tested.

Of particular interest are compounds that are metabolized anaerobically but not by bacteria or fungi when O_2 is present. For these molecules, only anaerobic transformations will result in destruction of the molecules and will be the basis of bioremediation. The substrates include highly chlorinated PCBs, hexachlorobenzene (Mohn and Tiedje, 1992), 2,6-dinitrotoluene, 3,5-dinitrobenzoic acid (Hallas and Alexander, 1983), RDX (McCormick et al., 1980), and DDT (Parr and Smith, 1974). It is likely that some of the substrates now believed to be degraded only anaerobically will ultimately be found to be metabolized by some aerobe, but that possible discovery will not change the conclusion that the process is clearly favored when no O_2 is available. In some instances, as with hexachlorocyclohexane (Ohisa and Yamaguchi, 1978), the conversion proceeds with or without O_2, but the anaerobic conversion is more rapid.

TABLE 13.2
Compounds Degraded under Anaerobic Conditions

Chloroalkanes and alkenes	Benzoates
Carbon tetrachloride	Benzoate
Chloroform	2-, 3-, and 4-Chlorobenzoate
Vinyl chloride	3,4- and 3,5-Dichlorobenzoate
1,2-Dichloroethane	
1,1,1-Trichloroethane	Aromatic hydrocarbons
Trichloroethylene	Toluene
1,1,2,2-Tetrachlorethane	Ethylbenzene
Tetrachloroethylene	o- and m-Xylene
Phenols	Others
Phenol	Highly chlorinated PCBs
2- and 3-Chlorophenol	Dimethyl phthalate
2,4- and 2,5-Dichlorophenol	Pyridine
Trichlorophenols	Quinoline
Tetrachlorophenols	m- and p-Cresol
Pentachlorophenol	2,4-D
2-, 3-, and 4-Nitrophenol	2,4,5-T
	Diuron
	Linuron

Although the initiation of anaerobic metabolism of some compounds is detected almost immediately after their introduction into a suitable environment, extremely long periods of acclimation are required before a detectable disappearance of others is evident. This acclimation may even be of several months' duration, and the reductive dechlorination of chlorinated benzoates or benzenes may have acclimation phases as long as 6 months (Mohn and Tiedje, 1992). In the case of benzoate or 2- or 3-hydroxybenzoates, an anaerobic enrichment acting on these compounds only developed after 18 months (Sahm et al., 1986).

Should an anaerobic bioremediation process lead to the accumulation of organic products, as is frequently true, it is likely that those products can be destroyed aerobically. Thus, monochlorobenzene that is generated anaerobically from hexachlorobenzene or the compounds that accumulate in the metabolism of PCBs under anoxic conditions can be transformed aerobically (Bédard et al., 1987; Mohn and Tiedje, 1992). Such two-stage processes involving an initial anaerobic phase followed by a final aerobic phase represent promising means for the mineralization of certain persistent pollutants.

Because of the mammalian toxicity and persistence of many highly chlorinated molecules and their susceptibility to anaerobic bacteria, considerable attention has been given to the reductive dechlorinations that frequently represent the first and critical metabolic steps. Reductive dechlorination or, more generically, reductive dehalogenation may lead to the replacement of one or two halogens with one or two hydrogens (Fig. 13.4). In the third equation in Fig. 13.4, the substrates are typically alkyl halides, and the halogens are removed from adjacent C atoms with the formation of a double bond between the two C atoms. The conversion requires microbial activity, but it is not certain whether microorganisms catalyze the reduction enzymatically or only generate a reductant that functions nonenzymatically to bring about halogen removal. Reductive dechlorination is evident in anoxic soils, sediments, and sludges and is responsible for initial phases in the metabolism of highly chlorinated PCBs, halogenated alkanes and ethylenes (e.g., chloroform, methyl chloride, and tri- and tetrachloroethylene), persistent chlorinated pesticides (including DDT, dieldrin, toxaphene, and lindane), and other halogenated molecules (Mohn and Tiedje, 1992).

An inadequate supply of electron acceptors often limits anaerobic conversions. However, some communities of anaerobic microorganisms are able to use nitrate, sulfate, or CO_2/bicarbonate as electron acceptors to destroy compounds they would not otherwise degrade. For example, a

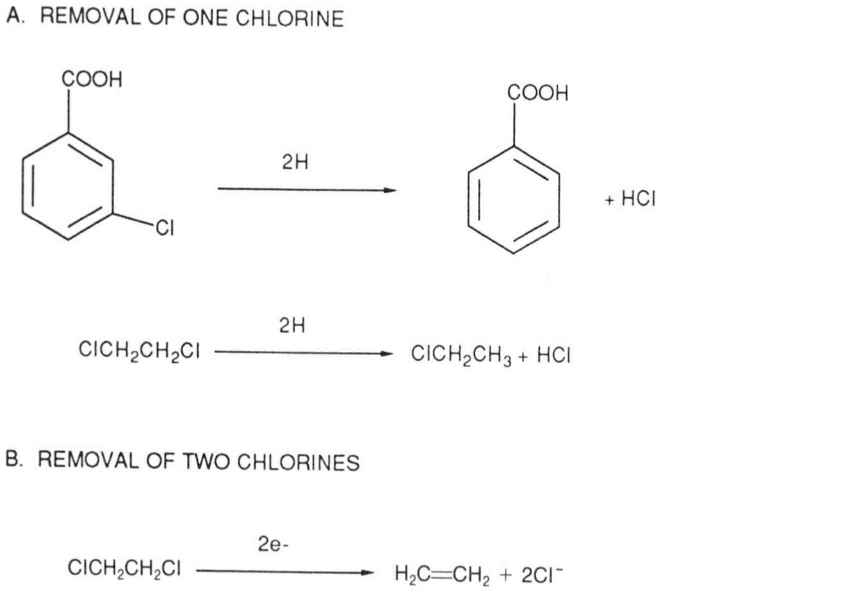

A. REMOVAL OF ONE CHLORINE

$$ClCH_2CH_2Cl \xrightarrow{2H} ClCH_2CH_3 + HCl$$

B. REMOVAL OF TWO CHLORINES

$$ClCH_2CH_2Cl \xrightarrow{2e-} H_2C{=}CH_2 + 2Cl^-$$

FIG. 13.4 Typical reductive dehalogenations.

mixture of microorganisms was found to be capable of degrading benzoate and 4-chlorobenzoate anaerobically but only if nitrate was present (Genthner *et al.,* 1989). These reactions not only destroy the organic molecules but reduce the electron acceptors nitrate, sulfate, and CO_2/bicarbonate to N_2 and N_2O, sulfide, and methane, respectively. If all three electron acceptors coexist, nitrate characteristically is metabolized first and disappears, sulfate reduction follows next, and finally methane is formed from CO_2. Ferric iron may serve as an electron acceptor, allowing some organisms to oxidize benzoate, phenol, and several other simple aromatic compounds anaerobically (Lovley *et al.,* 1989). On the other hand, sulfate may sometimes inhibit anaerobic conversions, for example, the reductive dehalogenation of aromatic compounds (Gibson and Suflita, 1986), and nitrate may also have harmful effects. Nevertheless, the supply of sulfate and nitrate in anaerobic environments is often limited, and the formation of additional amounts requires aerobic organisms; hence, if the supply is depleted, as would occur when the quantity of readily available C at a site is large, the anaerobic conversion may stop.

Recent research has thus disclosed the previously unrecognized potential of anaerobes for the decomposition of many pollutants, and these bacteria may be particularly important for compounds not metabolized by aerobes. Yet, many organic molecules persist in anaerobic environments, whether they be natural or polluted. Some of these are known to be potentially degradable or have actually been shown to be metabolized in nature or at some polluted site but, for reasons as yet unclear, the transformations are not ubiquitous—possibly because of the sparse distribution of organisms, the absence of suitable electron acceptors, toxins present at individual locations, or the need for O_2 not as an electron acceptor but rather because O_2 is a reactant in the actual oxidative step itself. Several of the compounds reported to be resistant to anaerobic degradation are listed in Table 13.3. Even for these examples, however,

TABLE 13.3
Compounds Reported to be Resistant to Anaerobic Degradation

Compounds	Reference
Anthracene, naphthalene	Bauer and Capone (1985)
2- and 4-Chlorobenzoate	Horowitz *et al.* (1983)
Benzene, chlorobenzene	Acton and Barker (1992)
Aniline, 4-toluidine	Hallas and Alexander (1983)
1- and 2-Naphthol, pyridine	Fox *et al.* (1988)
3,3'-Dichlorobenzidine	Boyd *et al.* (1984)
Saturated alkanes	Zehnder and Svensson (1986)

some transformation but possibly not mineralization may occur, or the reaction may proceed if periods long enough for extended acclimations are allowed to elapse. Still, the reactions are slow, incomplete, or not ubiquitous. Nevertheless, enrichments may be established, at least under laboratory conditions, that bring about the destruction of organic compounds that persist anaerobically in nature, as in the mineralization of benzene in sulfate-containing media (Edwards and Grbić-Galić, 1992). Studies of such enrichments may result in the development of practical bioremediation techniques that lead to the anaerobic destruction of otherwise long-lived pollutants.

REFERENCES

Acton, D. W., and Barker, J. F., *J. Contam. Hydrol.* **9**, 325–332 (1992).

Alvarez, P. J. J., and Vogel, T. M., *Appl. Environ. Microbiol.* **57**, 2981–2985 (1991).

Anderson, O. K., Goldman, J. C., Caron, D. A., and Dennett, M. R., *Mar. Ecol.: Prog. Ser.* **31**, 47–55 (1986).

Aprill, W., and Sims, R. C., *Chemosphere* **20**, 253–265 (1990).

Atlas, R. M., *Microbiol. Rev.* **45**, 180–209 (1981).

Atlas, R. M., and Bartha, R., *Biotechnol. Bioeng.* **14**, 309–318 (1972).

Bauer, J. E., and Capone, D. G., *Appl. Environ. Microbiol.* **50**, 81–90 (1985).

Beam, H. W., and Perry, J. J., *Arch. Microbiol.* **91**, 87–90 (1973).

Beam, H. W., and Perry, J. J., *J. Gen. Microbiol.* **82**, 163–169 (1974).

Bédard, D. L., Wagner, R. E., Brennan, M. J., Haberl, M. L., and Brown, J. F., Jr., *Appl. Environ. Microbiol.* **53**, 1094–1102 (1987).

Bordeleau, L. M., and Bartha, R., *Soil Biol. Biochem.* **3**, 281–284 (1971).

Bossert, I., and Bartha, R., *in* "Petroleum Microbiology" (R. M. Atlas, ed.), pp. 435–473. Macmillan, New York, 1984.

Bouwer, E. J., and McCarty, P. L., *Ground Water* **22**, 433–440 (1984).

Boyd, S. A., Kao, C.-W., and Suflita, J. M., *Environ. Toxicol. Chem.* **3**, 201–208 (1984).

Burkholder, P. R., *in* "Symposium on Marine Microbiology" (C. H. Oppenheimer, ed.), pp. 133–150. Thomas, Springfield, IL, 1963.

Cole, C. V., Elliott, E. T., Hunt, H. W., and Coleman, D. C., *Microb. Ecol.* **4**, 381–387 (1978).

Daughton, C. G., and Hsieh, D. P. H., *Appl. Environ. Microbiol.* **34**, 175–184 (1977).

Daughton, C. G., Cook, A. M., and Alexander, M., *Appl. Environ. Microbiol.* **37**, 605–609 (1979).

Dean-Raymond, D., and Alexander, M., *Appl. Environ. Microbiol.* **33**, 1037–1041 (1977).

deKlerk, H., and van der Linden, A. C., *Antonie van Leeuwenhoek* **40**, 7–15 (1974).

Dibble, J. T., and Bartha, R., *Appl. Environ. Microbiol.* **37**, 729–739 (1979).

DiGeronimo, M. J., Nikaido, M., and Alexander, M., *Appl. Environ. Microbiol.* **37**, 619–625 (1979).

Dolfing, J., and Tiedje, J. M., *FEMS Microbiol. Ecol.* **38**, 293–298 (1986).

Duah-Yentumi, S., and Kuwatsuka, S., *Soil Sci. Plant Nutr.* **26**, 541–549 (1980).

Edwards, E. A., and Grbić-Galić, D., *Appl. Environ. Microbiol.* **58**, 2663–2666 (1992).

Egli, T., *Antonie van Leeuwenhoek* **60**, 225–234 (1991).

Feinberg, E. L., Ramage, P. I. N., and Trudgill, P. W., *J. Gen. Microbiol.* **121**, 507–511 (1980).

Floodgate, G. D., *in* "Petroleum Microbiology" (R. M. Atlas, ed.), pp. 354–397. Macmillan, New York, 1984.

Focht, D. D., and Brunner, W., *Appl. Environ. Microbiol.* **50**, 1058–1063 (1985).

Fondén, R., *Oikos* **20**, 373–383 (1969).

Fournier, J.-C., *Chemosphere* **4**, 35–40 (1975).

Fox, P., Suidan, M. T., and Pfeffer, J. T., *J. Water Pollut. Control Fed.* **60**, 86–92 (1988).

Gaudy, A. F., Jr., Komolrit, K., and Gaudy, E. T., *Appl. Microbiol.* **12**, 280–286 (1964).

Genthner, B. R. S., Price, W. A., II, and Pritchard, P. H., *Appl. Environ. Microbiol.* **55**, 1472–1476 (1989).

Gibson, S. A., and Suflita, J. M., *Appl. Environ. Microbiol.* **52**, 681–688 (1986).

Guenzi, W. D., Beard, W. E., and Viets, F. G., Jr., *Soil Sci. Soc. Am. Proc.* **35**, 910–913 (1971).

Gunner, H. B., and Zuckerman, B. M., *Nature (London)* **217**, 1183–1184 (1968).

Hallas, L. E., and Alexander, M., *Appl. Environ. Microbiol.* **45**, 1234–1241 (1983).

Hallberg, K. B., Kelly, M. P., and Tuovinen, O. H., *Curr. Microbiol.* **23**, 65–69 (1991).

Harder, W., and Dijkhuizen, L., *Philos. Trans. R. Soc. London, Ser. B* **297**, 459–479 (1982).

Harder, W., Dijkhuizen, L., and Veldkamp, H., *in* "The Microbe" (D. P. Kelly and N. G. Carr, eds.), Part II, pp. 51–95. Cambridge Univ. Press, Cambridge, UK, 1984.

Hendriksen, H. V., Larsen, S., and Ahring, B. K., *Appl. Environ. Microbiol.* **58**, 365–370 (1992).

Hess, T. F., Schmidt, S. K., Silverstein, J., and Howe, B., *Appl. Environ. Microbiol.* **56**, 1551–1555 (1990).

Hoover, D. G., Borgonovi, G. E., Jones, S. H., and Alexander, M., *Appl. Environ. Microbiol.* **51**, 226–232 (1986).

Horowitz, A., Suflita, J. M., and Tiedje, J. M., *Appl. Environ. Microbiol.* **45**, 1459–1465 (1983).

Huang, T.-C., Chang, M.-C., and Alexander, M., *Appl. Environ. Microbiol.* **41**, 229–232 (1981).

Hutchinson, D. H., and Robinson, C. W., *Appl. Microbiol. Biotechnol.* **29**, 599–604 (1988).

Jamison, V. W., Raymond, R. L., and Hudson, J. O., Jr., *Dev. Ind. Microbiol.* **16**, 305–312 (1975).

Jensen, H. L., *Can. J. Microbiol.* **3**, 151–164 (1957).

Jobson, A., McLaughlin, M., Cook, F. D., and Westlake, D. W. S., *Appl. Microbiol.* **27**, 166–171 (1974).

Johannes, R. E., *in* "Advances in Microbiology of the Sea" (M. R. Droop and E. J. F. Wood, eds.), Vol. 1, pp. 203–213. Academic Press, London, 1968.

Jones, S. H., and Alexander, M., *FEMS Microbiol. Lett.* **52**, 121–126 (1988a).

Jones, S. H., and Alexander, M., *Appl. Environ. Microbiol.* **54**, 3177–3179 (1988b).

Kator, H., *in* "The Microbial Degradation of Oil Pollutants" (D. G. Ahearn and S. P. Meyers, eds.), pp. 47–65. Louisiana State University, Center for Wetland Resources, Baton Rouge, 1973.

Klečka, G. M., and Maier, W. J., *Biotechnol. Bioeng.* **31**, 328–333 (1988).

Knaebel, D. B., and Vestal, J. R., *Can. J. Microbiol.* **38**, 643–653 (1992).

Kuiper, J., and Hanstveit, A. O., *Ecotoxicol. Environ. Saf.* **8**, 15–33 (1984).

LaPat-Polasko, L. T., McCarty, P. L., and Zehnder, A. J. B., *Appl. Environ. Microbiol.* **47**, 825–830 (1984).

Laube, V. M., and Martin, S. M., *Appl. Environ. Microbiol.* **42**, 413–420 (1981).

Law, A. T., and Button, D. K., *J. Bacteriol.* **129**, 115–123 (1977).

LePetit, J., and N'Guyen, M.-H., *Can. J. Microbiol.* **22**, 1364–1373 (1976).

Lewis, D. L., Freeman, L. F., III, and Watwood, M. E., *Environ. Toxicol. Chem.* **5**, 791–796 (1986).

Lindley, N. D., and Heydeman, M. T., *Appl. Microbiol. Biotechnol.* **23**, 384–388 (1986).

Lochhead, A. G., *Bacteriol. Rev.* **22**, 145–153 (1958).

Lovley, D. R., Baedecker, M. J., Lonergan, D. J., Cozzarelli, I. M., Phillips, E. J. P., and Siegel, D. J., *Nature (London)* **339**, 297–300 (1989).

Mallory, L. M., Yuk, C.-S., Liang, L.-N., and Alexander, M., *Appl. Environ. Microbiol.* **46**, 1073–1079 (1983).

McCarty, P. L., and Smith, D. P., *Environ. Sci. Technol.* **20**, 1200–1206 (1986).

McCormick, N. G., Foster, D. M., Cornell, J. H., and Kaplan, A. M., *Abstr., Annu. Meet., Am. Soc. Microbiol.*, p. 196 (1980).

Mechalas, B. J., Meyers, T. J., and Kolpack, R. L., *in* "The Microbial Degradation of Oil Pollutants" (D. G. Ahearn and S. P. Meyers, eds.), pp. 67–79. Louisiana State University, Center for Wetland Resources, Baton Rouge, 1973.

Merkel, G. J., and Perry, J. J., *J. Agric. Food Chem.* **25**, 1011–1012 (1977).

Mohn, W. W., and Tiedje, J. M., *Microbiol. Rev.* **56**, 482–507 (1992).

Molin, G., *Appl. Environ. Microbiol.* **49**, 1442–1447 (1985).

Murakami, Y., and Alexander, M., *Biotechnol. Bioeng.* **33**, 832–838 (1989).

Nelson, M. J. K., Montgomery, S. O., O'Neill, E. J., and Pritchard, P. H., *Appl. Environ. Microbiol.* **52**, 383–384 (1986).

Nörtemann, B., Baumgarten, J., Rast, H. G., and Knackmuss, H.-J., *Appl. Environ. Microbiol.* **52**, 1195–1202 (1986).

Ohisa, N., and Yamaguchi, M., *Agric. Biol. Chem.* **42**, 1983–1987 (1978).

Olivieri, R., Robertiello, A., and Degen, L., *Mar. Pollut. Bull.* **9**, 217–220 (1978).

Ou, L. T., and Sikka, H. C., *J. Agric. Food Chem.* **25**, 1336–1339 (1977).

Owens, J. D., and Legan, J. D., *FEMS Microbiol. Rev.* **46**, 419–432 (1987).

Painter, H. A., Denton, R. S., and Quarmby, C., *Water Res.* **2**, 427–447 (1968).

Parr, J. F., and Smith, S., *Soil Sci.* **118**, 45–52 (1974).

Pfaender, F. K., and Alexander, M., *J. Agric. Food Chem.* **20**, 842–846 (1972).

Pfaender, F. K., and Alexander, M., *J. Agric. Food Chem.* **21**, 397–399 (1973).

Pritchard, P. H., Cripe, C. R., Walker, W. W., Spain, J. C., and Bourquin, A. W., *Chemosphere* **16**, 1509–1520 (1987).

Ramadan, M. A., El-Tayeb, O. M., and Alexander, M., *Appl. Environ. Microbiol.* **56**, 1392–1396 (1990).

Raymond, R. L., Hudson, J. O., and Jamison, V. W., *Appl. Environ. Microbiol.* **31**, 522–535 (1976).

Reddy, C. A., Bryant, M. P., and Wolin, M. J., *J. Bacteriol.* **110**, 126–132 (1972).

Robertson, B. K., and Alexander, M., *Appl. Environ. Microbiol.* **58**, 38–41 (1992).

Rogerson, A., and Berger, J., *J. Gen. Appl. Microbiol.* **29**, 41–50 (1983).

Rouatt, J. W., and Lochhead, A. G., *Soil Sci.* **80**, 147–154 (1955).

Rubin, H. E., and Alexander, M., *Environ. Sci. Technol.* **17**, 104–107 (1983).

Rubin, H. E., Subba-Rao, R. V., and Alexander, M., *Appl. Environ. Microbiol.* **43**, 1133–1138 (1982).

Sahm, H., Brunner, M., and Schoberth, S. M., *Microb. Ecol.* **12**, 147–153 (1986).

Shimp, R. J., and Pfaender, F. K., *Appl. Environ. Microbiol.* **49**, 394–401 (1985).

Skerman, T. M., *in* "Symposium on Marine Microbiology" (C. H. Oppenheimer, ed.), pp. 685–698. Thomas, Springfield, IL, 1963.

Smith, M. R., Ewing, M., and Rutledge, C., *Appl. Microbiol. Biotechnol.* **34**, 536–538 (1991).

Stumm-Zollinger, E., *Appl. Microbiol.* **14,** 654–664 (1966).

Subba-Rao, R. V., Rubin, H. E., and Alexander, M., *Appl. Environ. Microbiol.* **43,** 1139–1150 (1982).

Swindoll, C. M., Aelion, C. M., and Pfaender, F. K., *Appl. Environ. Microbiol.* **54,** 212–217 (1988).

Sylvestre, M., Massé, R., Ayotte, C., Messier, F., and Fauteux, J., *Appl. Microbiol. Biotechnol.* **21,** 192–195 (1985).

Uchiyama, H., Nakajima, T., Yagi, O., and Nakahara, T., *Appl. Environ. Microbiol.* **58,** 3067–3071 (1992).

Walker, A., *Proc. Br. Crop. Prot. Conf.–Weeds, 13th, 1976,* Vol. 2, 635–642 (1976).

Walton, B. T., and Anderson, T. A., *Appl. Environ. Microbiol.* **56,** 1012–1016 (1990).

Wang, Y.-S., Subba-Rao, R. V., and Alexander, M., *Appl. Environ. Microbiol.* **47,** 1195–1200 (1984).

Wang, Y.-S., Madsen, E. L., and Alexander, M., *J. Agric. Food Chem.* **33,** 495–499 (1985).

Watson, G. K., and Jones, N., *Soc. Gen. Microbiol. Q.* **6,** 78 (1979).

Wiggins, B. A., and Alexander, M., *Can. J. Microbiol.* **34,** 661–666 (1988).

Wiggins, B. A., Jones, S. H., and Alexander, M., *Appl. Environ. Microbiol.* **53,** 791–796 (1987).

Zaidi, B. R., Murakami, Y., and Alexander, M., *Environ. Sci. Technol.* **22,** 1419–1425 (1988).

Zaidi, B. R., Murakami, Y., and Alexander, M., *Environ. Sci. Technol.* **23,** 859–863 (1989).

Zehnder, A. J. B., and Svensson, B. H., *Experientia* **42,** 1197–1205 (1986).

14 INOCULATION

Microorganisms with a phenomenal array of catabolic activities are widespread. Soils, sediments, fresh and marine waters, and industrial and municipal waste-treatment systems possess large and often highly diverse microbial communities that potentially can exhibit many degradative capacities, and when these capacities are expressed fully and rapidly, organic chemicals are readily destroyed. Nevertheless, many synthetic compounds persist for some time in these same environments, even though these molecules are biodegradable, and the question has been asked whether inoculation might appreciably enhance the decomposition of these compounds. Such inoculation is sometimes called *bioaugmentation.*

In circumstances in which the period for destruction of the chemicals is not important, it is likely that inoculation is not warranted because the initially small population will multiply to destroy the unwanted chemical. However, when rapid destruction is important, it may not be appropriate to rely on the natural response of members of the indigenous community. For example, a slow biodegradation may result in the uptake by plants of toxicants present in soil, the movement of chemicals through soil to underlying groundwater, the transport of pollutants through a contaminated plume of groundwater to enter waters used for human consumption, or the dissemination of unwanted compounds through a biological treatment system to the receiving water and from that water to distant rivers or lakes, from which human, animal, or plant uptake may occur. The indigenous species may act, but often not sufficiently rapidly to prevent the spreading of a local problem.

It also is now clear that microorganisms acting on certain pollutants are absent from particular sites. A compound that is metabolized by many species will likely encounter one or several species in all microbial communities that can transform it. On the other hand, certain synthetic compounds are apparently transformed by very few species, and it is thus likely that not a single one of the very few species with the requisite enzymes may be present in a particular site. This view is in line with the frequent observation that some organic compounds are mineralized or otherwise metabolized in samples from one but not another environment and that active organisms can only be isolated from some environments.

Inoculation may also markedly reduce the acclimation period. If the time for the community to reach full activity is but a day or two, attempts to establish an organism probably would be pointless. However, if the acclimation period is weeks or months, as it often is, and the risk of human, animal, or plant exposure increases as the persistence of the toxicant increases, some form of intervention to enhance decomposition is called for.

Finally, inoculation may be necessary because conditions at the site preclude members of the resident community from functioning rapidly. Thus, when the unwanted chemical is present at a concentration high enough to suppress the native biodegrading species, when the temperature is too high, or the circumstances are otherwise stressful, the addition of a species able to destroy the chemical and also to tolerate the stress may be highly beneficial.

The approach to inoculation must be prudent. If there is an indigenous flora capable of carrying out the reaction, conditions favor their multiplication, and rapid destruction is not essential, additions of inocula are not needed. If these conditions do not pertain, intervention is called for. The lack of need for supplementation with microorganisms is well illustrated in waters and soils contaminated with oil. Such environments contain bacteria able to grow on and destroy a variety of hydrocarbons, and the persistence of components of oil is not a consequence of the absence of organisms but rather the absence of the full set of conditions necessary for the resident species to function rapidly (Atlas, 1977). In a typical study, the addition of a mixture of hydrocarbon-degrading bacteria to a marine-water microcosm did not enhance the degradation of crude oil polluting the seawater, and the indigenous microflora degraded the oil (Tagger et al., 1983). Similarly, the addition of soil with a large population of hydrocarbon degraders to soil freshly contaminated with oil reduced the acclimation period, but the indigenous population soon multiplied and carried out the desired transformation (DeBorger et al., 1978).

SUCCESSES

It is important to distinguish between microorganisms added to, or allowed to grow in, engineered systems and those added to natural environments. Aboveground bioreactors of many types have microorganisms added to them—either pure cultures, enrichments, or microbial mixtures—and these organisms usually develop and destroy the compounds on which they were grown or enriched. These engineered bioreactors, for example, may be industrial waste-treatment systems involving immobilized cells or biofilms designed for specific compounds or waste streams. The record of success in these instances is good.

In contrast, the record of success in enhancing biodegradation in soils, aquifers, and surface waters *in situ* is spotty. On the one hand, the initiation or enhancement of degradation has often been reported following the addition to natural environments (or, far more commonly, to samples of these environments brought to the laboratory) of bacteria and fungi that can metabolize and grow on specific organic compounds in culture. On the other hand, failures frequently have been reported also.

The usual way of obtaining a population for subsequent inoculation is to prepare an enrichment culture. This is typically done by adding a sample of soil, sewage, or natural water into a solution containing the organic compound and the variety of inorganic salts necessary for bacterial growth. The C source is usually added at a level far higher than exists in nature, so that a high cell yield is obtained. The pH is maintained near neutrality, and the mixture is incubated in the dark. When growth or chemical disappearance is evident, a subculture is added to a sterile portion of the same medium. This procedure may be repeated several times to increase the number of bacteria active on the test substrate relative to other organisms, and then the mixture usually is plated on an agar medium containing the test chemical as well as inorganic salts. The procedure provides isolates active on many compounds. However, the procedure favors organisms that grow well at high substrate concentrations, require no growth factors, multiply at pH values near 7.0, grow quickly, are not resistant to light inactivation, and multiply at levels of N and P in the enrichment medium that are far higher than prevail in the natural environment. Such approaches are based on other needs for isolates: to study the pathways of metabolism or the physiology of organisms catabolizing particular substrates. They are ideal for organisms active at high nutrient levels in the absence of biotic or abiotic stresses, but they do not fit in well with circumstances in natural ecosystems or polluted sites. It is unfortunately not common to make enrichments that are more likely to function well at the low levels of substrate, N, and P characteristic of the polluted environment or tolerate the pH, photochemical, or other stresses likely encountered in the environment in which rapid biodegradation is sought.

First, let us examine reports that inoculation enhances the destruction of pollutants in soil. These studies have centered on pesticides and either oil or specific hydrocarbon constituents of oil.

(a) **Parathion.** This insecticide is readily destroyed in soil inoculated with a mixture containing *Pseudomonas stutzeri* and *Pseudomonas aeruginosa*. The first bacterium converts parathion to 4-nitrophenol, and the second grows on and destroys 4-nitrophenol. In soil contaminated with 5.0 g of parathion per kilogram, more than 90% is destroyed within 3 weeks as a direct result of the inoculation. These tests were conducted using 10-g soil samples (Daughton and Hsieh, 1977). Addition of the same bacteria to soil in the field also results in the destruction of the insecticide. The latter study was done by adding the bacteria to soil contained within pipes (3.2 cm diameter) that were driven 10 cm into the ground (Barles *et al.*, 1979). The reason for stating the size of the containers or the amount of soil will be presented in the following.

(b) **IPC.** A mixture of bacteria added in a large volume of liquid to flats of soil (10 cm deep) in the greenhouse destroys this herbicide, as well as two related herbicides, when present initially at levels equivalent to 5, 10, and 15 kg/ha (McClure, 1972). Similarly, treating 57-g samples of soil in petri dishes with an *Arthrobacter* sp. able to grow in culture on IPC leads to inactivation of the herbicide (Clark and Wright, 1970). In these investigations, the action of the microorganisms was assessed by regular bioassay of the toxicity of the herbicide-treated soil to susceptible plants.

(c) **Chlorpropham.** The addition to 25-g quantities of soil of either of two chlorpropham-utilizing *Pseudomonas* species results in destruction of this herbicide (Milhomme *et al.*, 1989).

(d) **PCP.** Mixing an inoculum of *Rhodococcus chlorophenolicus* with 50-g quantities of soil brings about mineralization of PCP (Middledorp *et al.*, 1990). In earlier studies, it was found that introduction of a PCP-metabolizing strain of *Arthrobacter* into 250-g samples of PCP-contaminated soil that is mixed with the inoculum results in destruction of the chemical. The same bacterium also decomposes the chemical in soil placed 10 cm deep on a concrete floor, but more PCP is destroyed in soil that is mixed daily than if it is not mixed (Edgehill and Finn, 1983). Similar results are obtained when a PCP-degrading *Flavobacterium* is added to 100-g samples of soil supplemented with 100 mg of PCP per kilogram, although marked PCP mineralization in this soil is evident after 10 days as a result of the action of the indigenous community (Crawford and Mohn, 1985).

(e) **Oil.** Inoculation of 100 g of soil with *Candida guillermondii* promotes the rate of degradation of crude petroleum and hydrocarbons (Ismailov, 1985). In contrast, the addition of a mixture of oil-utilizing bacteria in field plots in an earlier study was found to cause only a slight enhancement in the rate of breakdown of *n*-alkane constituents with chain lengths of 20 to 25 carbons (Jobson *et al.*, 1974).

(f) 2,4-D. The typical acclimation phase prior to the onset of mineralization of the herbicide is almost wholly abolished if the soil is amended with a suspension containing organisms acting on this compound (Kunc and Rybarova, 1983).

(g) PCBs. Addition of *Acinetobacter* sp. stimulates the rate of PCB mineralization and favors the destruction of the more highly chlorinated PCBs (Focht and Brunner, 1985).

(h) Pyrazon. A gram negative coccus able to degrade this herbicide in liquid medium also enhances its breakdown in soil (Engvild and Jensen, 1969).

(i) 2,4,5-T. The introduction of a 2,4,5-T-utilizing strain of *Pseudomonas cepacia* into test tubes of soil containing 1.0 g of 2,4,5-T per kilogram of soil results in destruction of much of the herbicide as measured by bioassays with sensitive plants (Karns *et al.*, 1984). Tests involving addition of the fungus *Phanerochaete chrysosporium* to 1.0 g of soil amended with 4.0 g of ground corn cob show that 33% of the 2,4,5-T is mineralized in 30 days (Ryan and Bumpus, 1989).

(j) Lindane and chlordane. *Phanerochaete chrysosporium* mineralizes both of these chlorinated hydrocarbon insecticides when the fungus is inoculated into a mixture containing 1.0 g of sterile soil and 4.0 g of corn cobs (Kennedy *et al.*, 1990).

(k) Dicamba. Inoculation of soil with microorganisms degrading this herbicide results in destruction of the chemical and protection of seedlings from phytotoxicity (Krueger *et al.*, 1991).

(l) Nitrophenols. Inoculation of a flooded soil with a microbial mixture increases the rate of disappearance of 2-, 3-, and 4-nitrophenol and 2,4-dinitrophenol as compared to samples not so treated (Sudhakar-Barik and Sethunathan, 1978).

Soils are three-dimensional environments. From the point of inoculation, bacteria or fungi will have to be transported or migrate through an environment with pores ranging from macro- to microscopic in size. The cells or hyphae will not be able to penetrate into many micropores, and they will not be able to pass through the narrow necks between even some larger pores. Bacteria will also become sorbed to surfaces of the soil particles. From this viewpoint, experiments involving gram quantities or several centimeters of depth of soil do not serve as models for a three-dimensional environment composed of a solid matrix; they are merely two-dimensional trials. This is not to say that such inoculations will fail under more realistic conditions, rather it should be taken as an argument that realistic evaluations need to be conducted.

Inoculation has also been reported to result in the destruction of organic chemicals in natural waters. For example, a hydrocarbon-degrading bacterium obtained from an estuary enhances the biodegradation of oil spilled into a saline pond (Atlas and Busdosh, 1976). Similarly, data from laboratory studies indicate that oil-degrading bacteria added to seawater destroy a substantial part of the crude oil that has been added to the water (Miget et al., 1969). In both of the preceding tests, the water had been supplemented with inorganic nutrients. The effect of an inoculum is evident even when the concentration of the organic compound is low; thus, addition of a benzoate-utilizing bacterium to lake water stimulates the rate of biodegradation of benzoate present at initial concentrations of 5 and 50 μg/liter (Subba-Rao et al., 1982). In beach sands contaminated with oil from ocean pollution, an inoculum may also enhance the elimination of the unwanted oil (Ahlfeld and LaRock, 1973). On the other hand, a strain of a 4-nitrophenol-utilizing Corynebacterium had little activity in destroying that compound in lake water at 26 μg/liter although it was active at higher concentrations (Fig. 14.1). Biodegradation is also observed when a strain of Mycobacterium active on PAHs is added to a mixture of 180 ml of reservoir water with 20 g of sediment, and the bacterium is capable of mineralizing pyrene (Heitkamp and Cerniglia, 1989).

Again, the spatial issue should be raised. A small water sample is not a lake or a reasonable volume of the oceans. Although physical obstructions to microbial movement do not exist in open waters, the question can be raised about the environmental relevancy of very small scale tests that are not followed by pilot-scale experimentation, especially in a nonturbulent environment.

Activated-sludge systems for treating sewage are well mixed in actual practice, and here many of the spatial concerns are of less relevancy. Although possessing a large and metabolically diverse microbial community, sewage has long been known to exhibit an acclimation period prior to the initiation of rapid decomposition of many synthetic compounds. Thus, it is not surprising that addition of a 4-nitrophenol-metabolizing strain of Pseudomonas to samples of sewage enhances the mineralization of this compound (Goldstein et al., 1985).

Evidence does exist, however, of the successful outcome of inoculation to bring about pollutant destruction in the field. For example, either of two species of Phanerochaete added to a field site containing soil contaminated with PCP, which was used as a wood preservative, brings about destruction of about 90% of the PCP in less than 7 weeks, even under less than optimal conditions for these fungi. However, unidentified complexes are formed in the transformation (Lamar and Dietrich, 1990). PCP-contami-

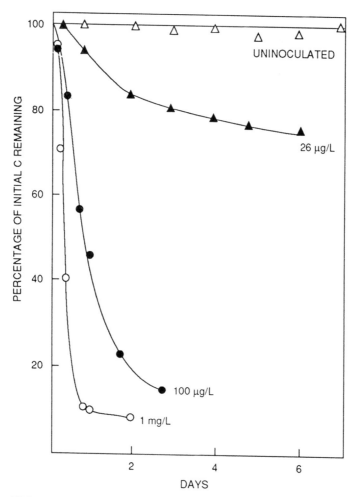

FIG. 14.1 Mineralization of 1.0 mg of 4-nitrophenol per liter in uninoculated lake water and of 26 μg, 100 μg, and 1.0 mg of 4-nitrophenol per liter in lake water inoculated with *Corynebacterium* sp. (From Zaidi *et al.*, 1988a. Reprinted with permission from the American Chemical Society.)

nated soil was also remediated at a field site by an inoculum containing a microbial mixture. In the latter instance, the pollutant was removed from the soil by a washing procedure, and the PCP in the wash solution was degraded by the added microorganisms in an aboveground bioreactor (Compeau *et al.*, 1991). The latter success is not the result of inoculation of the soil *in situ* but rather in an engineered system.

FAILURES

These findings of success have promoted considerable optimism among some investigators. All one needs to do, or so it seemed, is to set up an enrichment culture, isolate a bacterium or fungus able to use the unwanted chemical as a C source or cometabolize it, grow the organisms in culture to get a large cell biomass, and then add the organism to the natural environment containing the substance whose destruction is sought. That this unbridled optimism was premature soon became apparent as evidence became available that such inocula often failed to carry out in environmental samples the transformations they effected in laboratory media. For example, inoculation of alfalfa seeds with a 4-(2,4-DB)-utilizing *Flavobacterium* does not protect the plant from the toxic effects of this herbicide following sowing of the alfalfa in soil (MacRae and Alexander, 1965). Inoculation of soil with bacteria that can grow at the expense of aliphatic hydrocarbons does not enhance the degradation of fuel oil present in the soil (Lehtomäki and Niemelä, 1975). Similarly, a soil *Pseudomonas* sp. capable of mineralizing 2,4-dichlorophenol in liquid media does not mineralize this phenol when the organism is added to the surface of soil (Goldstein *et al.*, 1985), and a metolachlor-degrading *Streptomyces* added to soil does not enhance the biodegradation of this pesticide (Liu *et al.*, 1990). Failures have also been reported with samples of natural waters. Thus, a mixture of marine hydrocarbon-degrading bacteria added to a seawater microcosm does not stimulate the decomposition of crude oil (Tagger *et al.*, 1983), inoculation of lake water samples with a benzoate-utilizing bacterium does not enhance the rate of disappearance of benzoate initially present at 34 or 350 ng/liter (Subba-Rao *et al.*, 1982), a 4-nitrophenol-utilizing *Pseudomonas* sp. does not mineralize the compound in lake water samples, and a pseudomonad able to grow in culture using 2,4-dichlorophenol as its C and energy source does not mineralize the same molecule in samples of lake water or sewage (Goldstein *et al.*, 1985).

EXPLANATIONS FOR FAILURES

Such negative results come as no surprise to ecologists or agricultural scientists. The possession of one agronomically or ecologically useful trait is not sufficient to guarantee success. Without question, an organism having a substrate uniquely available to it has a distinct advantage, yet that advantage may not be sufficient to compensate for many other traits that are also necessary for survival, no less multiplication, in a natural ecosystem. Possessing the requisite enzymes to metabolize a novel compound is a *necessary* attribute for the organism to carry out that transformation in a natural ecosystem, but it is not *sufficient* for the organism to succeed. Populations of microorganisms are subject to a variety of abiotic and biotic stresses, and these must be overcome for an introduced organism to be able to express its beneficial traits. A simple analogy is evident in agriculture: the development by plant breeders of a new crop variety that has desirable characteristics and shows vigor and outstanding growth potential in the greenhouse. The novice might think that the plant will be successful because it has these beneficial traits, but both the agricultural scientist and any good farmer realizes that there are a variety of soil fertility and soil structural problems, plant pathogens, weeds, and insects that must be controlled in order for the introduced species to be successful. In the same way, a microorganism that has beneficial traits for biodegradation must also be able to overcome the biotic and nonbiological stresses in the environment in which it is to be introduced.

The reasons for the frequent failures of inocula to function in nature are many, even for species selected because they can rapidly grow on and mineralize pollutants present at a site. These reasons for failure often reflect ecological constraints on the introduced organism. These constraints are of several types, and to be able to colonize a particular environment, the added species must be able to cope with each. Some are immediately obvious to the laboratory scientist, others are not.

a. Limiting nutrients. The added population has a nutrient uniquely available to it, namely, the organic compound whose destruction is sought, but it must also obtain N, P, O_2, other inorganic nutrients, and possibly growth factors from the environment into which it is introduced. The supply of these nutrients is frequently less than the demand, particularly where organic pollution is extensive, so that the members of the microbial community compete for the limiting inorganic nutrients. A species that grows slowly will not be as effective a competitor as its rapidly growing neighbors, and in instances where the chemical of interest is at a concentration below the K_s value, the added organism will probably grow slowly. Obviously, the addition of N, P, aeration to supply O_2, or other nutrient elements will reduce or eliminate the stress of competition for these requisites. Competition for inorganic nutrients may explain why an inoculum destroys synthetic chemicals in sterile but not in identical but nonsterile environmental samples, as has been observed in studies of *Streptomyces* introduced into soil to degrade metolachlor (Liu *et al.*, 1990), *Flavobacterium* sp. added to soil to destroy 4-(2,4-DB) (MacRae and Alexander, 1965), or *Pseudomonas* sp. added to mineralize 2,4-dichlorophenol (Goldstein *et al.*, 1985). These bacteria decomposed the chemicals in sterile soil.

Even apart from the competition for inorganic nutrients in insufficient supply, a problem may exist because of the low concentration of inorganic nutrients. Probably for every inorganic nutrient, one can describe the growth rate of a microorganism as a function of nutrient concentration, as previously described for Monod kinetics applied to the organic nutrient. At high levels of the inorganic nutrient, growth rate is independent of its concentration. At low levels, the growth rate is directly dependent on its concentration and would be slower at progressively lower concentration. A threshold probably also exists below which growth does not occur. Thus, a strain to be used for inoculation would function slowly if its K_s value for a particular nutrient is higher than the prevailing level in the environment of interest, and it would not function at all if the prevailing concentration is below its threshold. Clearly, the level of the nutrient would be reduced owing to competition with other species. Alternatively, the effect of low nutrient concentration may be compounded by grazing protozoa; that is, at concentrations of the nutrient near or below K_s, the inoculum strain would grow slowly, and the bacteria may not multiply fast enough to replace the cells consumed by predators. Low concentrations of P, N, and possibly other elements may in this way be the cause of failure of inoculation in natural ecosystems.

b. Suppression by predators and parasites.   Soils, natural waters, sewage, and sediments contain predators and parasites that may suppress not only the indigenous bacteria but also an added species. Especially prominent among the predators are the protozoa, which often are abundant in these environments. *Bdellovibrio,* bacteriophages, myxobacteria, cellular slime molds, and species producing lytic enzymes may also be numerous, although little evidence exists that they markedly suppress particular species or control the activities of bacterial communities in such environments. Protozoa feed on many bacterial genera, and although their grazing requires a bacterial cell density greater than about 10^6 per milliliter or gram (Alexander, 1981), such densities are common in nature. Should a bacterial species be multiplying rapidly, the cells lost to protozoan grazing, especially if grazing is not intense, may be replaced as new cells are formed. However, if a bacterium is multiplying slowly, as is likely the case for a species added to destroy a chemical at concentrations near or below the K_s value, the cells eliminated by protozoan grazing will not be replaced. Hence, that species is suppressed or eliminated even as the total bacterial community is maintained (Mallory *et al.,* 1983; Wiggins and Alexander, 1988).

Direct evidence that protozoa affect the ability of inocula to carry out biodegradation comes from studies of the biodegradation of 4-nitrophenol by nitrophenol-utilizing bacteria. A *Corynebacterium* with this capacity in culture mineralizes little of the compound present at low concentrations in lake water, and its population declines; however, the bacterium grows and the compound is mineralized in lake water amended with cycloheximide, an inhibitor of protozoa and other eucaryotes but not of the bacterium (Zaidi *et al.*, 1989). An investigation using low but environmentally realistic levels of inoculation by another 4-nitrophenol-degrading bacterium revealed that the organism fails to cause appreciable mineralization of the nitro compound unless the protozoa are suppressed by eucaryotic inhibitors (Fig. 14.2). Large inocula do effect mineralization, but population estimates show that some survivors remain after protozoan feeding on the cells in the large inoculum. These survivors then multiply and metabolize 4-nitrophenol when grazing pressure by the protozoa is reduced. Nevertheless, inhibition of protozoa also enhances activity by the large inoculum (Ramadan *et al.*, 1990). Hence, protozoa represent a major deterrent to successful remediation by introduced bacteria in natural environments containing active protozoan communities.

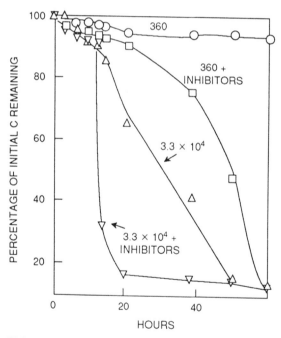

FIG. 14.2 Effect of cycloheximide and nystatin, two eucaryotic inhibitors, on the mineralization by *Pseudomonas cepacia* of 1.0 mg of 4-nitrophenol per liter of lake water. The inoculum was added to give 360 or 3.3 × 10⁴ cells per milliliter. (From Ramadan *et al.*, 1990. Reprinted with permission from the American Society for Microbiology.)

Protozoa are also active in suppressing the development of bacteria introduced into soil (Acea *et al.*, 1988). Therefore, they may have a similar role in soil as in lake water in determining the outcome of inoculation to effect biodegradation.

c. Inability of bacteria to move appreciably through soil.

The finding of both successes and failures as a result of inoculation of soil might seem to be anomalous. However, in studies in which the introduced bacteria or fungi brought about biodegradation, the test system usually contained only 1 to 250 g of soil in centrifuge tubes, test tubes, petri dishes, or occasionally pots, the soil was in depths of 10 cm, or the soil sample was mixed with the added bacteria. These procedures probably result in penetration or mixing of the bacteria with the body of soil to an extent that the organisms encounter much of their substrate. However, even when a bacterium able to grow on 2,4-dichlorophenol or another able to grow on 4-nitrophenol is added to the surface of sterile soil (in which there is no competition, predation, or parasitism), the bacteria mineralize little of either of the two phenols, although they are active if mixed with the sterile soil (Fig. 14.3). The latter findings probably result from the lack of appreciable movement of the microorganisms through the soil matrix to degrade the chemical at a distance from the point of inoculation. Presumably, only the chemical very near the sites of bacterial introduction is destroyed, as would occur in shallow soil depths or if bacteria are well mixed with the soil.

Such findings are not surprising in view of the many observations, in both the laboratory and field, that bacteria do not move appreciably through soil. It is not enough for individual cells to move through macropores, channels, or openings previously made by roots, root hairs, earthworms, or other soil animals because much of the chemical in soil is at a distance from these channels. The inoculated bacterium must be distributed to nearly all sites immediately adjacent to the chemical. The inability of bacteria, even in sterile soil with no competition or predation, to destroy a chemical only a short distance from the point of their addition presumably results from the lack of movement of the cells through the porous matrix (Goldstein *et al.*, 1985). If the soil is made into a slurry with water or if an economically feasible way exists to mix soil intimately with an inoculum in the field, the failures to achieve remediation because of poor movement of the inoculum may be overcome.

Limited movement through soil in the field or laboratory has been frequently demonstrated. Thus, it has been shown that few of the bacteria applied to the surface penetrate past the first 5 cm (Edmonds, 1976), and most fecal coliforms, even when applied in liquid derived from a sewage lagoon effluent, do not pass beyond the surface 8 cm (Bell and Bole,

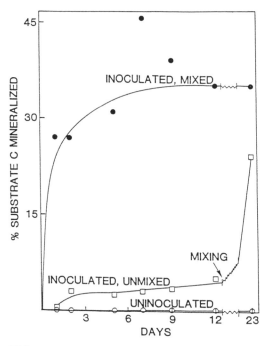

FIG. 14.3 Mineralization of 5.0 μg of 4-nitrophenol per gram in sterile soil that received no inoculum or that was inoculated at the surface with a 4-nitrophenol-utilizing *Pseudomonas* and either mixed or left unmixed. The arrow indicates when the unmixed soil was shaken vigorously. (From Goldstein *et al.*, 1985. Reprinted with permission from the American Society for Microbiology.)

1978). Although the extent of movement is probably greater in sandy soils because of their larger pores, bacterial transport with water is probably very limited in nonsandy soils. For example, the movement of *Pseudomonas putida* and *Bradyrhizobium japonicum* through a soil does not exceed 3 cm; although earthworm activity or the presence of growing roots increases vertical transport, the bacteria still did not move far (Madsen and Alexander, 1982). With enormous amounts of water and considerable time, good movement does occur through sand (Robeck *et al.*, 1962), but the amount of time required and the fact that soils do not have as large pores as sand make extensive movement highly unlikely. On the other hand, because the degree of movement does vary among bacteria (Wong and Griffin, 1976; Gannon *et al.*, 1991) and among the spores of certain fungi (Hepple, 1960), it may be possible to find species that are susceptible to transport with water and hence that are more suitable for soil inoculation. Nevertheless, one must consider that dispersal of cells through soil to destroy chemicals more than a few centimeters from the point of inoculation is a major obstacle to success.

It is not clear whether bacteria that are deliberately injected into groundwater will be dispersed sufficiently or will move with a contaminant plume to an extent that they will destroy contaminants in aquifers. Most available evidence, largely obtained with fecal coliforms but sometimes with other bacteria, indicates that bacteria do not move far through septic fields, but successful transport for somewhat more than 10 m is possible (Reneau and Peittry, 1975; Viraraghavan, 1978). Conversely, distant dissemination of bacteria does occur through fractured bedrock, at least for 30 m (Morrison and Allen, 1972), and distant transport also probably occurs through a variety of underground cracks and fissures. Nevertheless, because appreciable biodegradation in groundwaters requires extensive dispersal of the organisms added at discrete points of belowground injection, it is not yet clear whether such approaches are feasible.

d. Use of other C sources. An organism isolated because of its capacity to metabolize an unwanted chemical is able to grow on a number of other organic substrates, some of which may be present in the environment of interest. Following its addition to that environment, the organism may multiply by preferentially using one or more of these other C sources, leaving unmetabolized the unwanted compound (Goldstein et al., 1985).

e. Concentration of organic substrate too low to support multiplication. For many practical problems, the initial population density of the organisms added for the purposes of biodegradation will be small because the volume of material to be freed of the compound will be large. Hence, to effect a significant loss of the compound, the organism will have to multiply. However, because microorganisms will not grow on organic molecules below a threshold concentration and the selective advantage of an introduced species is its ability to grow on the unwanted molecule, it cannot degrade that substrate to a significant degree if its selective advantage is lost. Evidence for such a problem is found in the reports that a benzoate-utilizing bacterium fails to mineralize 34 and 350 ng of benzoate per liter although it destroys benzoate at higher levels following inoculation into lake water (Subba-Rao et al., 1982). It is possible that the inability to function at low substrate concentrations may sometimes not be a consequence of concentrations below the threshold for growth but rather reflect the inability of the inoculated organism to grow sufficiently rapidly at the low substrate concentration to replace cells consumed by protozoan grazing.

f. Need for C source to support growth. A species that acts on an organic substrate by cometabolism must have a C source for growth. It is unlikely that the site of interest will have a sufficient supply of such a C source to support the introduced cells because, even if present, the indigenous species would likely use it more readily than an introduced organism. Hence, without additions of such an organic nutrient, the inoculum will not carry out the desired transformation. This possible cause of failure is illustrated by studies of a strain of *Acinetobacter* that cometabolizes a number of PCBs. Inoculation of soil with this bacterium does not enhance PCB mineralization unless the soil is also amended with biphenyl, a nonchlorinated analogue of PCBs that supports proliferation of the bacterium. In this instance, *Acinetobacter* only carries out the first step in the mineralization, and indigenous populations convert the products of PCB cometabolism to CO_2 (Brunner *et al.*, 1985). However, some bioremediation strategies are based on additions of growth substrates to support the cometabolizing species.

g. Temperature. It is common to isolate bacteria using enrichment cultures incubated at 25 to 37°C. Such temperatures are characteristic of some natural and man-made environments, but the temperatures of most are significantly lower. Many species that multiply well at these common laboratory temperatures do not multiply at lower temperatures. On the other hand, enrichments could just as well be established and organisms isolated at temperatures similar to those that prevail at the sites of interest.

h. pH. It is also common to maintain enrichment cultures near neutrality and thus to isolate microorganisms that grow well or that only multiply at pH values close to 7.0. Nevertheless, the oceans and many inland waters are at higher pH values, and many soils have lower values. Thus, an organism may fail to destroy an organic compound in nature simply because its pH range for growth does not include the pH at the site of interest. Overcoming or preventing such failures is simple: the enrichment should be maintained at pH values similar to those at the site of concern, and the isolate thus will be capable of development at those values (Zaidi *et al.*, 1988b).

i. Salinity. Some waters and soils have modest to occasionally high levels of salts. The oceans and estuaries are more saline than nearly all inland waters, and some soils are also reasonably rich in salts. As with temperature and pH, the inoculum strain must be able to multiply at the prevailing levels of salts, and its inability to do so will result in the failure of the organism to function. Conversely, an organism isolated from a salt-rich habitat may not degrade a chemical in fresh water; for example, Atlas and Busdosh (1976) found that a hydrocarbon-utilizing bacterium isolated from an estuary enhanced degradation of oil spilled in a saline but not in a freshwater pond.

j. Toxins. Natural inhibitors affecting bacteria are present in some unpolluted waters and soils. Although their identities are largely unknown, these toxins prevent the growth and may affect the survival of a species introduced into an environment in which it is not native. Polluted soils and surface and ground waters usually contain many inhibitors harmful to microorganisms. To function at a site in which natural or synthetic inhibitors are present at injurious levels, the organism to be used must be resistant to the toxins.

Overcoming some of these constraints is easy. Overcoming others will be difficult or, in some instances, impossible. Bacteria or fungi able to cope with many of these stresses can be isolated, or the site containing the unwanted chemical could be modified by aeration, mixing, nutrient supplementation, etc. Frequent success will then be attained when the identities of the constraints are defined and means are devised to overcome or minimize their importance.

GENETICALLY ENGINEERED MICROORGANISMS

Molecular biology has provided highly useful techniques to modify the genetic composition of microorganisms and thus to allow for the potential construction of new organisms having the capacity to carry out catabolic sequences not possible in existing organisms or under conditions not suitable for existing organisms. These genetic modifications thus provide a highly important and potentially very useful approach to effect bioremediation of compounds not otherwise destroyed rapidly by microbiological means or under conditions in which microbial transformations would otherwise be too slow to be practical. The constraints affecting introduced organisms, to be sure, are the same as those that apply to existing organisms, and an inoculated genetically engineered microorganism must be able to cope with the ecological and environmental stresses that apply to any nonindigenous species. However, the constructed microorganism still is of special significance because of its new characteristics.

A variety of problems in the future may be solved by use of genetically engineered microorganisms. (a) Constructing microorganisms able to grow on and mineralize pollutants that presently are only cometabolized. Substrates that are transformed solely by cometabolism are biodegraded slowly and thus persist, and they yield products that are often toxic, long-lived, or both. However, by combining in one organism the genes encoding for the enzymes that cometabolize the compound of concern with the genes encoding for enzymes that allow an organism to grow on and mineralize what otherwise would be the end-product of cometabolism, the engineered organism would use the parent molecule as C source and bring about its mineralization. The new organism would thus have two catabolic sequences that complement one another, and such complementary catabolic sequences would then effect a conversion not otherwise possible. (b) Creation of new catabolic pathways to effect transformations not presently carried out efficiently or rapidly, such as by altering the range of substrates used by a particular microorganism. (c) Increasing the amount or activity of specific enzymes in a microorganism. This increase might be useful in enhancing the rate of degradation brought about by a microorganism deliberately added to a polluted site or in providing highly active bacteria for use as immobilized cells or for the preparation of immobilized enzymes. (d) Construction of microorganisms that not only can destroy target pollutants but also are resistant to inhibitors at the site that prevent degradation by indigenous microorganisms. Many industrial sites, possibly most, that have high levels of synthetic organic compounds also contain high concentrations of heavy metals or other substances that suppress microbial development.

Most of the genetic determinants of bacteria are on a single, circular chromosome. In addition, many bacteria have far smaller genetic elements known as plasmids that bear some of the genetic determinants of the organisms; the plasmids are considered to be not crucial for the survival of the cell, but they do have special significance in many catabolic sequences. The enzymes catalyzing the degradation of a particular compound may be encoded by chromosomal genes, by genes on a plasmid, or partly by chromosomal and partly by plasmid genes.

Bacteria may exchange genetic material in three ways: transduction, transformation, or conjugation. Gene exchange in transduction is mediated by a bacteriophage, whereas transformation entails the release of DNA by lysis of one bacterium and its uptake by a second. In conjugation, the DNA is transferred from one cell to another through a conjugal tube joining the two cells. Construction of novel bacterial strains by transformation, transduction, or conjugation is termed *in vivo* genetic engineering, that is, the genetic rearrangement occurs in living organisms. *In vitro* genetic engineering, in contrast, may involve separation of DNA from the cell, its treatment with a specific restriction endonuclease to cleave the DNA molecule, the rejoining of DNA fragments with DNA ligase to give a new sequence of nucleotide bases, and the reintroduction of this hybrid molecule into a suitable bacterial cell in which it will replicate and be expressed. Protoplast fusion and transposon-mediated gene manipulation may also be used to construct organisms with new characteristics. Bacterial protoplasts are cells with the rigid, outer peptidoglycan layer removed enzymatically, and fusion of such protoplasts may lead to genetic rearrangements. Transposons are short sequences of DNA bases that can be inserted *in vivo* into many sites in replicating DNA molecules.

Plasmids have been of great attraction as a means to construct novel bacteria. This attraction results from the many plasmid-borne genes that encode enzymes important in biodegradation. These are known as catabolic plasmids, and they give the bacterium containing them the ability to degrade certain compounds. The type of plasmid that is of particular interest is the one that can be transferred from one organism to another; some of these can only be transferred between closely related strains (narrow host-range plasmids), but others are transferred freely between different species and genera. The latter, the broad host-range plasmids, replicate in the cells in which they have been introduced, and genetic information thereby may be transmitted to quite dissimilar species. Catabolic plasmids have been discovered that encode enzymes catalyzing the degradation of ABSs, benzoate and chlorobenzoates, biphenyl and 4-chlorobiphenyl, chloroacetate, *p*-cresol, 2,4-D, naphthalene, octane, parathion, phenanthrene, styrene, toluene, and other compounds, and they have been found in species of *Pseudomonas, Alcaligenes, Acinetobacter, Flavobacterium, Beijerinckia, Klebsiella, Moraxella,* and *Arthrobacter* (Sayler *et al.,* 1990). Some of the plasmid-encoded degradative activities that have been transferred from one bacterial species to a second are listed in Table 14.1. The recipient of the plasmid in some instances is a different species in the same genus as the source of the plasmid, but sometimes the transfer involves different genera. In several instances, the plasmid confers on the new host the capacity to metabolize the compound but not use it as a C source for growth. In other instances, however, the recipient acquires the ability to grow on the molecule it previously was unable to metabolize.

By means of transmissible plasmids or by the use of other genetic techniques, bacteria have been constructed that have activities different from those of the original organisms. One of the first of such new organisms was a bacterium that had acquired the ability to destroy the herbicide 2,4,5-T (Kellogg *et al.,* 1981). The fusion of protoplasts prepared from two species of *Streptomyces* serves as a means of obtaining an actinomycete that more readily converts lignocellulose to a water-soluble polymer than the original actinomycetes (Pettey and Crawford, 1984). The issue

TABLE 14.1
Plasmid-Borne Genes That Have Been Transferred from One Bacterium to Another

Activity encoded by plasmid that is transferred	Activity of bacterial recipient of plasmid	Reference
2,4-D Degradation	Metabolism but not growth on 2,4-D	Friedrich et al. (1983)
Benzene metabolism	Growth on benzene	Irie et al. (1987)
Haloacetate dehalogenase	Metabolism but not growth on chloroacetate	Kawasuki et al. (1981)
Naphthalene metabolism	Metabolism but not growth on naphthalene	Oh et al. (1985)
3-Chlorobenzoate metabolism	Growth on chlorophenols	Reineke et al. (1982)
Five catabolic pathways	Growth on chlorobenzoates	Rojo et al. (1987)
TCE cometabolism	Degradation of TCE	Winter et al. (1989)

of obtaining bacteria active in biodegradation and also tolerant to the heavy metals found at many hazardous waste sites was addressed in a study in which a strain of *Alcaligenes eutrophus* was constructed that degraded 2,4-D and di- and trichlorobiphenyl isomers and also tolerated high concentrations of Ni and Zn (Springael *et al.*, 1993). Of special

TABLE 14.2
Degradative Activity of Genetically Engineered Bacteria

Activity of parent cultures	Compounds metabolized by constructed bacteria	Reference
Biphenyl-grown *Acinetobacter*, 3-chlorobenzoate-grown *Pseudomonas*	3-Chlorobiphenyl	Adams et al. (1992)
Pseudomonas using 4-chloro-2-nitrophenol for N, *Alcaligenes* using haloaromatics	4-Chloro-2-nitrophenol used as C source	Bruhn et al. (1988)
Toluene-grown *Pseudomonas putida*, benzoate-grown *P. alcaligenes*	1,4-Dichlorobenzene	Kröckel and Focht (1987)
Aniline-degrading *Pseudomonas*, chlorocatechol-degrading *Pseudomonas*	Chloroanilines	Latorre et al. (1984)
4-Chlorophenol-utilizing *Pseudomonas*, phenol-utilizing *Alcaligenes*	2- and 3-Chlorophenols	Schwien and Schmidt (1982)
Biphenyl-grown *Pseudomonas putida*, 4-chlorobenzoate-grown *P. cepacia*	Dichlorobiphenyls	Havel and Reineke (1991)

significance is the finding that genes encoding for a metabolic activity (the conversion of halogen-containing aromatic compounds to halogenated catechols) can be cloned in a bacterium that is able to grow on the product of cometabolism by the first species, the result being a new organism that is able to mineralize the original halogenated aromatic compound (Reineke, 1986). Other constructed bacteria are listed in Table 14.2.

Thus, genetic engineering does indeed promise to provide organisms with novel and presently nonexisting biodegradative activities. However, the possession of these new catabolic traits will only lead to successful bioremediations if these unique organisms can cope with the stresses in the environments in which they will be introduced. Those that are successful ecologically and catabolically should greatly enhance society's ability to destroy pollutants in natural environments. In bioreactors, in which such stresses are less important, on the other hand, the availability of constructed organisms tailor-made for the compounds of concern should have a more immediate impact.

REFERENCES

Acea, M., Moore, C. R., and Alexander, M., *Soil Biol. Biochem.* **20**, 509–515 (1988).

Adams, R. H., Huang, C.-M., Higson, F. K., Brenner, V., and Focht, D. D., *Appl. Environ. Microbiol.* **58**, 647–654 (1992).

Ahlfeld, T. E., and LaRock, P. A., in "The Microbial Degradation of Oil Pollutants" (D. G. Ahearn and S. P. Meyers, eds.), pp. 199–203. Louisiana State University, Center for Wetland Studies, Baton Rouge, 1973.

Alexander, M., *Annu. Rev. Microbiol.* **35**, 113–133 (1981).

Atlas, R. M., *CRC Crit. Rev. Microbiol.* **5**, 371–386 (1977).

Atlas, R. M., and Busdosh, M., in "Proceedings of the Third International Biodegradation Symposium" (J. M. Sharpley and A. M. Kaplan, eds.), pp. 79–85. Applied Science Publishers, London, 1976.

Barles, R. W., Daughton, C. G., and Hsieh, D. P. H., *Arch. Environ. Contam. Toxicol.* **8**, 647–660 (1979).

Bell, R. G., and Bole, J. B., *J. Environ. Qual.* **7**, 193–196 (1978).

Bruhn, C., Bayly, R. C., and Knackmuss, H.-J., *Arch. Microbiol.* **150**, 171–177 (1988).

Brunner, W., Sutherland, F. H., and Focht, D. D., *J. Environ. Qual.* **14**, 324–328 (1985).

Clark, C. G., and Wright, S. J. L., *Soil Biol. Biochem.* **2**, 19–26 (1970).

Compeau, G. C., Mahaffey, W. D., and Patras, L., in "Environmental Biotechnology for Waste Treatment" (G. S. Sayler, R. Fox, and J. W. Blackburn, eds.), pp. 91–109. Plenum, New York, 1991.

Crawford, R. L., and Mohn, W. W., *Enzyme Microb. Technol.* **7**, 617–620 (1985).

Daughton, C. G., and Hsieh, D. P. H., *Bull. Environ. Contam. Toxicol.* **18**, 48–56 (1977).

DeBorger, R., Vanloocke, R., Verlinde, A., and Verstraete, W., *Rev. Ecol. Biol. Sol* **15**, 445–452 (1978).

Edgehill, R. U., and Finn, R. K., *Appl. Environ. Microbiol.* **45**, 1122–1125 (1983).

Edmonds, R. L., *Appl. Environ. Microbiol.* **32**, 537–546 (1976).

Engvild, K. C., and Jensen, H. L., *Soil Biol. Biochem.* **1**, 295–300 (1969).
Focht, D. D., and Brunner, W., *Appl. Environ. Microbiol.* **50**, 1058–1063 (1985).
Friedrich, B., Meyer, M., and Schlegel, H. G., *Arch. Microbiol.* **134**, 92–97 (1983).
Gannon, J. T., Mingelgrin, U., Alexander, M., and Wagenet, R. J., *Soil Biol. Biochem.* **23**, 1155–1160 (1991).
Goldstein, R. M., Mallory, L. M., and Alexander, M., *Appl. Environ. Microbiol.* **50**, 977–983 (1985).
Havel, J., and Reineke, W., *FEMS Microbiol. Lett.* **78**, 163–169 (1991).
Heitkamp, M. A., and Cerniglia, C. E., *Appl. Environ. Microbiol.* **55**, 1968–1973 (1989).
Hepple, S., *Trans. Br. Mycol. Soc.* **43**, 73–79 (1960).
Irie, S., Shirai, K., Doi, S., and Yorifuji, T., *Agric. Biol. Chem.* **51**, 1489–1493 (1987).
Ismailov, N. M., *Mikrobiologiya* **54**, 835–841 (1985).
Jobson, A., McLaughlin, M., Cook, F. D., and Westlake, D. W. S., *Appl. Microbiol.* **27**, 166–171 (1974).
Karns, J. S., Kilbane, J. J., Chatterjee, D. K., and Chakrabarty, A. M., *in* "Genetic Control of Environmental Pollutants" (G. S. Omenn and A. Hollaender, eds.), pp. 3–21. Plenum, New York, 1984.
Kawasuki, H., Tone, N., and Tonomura, K., *Agric. Biol. Chem.* **45**, 29–34 (1981).
Kellogg, S. T., Chatterjee, D. K., and Chakrabarty, A. M., *Science* **214**, 1133 (1981).
Kennedy, D. W., Aust, S. D., and Bumpus, J. A., *Appl. Environ. Microbiol.* **56**, 2347–2353 (1990).
Kröckel, L., and Focht, D. D., *Appl. Environ. Microbiol.* **53**, 2470–2475 (1987).
Krueger, J. P., Butz, R. G., and Cork, D. J., *J. Agric. Food Chem.* **39**, 1000–1003 (1991).
Kunc, F., and Rybarova, J., *Soil Biol. Biochem.* **15**, 141–144 (1983).
Lamar, R. T., and Dietrich, D. M., *Appl. Environ. Microbiol.* **56**, 3093–3100 (1990).
Latorre, J., Reineke, W., and Knackmuss, H.-J., *Arch. Microbiol.* **140**, 159–165 (1984).
Lehtomäki, M., and Niemelä, S., *Ambio* **4**, 126–129 (1975).
Liu, S.-Y., Lu, M.-H., and Bollag, J.-M., *Biodegradation* **1**, 9–17 (1990).
MacRae, I. C., and Alexander, M., *J. Agric. Food Chem.* **13**, 72–76 (1965).
Madsen, E. L., and Alexander, M., *Soil Sci. Soc. Am. J.* **46**, 557–560 (1982).
Mallory, L. M., Yuk, C.-S., Liang, L.-N., and Alexander, M., *Appl. Environ. Microbiol.* **46**, 1073–1079 (1983).
McClure, G. W., *J. Environ. Qual.* **1**, 177–180 (1972).
Middledorp, P. J. M., Briglia, M., and Salkinoja-Salonen, M. S., *Microb. Ecol.* **20**, 123–139 (1990).
Miget, R., Oppenheimer, C. H., Kator, H. I., and LaRock, P. A., "Proceedings of the Joint Conference on Prevention and Control of Oil Spills" pp. 327–331. American Petroleum Institute, New York, 1969.
Milhomme, H., Vega, D., Marty, J.-L., and Bastide, J., *Soil Biol. Biochem.* **21**, 307–311 (1989).
Morrison, S. M., and Allen, M. J., "Bacterial Movement through Fractured Rock." Environmental Resources Center, Colorado State University, Fort Collins, 1972.
Oh, S., Quensen, J., Matsumura, F., and Momose, H., *Environ. Toxicol. Chem.* **4**, 21–27 (1985).
Pettey, T. M., and Crawford, D. L., *Appl. Environ. Microbiol.* **47**, 439–440 (1984).
Ramadan, M. A., El-Tayeb, O. M., and Alexander, M., *Appl. Environ. Microbiol.* **56**, 1392–1396 (1990).
Reineke, W., *J. Basic Microbiol.* **26**, 551–567 (1986).
Reineke, W., Wessels, S. W., Rubio, M. A., Latorre, J., Schwien, U., Schmidt, E., Schlömann, M., and Knackmuss, H.-J., *FEMS Microbiol. Lett.* **14**, 291–294 (1982).

Reneau, R. B., Jr., and Peittry, D. E., *J. Environ. Qual.* **4,** 41–44 (1975).

Robeck, G. G., Bryant, A. R., and Woodward, R. L., *J. Am. Water Works Assoc.* **54,** 75–82 (1962).

Rojo, F., Pieper, D. H., Engesser, K.-H., Knackmuss, H.-J., and Timmis, K. N., *Science* **238,** 1395–1398 (1987).

Ryan, T. P., and Bumpus, J. A., *Appl. Microbiol. Biotechnol.* **31,** 302–307 (1989).

Sayler, G. S., Hooper, S. W., Layton, A. C., and King, J. M. H., *Microb. Ecol.* **19,** 1–20 (1990).

Schwien, U., and Schmidt, E., *Appl. Environ. Microbiol.* **44,** 33–39 (1982).

Springael, D., Diels, L., Hooyberghs, L., Kreps, S., and Mergeay, M., *Appl. Environ. Microbiol.* **59,** 334–339 (1993).

Subba-Rao, R. V., Rubin, H. E., and Alexander, M., *Appl. Environ. Microbiol.* **43,** 1139–1150 (1982).

Sudhakar-Barik and Sethunathan, N., *J. Environ. Qual.* **7,** 349–352 (1978).

Tagger, S., Bianchi, A., Julliard, M., LePetit, J., and Roux, B., *Mar. Biol. (Berlin)* **78,** 13–20 (1983).

Viraraghavan, T., *Water, Air, Soil Pollut.* **9,** 355–362 (1978).

Wiggins, B. A., and Alexander, M., *Can. J. Microbiol.* **34,** 661–666 (1988).

Winter, R. B., Yen, K.-M., and Ensley, B. D., *Bio/Technology* **7,** 282–285 (1989).

Wong, P. T. W., and Griffin, D. M., *Soil Biol. Biochem.* **8,** 215–218 (1976).

Zaidi, B. R., Murakami, Y., and Alexander, M., *Environ. Sci. Technol.* **22,** 1419–1425 (1988a).

Zaidi, B. R., Stucki, G., and Alexander, M., *Environ. Toxicol. Chem.* **7,** 143–151 (1988b).

Zaidi, B. R., Murakami, Y., and Alexander, M., *Environ. Sci. Technol.* **23,** 859–863 (1989).

15 BIOREMEDIATION TECHNOLOGIES

Recent years have witnessed an enormous growth in the controlled, practical use of microorganisms for the destruction of chemical pollutants. These various technologies rely on the biodegradative activities of microorganisms, and they focus on enhancing existent but slow biodegradation processes in nature or technologies that bring chemicals into contact with microorganisms in some type of reactor that allows for rapid transformation. In many instances, the focus of attention is on existing sites of pollution, and such technologies are encompassed by the term *bioremediation*. The term is an apt one because a remedy is being applied to a problem. Approaches that deal with waste streams from industry are often considered under the purview of bioremediation, although they might be more appropriately termed bioprophylaxis.

Bioremediation of contaminated sites is a new field of endeavor, and many new or altered technologies are appearing; nevertheless, the utilization of microbial processes to destroy chemicals is neither a novel idea nor a new technology. Such processes have been used for decades for the elimination of chemicals from waste streams of industries that had biological treatment systems for their effluents and, knowingly or unknowingly, for the breakdown of chemicals from households or industries serviced by municipal waste-treatment systems. The fact that many compounds were not so destroyed was not necessarily the result of the absence of a biodegradative microflora but rather that the systems were optimized for different purposes.

The goal of bioremediation is to degrade organic pollutants to concentrations that are either undetectable or, if detectable, to concentrations below the limits established as safe or acceptable by regulatory agencies. Bioremediation is being used for the destruction of chemicals in soils, groundwater, wastewater, sludges, industrial-waste systems, and gases. The list of compounds that may be subject to biological destruction by one or another bioremediation system is long. However, because they are widespread, represent health or ecological hazards, and are susceptible to microbial detoxication, most interest has been directed to oil and oil products, gasoline and its constituents, polycyclic aromatic hydrocarbons, chlorinated aliphatics such as TCE and tetrachloroethylene (also called perchloroethylene or PCE), and chlorinated aromatic hydrocarbons. Although they are not biodegraded, metals are of interest in bioremediation because they can be altered and rendered less harmful by microorganisms.

Certain criteria must be met for bioremediation to be seriously considered as a practical means for treatment. (a) Microorganisms must exist that have the needed catabolic activity. (b) Those organisms must have the capacity to transform the compound at reasonable rates and bring the concentration to levels that meet regulatory standards. (c) They must not generate products that are toxic at the concentrations likely to be achieved during the remediation. (d) The site must not contain concentrations or combinations of chemicals that are markedly inhibitory to the biodegrading species, or means must exist to dilute or otherwise render innocuous the inhibitors. (e) The target compound(s) must be available to the microorganisms. (f) Conditions at the site or in a bioreactor must be made conducive to microbial growth or activity, for example, an adequate supply of inorganic nutrients, sufficient O_2 or some other electron acceptor, favorable moisture content, suitable temperature, and a source of C and energy for growth if the pollutant is to be cometabolized. (g) The cost of the technology must be less or, at worst, no more expensive than other technologies that can also destroy the chemical. None of these criteria is trivial, and none is platitudinous. The failure to meet any one has resulted in a rejection of a biodegradative approach or the inability to meet the cleanup goals that were established.

An appreciation of current activities, in the United States at least, can be obtained from the results of a survey conducted by the U.S. Environmental Protection Agency (Devine, 1992). That agency received information on 240 cases of biomediation. Undoubtedly, many more activities existed, but the information was not transmitted, in part because there was no obligation to do so, in part probably because of proprietary concerns. Of the 240 cases reported, many were not included in the survey, often because they contained too little information. Nevertheless, 132 cases had sufficient information to be included in the report of the survey. Of the 132, 75 dealt with petroleum and related materials, 13 with wood preservatives, 10 with solvents as contaminants, 7 with agricultural chemicals (presumably pesticides), 5 with tars, and 4 with munitions. Most of the bioremediations dealt with soil or groundwater as media, and a few were cleanups of sludge. Such data indicate the types of field activities and the most common environments being treated by biological means.

A variety of different technologies and procedures are currently being used, and a number of new and promising approaches have been suggested or have reached advanced stages of development. Some of these technologies are *in situ* treatments, in which soil is not removed from the field or groundwater is not pumped for aboveground treatment. *In situ* biomediation has the advantage of relatively low cost but the disadvantage of being less subject to rigorous control. Other bioremediation technologies require removal of the contaminated material in some manner from its original location. Such removals increase the costs modestly or appreciably, but the processes are more subject to control.

Microorganisms in soil have a broad array of catabolic activities, and a simple way of destroying pollutants is to add the compounds or materials containing them to the soil and rely on the indigenous microflora. This procedure, often called *land farming* or land treatment, has been frequently used by the oil industry to destroy oily wastes, and it is a procedure that has been utilized for many years. It is also employed if oily or hydrocarbon-rich materials are spilled on soil. The considerable amounts of C added in these wastes have the potential to support a large biomass, but the soil has too little N and P—and possibly other inorganic nutrients—to support such large biomasses, so N and P are added to the soil, often in the form of commercial fertilizers. Furthermore, the O_2 demand of the microflora increases with the added organic C, and the rate of diffusion of O_2 from the overlying air into the soil is too slow to sustain the aerobic bacteria that are chiefly, or solely, responsible. The need for supplemental O_2 is satisfied by mixing the soil in some way, sometimes by simple plowing, sometimes by more thorough mixing. Another common limiting factor for rapid microbial transformation is moisture because surface soil often dries out, so arrangements are made to provide water to maintain optimum moisture levels for aerobic organisms. The effectiveness of this type of remediation in a soil containing diesel oil is shown in Fig. 15.1. Remediation by these means is limited to times of year when the soil temperature is in a range that permits reasonably rapid microbial growth and activity; little or no biodegradation occurs during the cold parts of the year in the temperate zone. Sometimes, microorganisms are added to the soil by some form of inoculation, but such inoculations usually, or possibly always, are ineffective—or at least convincing data do not exist to show any benefit.

The efficacy of land treatment for spills of oil and oil products has been confirmed in carefully controlled experiments in the laboratory and in the field. Thus, the hydrocarbons in gasoline, jet fuel, and heating oil were found to be extensively degraded in soils in the laboratory that were treated with fertilizer, lime, and simulated tilling (Song *et al.*, 1990), and crude oil, crankcase oil, jet fuel, heating oil, and diesel oil disappeared

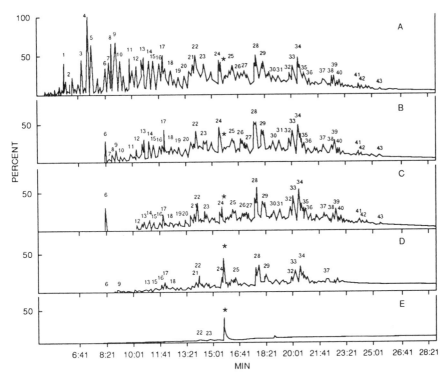

FIG. 15.1 Gas chromatograms of the aromatic fraction of diesel oil residue in contaminated soil. Zero time (A); 2 weeks, without (B) and with (C) bioremediation; 12 weeks, without (D) and with (E) bioremediation. The numbers over the peaks designate different PAHs, and the asterisk is a compound added as a standard for analytical purposes. (From Wang, et al., 1990. Reprinted with permission from the American Chemical Society.)

at an enhanced rate in field plots receiving fertilizer, lime, and simulated tilling as compared to soil not receiving these treatments (Raymond *et al.*, 1976; Wang and Bartha, 1990). Such experiments are worth citing because they contain untreated controls and are thus scientifically more convincing. Controls are not included in actual remediations because the issue is not to convince scientists of the response to the treatments but rather to remove the unwanted materials. However, the efficacy if not the scientific rigor is evident in analyses conducted in a field inundated with approximately 1.9 million liters of kerosene. After the initial emergency cleanup operation, the oil content of the soil was 0.87% in the top 30 cm and about 0.7% at 30–45 cm. However, in soil that received 200 kg N and 20 kg P as well as lime, the oil content had declined to <0.1% in the upper 30 cm, although only to 0.3% at the lower depth (Dibble and Bartha, 1979). On the other hand, some hydrocarbon-rich materials are not readily destroyed by land farming, for example, sediment contaminated with PCP, creosote (Mueller *et al.*, 1991a), and bunker C oil (Song *et al.*, 1990).

In a similar technology, more engineering controls are included. The additions include systems to provide irrigation water and nutrients, a liner at the bottom of the soil, and a means to collect leachate. This is termed a *prepared bed reactor* (Fig. 15.2). Either clay or a synthetic material acts as the liner. These reactors are used at many Superfund sites in which bioremediation is being used, and often the contaminants are polycyclic aromatic hydrocarbons, BTEX (benzene, toluene, ethylbenzene, and xylene), or both (Ryan *et al.*, 1991). The liner and a system to collect leachate are included because of concern that conventional land treatment may result in contamination of the underlying groundwater with the parent compounds or products of microbial transformation that are carried downward with percolating water. The level of sophistication, and consequently the cost, will vary enormously. In some instances, perforated pipes are placed above the liner to collect the leachate, and sand is placed on the liner and over the pipes to improve drainage and hence the ultimate collection of the leachate. The leachate is removed for subsequent treatment, which may be in an adjacent bioreactor. Water and nutrients may be dispensed through an overhead spray-irrigation system, and the entire operation may be enclosed in a plastic greenhouse if volatile hazardous products may be emitted.

This is essentially the method that was used for the treatment of 115,000 m³ of soil contaminated with bunker C fuel oil (Compeau *et al.*, 1991) and

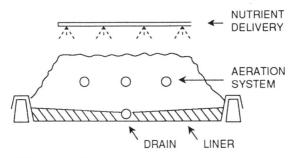

FIG. 15.2 Diagram of prepared bed reactor for treatment of excavated contaminated soil. [Reprinted with permission from Fogel, S., Findlay, M., and Moore, A. *In* "Petroleum Contaminated Soils" (E. J. Calabrese and P. T. Kostecki, eds.), Vol. 2. Copyright CRC Press, Inc., Boca Raton, Florida.]

23,000 m³ of soil contaminated with gasoline and fuel oil (Block *et al.,* 1990). The treatments were fertilizer, lime, irrigation, and mixing the soil to provide O_2 to the aerobic populations able to degrade the unwanted material. Sometimes the soil is placed in piles that extend laterally for some distance (Hildebrandt and Wilson, 1991).

Land treatment is also a means to dispose of contaminated water. This is illustrated in a system in which contaminated water from a facility originally designed for treating wood with creosote was introduced into a soil to destroy PAHs. The soil was placed on a polyethylene liner, and the treatment unit was provided with a means to collect leachate. In the first year of operation in the field, 60% of the extractable hydrocarbons, more than 95% of the 2- and 3-ring PAHs, and more than 70% of the 4- and 5-ring PAHs disappeared, this destruction occurring chiefly in the first 90 days when the temperatures were still warm (Lynch and Genes, 1989).

A high percentage of the bioremediations that have been completed recently are land treatments by one of these means (Devine, 1992). These successes supplement the several decades of use of conventional land treatment for the destruction of oily and petroleum wastes.

The addition of nutrients to soil to enhance biodegradation, which is sometimes called *biostimulation,* is not always beneficial. On occasion, in laboratory tests, the addition of N inhibits the mineralization of aromatic and aliphatic hydrocarbons (Manilal and Alexander, 1991; Morgan and Watkinson, 1990). The frequency and the explanations for the reduced rate of mineralization are unknown. However, it is possible that more substrate-C is incorporated into biomass in the presence of higher levels of N and hence less CO_2 is produced; if this explanation is correct, the rate of loss of the substrate may not have diminished but rather the flow of C from substrate may be changed to yield more cells and less CO_2.

Recent studies suggest that certain surfactants, specifically some non-ionic alcohol ethoxylates, at low concentration also stimulate biodegradation of hydrocarbons sorbed in soil. This promoting effect occurs even though little of the compounds are desorbed from the soil by the surfactants (Aronstein *et al.,* 1991).

Another solid-phase treatment involves the same approach to enhancing microbial activity but relies on a different way of providing O_2. Additional air is provided by vacuum extraction of soil above the water table (the *vadose* zone or the unsaturated soil layer), thereby supplying the terminal electron acceptor needed by the aerobic bacteria. This process, which is designed for hydrocarbon-contaminated sites, is termed *bioventing* or simply *venting* (Hinchee *et al.,* 1991).

SOLID-PHASE TREATMENT: COMPOSTING

In composting as a treatment procedure, the polluted material is mixed together in a pile with a solid organic substance that is itself reasonably readily degraded, such as fresh straw, wood chips, wood bark, or straw that had been used for livestock bedding. The pile is often supplemented with N, P, and possibly other inorganic nutrients. The material is placed in a simple heap, it is formed in long rows known as windrows, or it is introduced into a large vessel equipped with some means of aeration. Moisture must be maintained, and aeration is provided either by mechanical mixing or by some aeration device. A contained vessel is desirable when the compost contains hazardous chemicals. Heat released during microbial growth on the solid organic material is not adequately dissipated, and hence the temperature rises. The higher temperatures (50–60°C) are often more favorable to biodegradation than the lower temperatures that are maintained in some composts.

Composting has been used as a means of treating soil contaminated with chlorophenols. The various chlorophenols present in the composted material, which is placed in windrows in the field, decline markedly during the summer months, when the temperature in the compost is high, but the conversion is slow during the cold part of the year (Valo and Salkinoja-Salonen, 1986). A field-scale demonstration has also shown that the concentrations of TNT, RDX, and HMX in contaminated sediments placed into composts decline to a marked extent as these three explosives are biodegraded (Ziegenfuss et al., 1991).

SLURRY REACTORS

Bioremediation can be effected by a variety of procedures in which contaminated solids are mixed constantly with a liquid in a *slurry-phase treatment*. The system may be reasonably unsophisticated and entail introduction of the contaminated soil, sludge, or sediment into a lagoon that has been constructed with a liner, or it may be a sophisticated reactor in which the contaminated materials are mixed (Fig. 15.3). The operation in many ways resembles the activated-sludge procedure that is common for treatment of municipal wastes, and it allows for aeration, adequate mixing, and control of many of the factors affecting biodegradation. Some designs allow for the capture of volatile organic products that may be generated. The level of dissolved O_2, the pH, and the concentration of inorganic nutrients may be monitored and controlled. Some bioreactors are inocu-

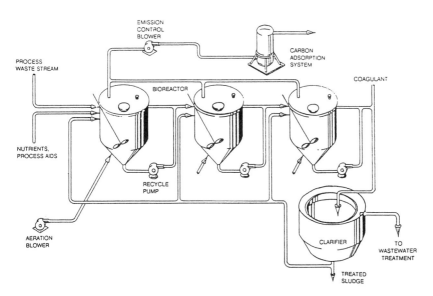

FIG. 15.3 A system for bioremediation by slurry-phase treatment. (From Ryan *et al.*, 1991. Reproduced with permission from Elsevier Science Publishers.)

lated with a single species or a mixture of microorganisms able to function effectively under the controlled conditions. In locations where biodegradation slows and sometimes ceases because of low temperatures during part of the year, as is common in many land-treatment systems, the temperature in slurry reactors is maintained in ranges suitable for rapid biodegradation.

Carefully controlled laboratory experiments show that many PAHs, heterocycles, and phenols in creosotes are quickly destroyed, and more than 50% of the lower-molecular-weight PAHs may be destroyed in 3 to 5 days. On the other hand, some of the higher-molecular-weight PAHs are only slowly destroyed (Mueller *et al.*, 1991b). The rate of destruction of compounds sorbed to soil or of oil in soil may be increased markedly by the addition of appropriate surfactants or dispersants to the slurries of soil (Aronstein and Alexander, 1992; Rittmann and Johnson, 1989). Laboratory studies also suggest that the addition of relatively inexpensive materials that are quickly degraded may result in anaerobic conditions that enhance the degradation of specific compounds, such as dinoseb (Kaake *et al.*, 1992).

Slurry-phase procedures may be combined with a washing technique to remove the contaminant from soil. Such a soil washing was used for a field-scale bioremediation at a wood-treating facility that contained up

TABLE 15.1

Bioremediation of Sludge Rich in Oil and Grease in
an Aerated Lagoon[a]

Compound	Concentration (mg/kg)	
	Initially	At end
Benzene	64.4	1.19
Toluene	19.4	1.14
Ethylbenzene	32.4	0.32
Naphthalene	290	ND[b]
Phenanthrene	150	ND
Pyrene	540	0.03
Anthracene	20	0.02
Benzoanthracene	91	ND
Chrysene	20	ND
Benzopyrene	100	<0.01

[a]From Vail (1991).
[b]Not detected.

to 8000 mg of PCP per kilogram of soil. The soil was initially washed to remove the PCP, and then the wash solution was introduced into a slurry-phase reactor. This two-phase procedure reduced the PCP concentration in the soil to the target cleanup level at the site, namely, less than 0.5 mg/kg of soil (Compeau *et al.*, 1991).

Aerated lagoons have been employed to bioremediate refinery wastes and other materials containing petrochemicals. One such cleanup was designed for a wastewater–sludge mixture from a refinery. Aeration, N, and P were provided, but the temperature was not controlled. The 7000 m^3 of oil- and grease-rich sludge contained 2000 mg of PAHs and 200 mg of BTEX per kilogram. The extent of biodegradation is depicted in Table 15.1. Wastes from refineries and petrochemical manufacturers were also successfully treated in a lagoon that was aerated and amended with phosphate and nitrate (Woodward and Ramsden, 1990).

IN SITU GROUNDWATER BIORESTORATION

A common procedure for *in situ* bioremediation entails the introduction of nutrients and O_2 into subsurface aquifers, relying on the indigenous microflora to destroy the unwanted molecules. This process is sometimes called *biorestoration*. Most of the contaminated sites treated so far contain petroleum hydrocarbons as the contaminants. Leakages from underground storage tanks containing gasoline result in the appearance of benzene, toluene, ethylbenzene, and xylenes. Although these BTEX compounds are initially in the gasoline phase, particular attention is given to them because they are toxic and because they enter the aqueous phase in the form of a sustained release. This sustained release and the amounts present in the aqueous phase are consequences of their reasonable water solubilities and the constant partitioning of these compounds from the gasoline to the aqueous phases. Groundwaters contaminated with diesel fuel and JP-4 aviation fuel are also treated in similar manners.

Initially, as much of the free oil or hydrocarbon as possible is removed by one of several physical means. Bioremediation without such removal is pointless because the bulk source would continue adding new chemical to the groundwater. Laboratory tests are also conducted to determine the optimal amount of nutrients to add; this is especially important to avoid too little or too much being supplied. Too little will result in a slow transformation, and too much might clog the aquifer because of the large biomass that is formed, thereby causing cessation of the remediation. The three nutrients that are commonly required for optimal activity are N, P, and O_2, which are typically the factors that limit the activity of the indigenous microflora. The N and P salts are usually dissolved in the groundwater that is circulated through the contaminated site. A common procedure is to add the nutrients in solution through injection wells into the saturated zone or through infiltration galleries into the unsaturated or surface-soil zone (Fig. 15.4). The water is recovered from production wells, and that water is again amended with nutrients and recirculated. The concentrations of contaminants and nutrients are often measured on a regular basis by taking samples from wells that are installed between the points of injection and removal. In some instances, the water is not recirculated but, instead, is disposed at the surface (Thomas and Ward, 1989).

Rapid biodegradation of hydrocarbons typically is carried out by aerobic bacteria, and their activity must be sustained. This poses major problems because even under the best conditions, little O_2 is present in groundwater and natural sources provide O_2 at exceedingly slow rates. The same problem of poor O_2 solubility applies to the water added during the remediation itself. For example, for the degradation of a 4000-liter spill of hydrocarbons, 5000 kg of O_2 is needed. An enormous volume of water mixed with air to give up to 8 mg O_2/liter would have to be pumped through the aquifer to provide the requisite O_2. Even if the introduced water is saturated with pure O_2 (to give 40 mg O_2/liter) rather than air, approximately 110 million liters of water would be needed for the 4000-liter spill. As a result, H_2O_2

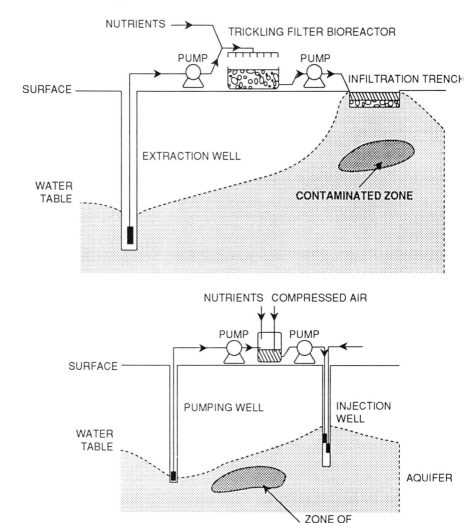

FIG. 15.4 *In situ* groundwater treatment using an infiltration trench (top) and an injection well (bottom). Note the use of an aboveground trickling-filter bioreactor. (From Morgan and Watkinson, 1989. Reproduced with permission from Elsevier Science Publishers.)

is sometimes added to the nutrient solution as a source of O_2 (McCarty, 1988). H_2O_2 is highly water soluble, and it slowly breaks down in the aquifer to give free O_2. However, care needs to be taken because H_2O_2 is toxic to some species at 100 to 200 mg/liter. The problem of toxicity can be minimized or avoided if a low H_2O_2 concentration is added initially, possibly 50 mg/liter, and then the amounts can be stepwise increased to give 1000 mg/liter (Thomas and Ward, 1989; Wilson and Ward, 1987). The success of biorestoration depends on the hydrogeology of the site. If the hydrogeology is complex, success is problematic, and bioremediation sometimes will be of dubious value. Adequate procedures to characterize many sites are not currently available, and thus the likelihood of success at complex sites is questionable. Moreover, the subsurface environment must be sufficiently permeable to permit the transport of the added N, P, and O_2 to the microorganisms situated at the various subsurface sites containing the contaminants. This water movement, referred to as hydraulic conductivity, is critical to a positive outcome (McCarty, 1991; Thomas et al., 1992).

Such procedures have successfully destroyed a number of pollutants at a variety of sites. For example, the groundwater at a former retail gasoline facility was essentially freed of petroleum hydrocarbons after 10 months of treatment with solutions containing H_2O_2 and ammonium and phosphate salts (Fogel et al., 1989). A site contaminated with ethylene glycol was freed of detectable levels in 26 days (Jerger and Flatham, 1990), and a 94% or greater decline in PCP concentration in the top 10 m was achieved in a pilot-scale test for the remediation of groundwater at a site where PCP had previously been used to treat wood and wood products (Fu and O'Toole, 1990). The microbial populations and their activity may remain high in treated areas even 2 years after the in situ bioremediation has ended (Thomas et al., 1990).

A similar procedure has been used to enhance the biodegradation of TCE in a field demonstration of cometabolism. Methane and O_2 were added alternatively at concentrations of 20 and 32 mg/liter, respectively, the two additions serving to stimulate indigenous CH_4-oxidizing bacteria (methanotrophs) that cometabolize TCE. Other nutrients were not added. From 20 to 30% of the TCE was destroyed in the test period (P. L. McCarty, cited by Thomas and Ward, 1989).

In situ treatments involving nutrients and one or another source of O_2 are utilized in many field operations (Devine, 1992).

Some bacteria are able to use, in addition to O_2, nitrate as an electron acceptor, and they thus grow and degrade a number of substrates in anoxic waters provided with nitrate. Nitrate is attractive because of its high solubility in water and low cost, although caution must be exercised because nitrate, if present in drinking water at levels in excess of 10 mg/ liter (as N), is a pollutant itself. A field test has been conducted at a location in Michigan contaminated with JP-4 jet fuel to determine the efficacy of nitrate. Although the concentration of each constituent of

BTEX was reduced from 760, 4500, and 840 to <1, <1, and 6 μg/liter for benzene, toluene, and ethylbenzene but only to 20–40 μg/liter for total xylene isomers, the design of the test did not permit separating the degree of biodegradation associated with O_2 and nitrate as electron acceptors (Hutchins et al., 1991).

MARINE OIL SPILLS

Interest in the cleanup of oil spilled in marine, estuarine, and fresh waters and in the use of microorganisms for freeing the adjacent shorelines of oil has existed for many years. Early studies showed that hydrocarbon-oxidizing bacteria were widespread, that they were limited by N and P when oil was introduced into the water, and that formulations containing oleophilic fertilizers were particularly beneficial. However, it was not until the tanker Exxon Valdez was grounded on a reef in Prince William Sound, Alaska, that a major bioremediation of oil in surface waters was undertaken. The March 24, 1989, grounding of this tanker released 42 million liters of oil into the sound, contaminating both the water and adjacent beaches. As a result, a program was immediately mounted to determine, both in the laboratory and in the field, how to promote the biodegradation. The subsequent cleanup represents the largest field bioremediation ever undertaken.

Laboratory investigations conducted shortly after the spill confirmed the early studies that N and P were limiting, and that almost all the alkanes in the Alaskan oil and an appreciable amount of the PAHs were metabolized in 6 weeks with an inorganic salts solution or an oleophilic fertilizer containing N and P. Far less hydrocarbon loss was evident with no added N and P (Tabak et al., 1991). Field tests confirmed the abundance of hydrocarbon-degrading bacteria. Specific N and P fertilizers were chosen to be added to the beaches because they would remain associated with the oil, a critical issue in Prince William Sound owing to water movement associated with tidal activity and the occasional strong storms. The oleophilic fertilizer chosen was a liquid containing urea in oleic acid as the N source and tri(laureth-4)-phosphate as the P source. Although this oleophilic preparation remained with the oil on the beach surface, little penetrated to the underlying sediment; therefore, an encapsulated formulation of N and P was added to stimulate the subsurface microflora. Within 2 weeks, differences in the quantities of oil were visually evident between fertilizer-treated and untreated beaches, and subsequent quantitative measurements revealed that 60–70% of the oil was degraded within

16 months of the spill. Even in the absence of the fertilizer, biodegradation occurred, but the added N and P promoted the process. As a result of the positive response to N and P, more than 117 km of shoreline was treated in 1989, additional locations were fertilized in 1990, and those isolated areas still containing oil were treated in 1991. Although oil constituents were degraded with no added N and P as the indigenous bacteria used the N and P initially present and that subsequently made available as the initial populations died, an enhancement in rate is especially important in a region, such as Alaska, in which only a few months are warm enough for appreciable microbial decomposition to occur (Prince, 1992; Pritchard and Costa, 1991). On the beaches that were treated, O_2 deficiencies were not deemed to be of practical significance because of wave action with aerated water in the sound.

The response of microorganisms to N, P, and O_2 or the effectiveness of biodegradation is simple to determine in the laboratory. It is also easy to assess in a bioreactor in which measurements can be made of the inflow to the reactor and the outflow from it. The same is not true in an *in situ* bioremediation. The pollutants may disappear from the site as a result of volatilization from the soil or water or merely by dilution in flowing water. A number of methods have been proposed to show that biodegradation has occurred or that a deliberate bioremediation has been successful in the field (Madsen, 1991). The issue of confirming effectiveness of a bioremediation was particularly acute at Prince William Sound because of the enormous cost for the cleanup of the large affected area.

One way of assessing microbial activity in the field is to determine the changes in concentration with time of two substances that have similar physical and chemical behaviors in nature but only one of which is degraded reasonably quickly. Both might be subject to abiotic degradation or volatilization, but a finding that the more biodegradable compound disappears more readily strongly suggests microbial degradation. This assessment may be accomplished by comparing the rate of disappearance of linear alkanes having 17 or 18 C atoms with highly branched alkanes of similar molecular weights, such as pristane and phytane. The linear alkanes usually disappear far more quickly than the highly branched hydrocarbons. Tests following the Exxon Valdez spill showed that the ratio of the straight-chain 18-C alkane to phytane did indeed change with time (Fig. 15.5). However, once the extent of biodegradation is extensive, even most of the branched alkanes are destroyed by the microflora. Hence, another internal standard for bioremediation is necessary. For this purpose, a still less degradable standard is needed, and a highly resistant five-ring saturated hydrocarbon was chosen, namely, $17\alpha(H),21\beta(4)$-hopane.

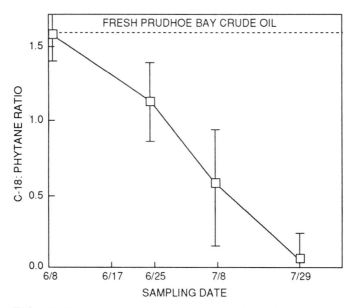

FIG. 15.5 Change with time in the ratio of octadecane to phytane (on a mg: g weight basis) following the application of an oleophilic fertilizer on June 8, 1989. (Reproduced with permission from Pritchard and Costa, 1991.)

Hopane is not one of the PAHs, which are, by definition, unsaturated. Tests with hopane as an internal standard confirmed biodegradation of the oil from the spill in Prince William Sound (Prince, 1992).

Other means have been proposed to confirm *in situ* biodegradation. One makes use of the facts that the ratio of stable C isotopes (^{13}C and ^{12}C) in the CO_2 produced biologically in mineralization differs from that associated with abiotic processes and that the relative rates of utilization of ^{12}C and ^{13}C in organic compounds differ in biotic and abiotic processes (Aggarwal and Hinchee, 1991). Field use has not been made of the method employing stable C isotopes.

Nearly all the successful biomediations using the techniques described here rely on the actions of indigenous microorganisms. Although proprietary concerns in commercial practice usually preclude the release of information on procedures and sometimes of results, little information exists in the peer-reviewed literature suggesting that inoculation (or bioaugmentation) is needed for the previously discussed technologies. The claims by some entrepreneurs of remarkable successes by undescribed organisms under undefined operating conditions must be taken as just that—claims. Undoubtedly, some such inoculations are indeed beneficial, but better information is needed before conclusions can be reached. On the other hand, there is no question that some of the treatments carried out in the reactors described next benefit, modestly or enormously, from deliberate introduction of specific microorganisms.

ABOVEGROUND BIOREACTORS

A variety of techniques have been described in which pollutants are treated above ground. Some are old practices, such as activated-sludge treatment of waste streams containing synthetic chemicals; others have been introduced into practice quite recently; and still others exist only in laboratory or sometimes in pilot-plant treatment units. As indicated earlier, the semantic purist might object to designating an aboveground bioreactor as a means of remediation since a remedy is sought only after a problem is found, and many of these reactors are designed to destroy chemicals in industrial waste streams before those compounds reach natural environments.

When coupled with an *in situ* treatment, however, the concept of an aboveground bioreactor is completely appropriate as a technology for bioremediation. These reactors may be mobile units taken to field sites and used to treat groundwater pumped out of a contaminated aquifer or wash water used to remove chemicals from a contaminated soil. Such processes may be continuous, or they may be conducted in batches with contaminated material added and treated material removed intermittently.

A typical example is the mobile unit used to degrade PCP-containing groundwater at a facility in which preservatives were once added to wood. Groundwater was pumped up from a well, the pH was adjusted, and nutrients were added to the liquid. The fixed-film bioreactor, which had a means for temperature control, brought about extensive biodegradation of PCP (Stinson *et al.*, 1991). Aboveground reactors have also been used for other compounds and for treatment of slurries (Litchfield, 1991).

BIOFILTERS

Microorganisms are also being used to destroy a variety of volatile compounds. In such technologies, the microorganisms are allowed to grow on some solid support, and a stream of gas containing the unwanted molecules is passed through the solid support. The resulting microbial action leads to destruction of the contaminants. This process, which is termed *biofiltration,* is common in Germany, the Netherlands, Japan, and a few other countries. The *trickle-bed reactor,* in which the bacteria are fixed on a column, may be designed for volatile compounds, the chemical to be degraded (or "scrubbed") being passed through and dissolved in a solution. Somewhat similar to biofilters are *bioscrubbers,* in which the gases and O_2 are usually first passed into a unit in which the volatiles dissolve in water, and then the solution is introduced into a system, usually an activated-sludge system, in which the organic compounds are degraded by microorganisms dispersed in the aqueous phase. These treatment systems are attractive because of their low cost compared to physical and chemical methods and the ability to destroy compounds at low concentrations.

The solid phase of the biofilter may be peat, soil, composted organic matter, sawdust, bark chips, activated carbon, clay particles, or porous glass. Soil and compost are popular solid phases, and a volatile organic compound passing through a soil or compost bed is first sorbed by the solid phase, which must be kept moist to maintain biological activity, and then the sorbed molecule is metabolized by the film of microorganisms adhering to the solid. An unsophisticated but effective biofilter is simply a bed of soil placed on top of a system of pipes through which the volatiles pass.

A large number of volatile compounds can be degraded in biofilters, including naphthalene, acetone, propionaldehyde, volatile S compounds, toluene, benzene, dichloromethane, and vinyl chloride. Particular attention has been given to volatiles with offensive odors, and a number of systems have been designed to remove H_2S, methane thiol, dimethyl disulfide (Shoda, 1991), SO_2 (Dasu and Sublette, 1989), and dimethyl sulfide (Hirai *et al.,* 1990). Among the organic pollutants, waste air containing the components of BTEX can be treated to destroy the aromatic compounds in the vapor phase by passing the air through soil. Such an approach offers promise for destruction of the organic volatiles generated when gasoline-contaminated groundwaters are treated by air stripping or when soil venting is employed to eliminate volatiles emitted from the vadose zone. Alternative treatments to remove the volatiles, such as incineration or sorption by activated carbon, are costly (Canter *et al.,* 1989; Miller and Canter, 1991). Propane and *n-* and isobutane are also readily degraded when exposed to the microorganisms present in soil (Kampbell *et al.,* 1987), as are volatile organics present in fumes from aviation gasoline (Kampbell and Wilson, 1991) and carbon monoxide (Frye *et al.,* 1992).

Biofilters or bioscrubbers can also be operated to destroy volatile chlorinated compounds. This may be accomplished by organisms cometabolizing the pollutants, for example, in systems in which gaseous methane or propane is passed through a column of an aquifer material. Perfusing a liquid containing TCE and 1,1,1-trichloroethane leads to their cometabolism by populations growing on the methane or propane (Wilson *et al.*, 1987). A similar type of degradation of TCE occurs with a muck soil as the solid phase (Kampbell *et al.*, 1987). Dichloromethane and vinyl chloride in a gas stream also can be metabolized, and most of the compounds are destroyed in trickle-bed filters inoculated with species of *Mycobacterium* or *Hyphomicrobium*, respectively (Diks and Ottengraf, 1991; Hartmans *et al.*, 1985).

FIXED FILMS AND IMMOBILIZED CELLS

A common way of bringing about biodegradation is to use reactors in which microbial cells become attached as a film to some matrix. A solution containing the chemicals is passed over the resulting biofilm, which brings about a rapid biodegradation because of the high cell density. A modification of fixed-film treatment employs immobilized or strongly sorbed cells. The cells are immobilized by firmly attaching the organisms or physically embedding them in the solid matrix. The cells may thus be immobilized in or on a variety of materials, including alginate beads, diatomaceous earth, hollow glass fibers, polyurethane foam, activated C, and polyacrylamide beads. Common to many of these systems is the greater tolerance to high chemical concentrations of the cells that are in the films or that are immobilized than cells in suspension. The greater resistance may be associated with sorption of the substrate to the solid or immobilizing material, thereby reducing the amount available to suppress the microorganisms, or to some other mechanism.

Several bacteria and fungi as well as microbial mixtures have been used in immobilized cell systems, and a number of compounds are readily biodegraded by these procedures (Table 15.2). The immobilized cells are contained in one of several different types of reactors to facilitate rapid biodegradation. Such technologies are particularly useful for waste streams from chemical-manufacturing facilities, as illustrated by a pilot plant used for the high-volume treatment of low concentrations of glyphosate. Wastewater containing this compound was introduced at 45 liter/min into a column containing immobilized cells configured as an upflow reactor. The immobilized microorganisms were found to bring about rapid and extensive degradation (Hallas *et al.*, 1992).

TABLE 15.2
Biodegradation of Organic Compounds by Immobilized Cells or Strongly
Sorbed Cells

Compound	Immobilizing material or solid phase	Microorganism	Reference
Acrylonitrile	Alginate	Brevibacterium sp.	Hwang and Chang (1989)
Aniline, chloroanilines	Diatomaceous earth	Mixed culture	Livingston and Willacy (1991)
Anthracene	Alginate beads	Trichoderma harzianum	Ermisch and Rehm (1989)
2-Chlorophenol	Alginate beads	Phanerochaete chrysosporium	Lewandowski et al. (1990)
Glyphosate	Diatomaceous earth	Mixed culture	Hallas et al. (1992)
4-Nitrophenol	Diatomaceous earth	Pseudomonas sp.	Heitkamp et al. (1990)
PCP	Alginate beads	Arthrobacter sp.	Lin and Wang (1991)
Phenol	Activated carbon	Pseudomonas putida	Ehrhardt and Rehm (1989)

COMETABOLISM

Few engineered systems have been designed to exploit biodegradation by cometabolizing microorganisms. Any such technology must provide not only the specific compound of concern but also a substrate that would provide C and energy to support growth. However, bioremediation technologies for TCE cometabolism are being developed. The potential use of methane and indigenous methane-oxidizing (methanotrophic) bacteria in an aquifer has already been considered. One procedure adapted for controlled treatment involves a two-stage reactor, the first being a conventional fermenter in which methanotrophic bacteria are grown on a mixture of methane and O_2. The resulting cells are then transferred to a plug-flow reactor in which TCE is transformed (Alvarez-Cohen and McCarty, 1991). Another procedure involves an expanded bed consisting of crushed glass on which the bacteria grow. The reactor is supplied with a solution of inorganic nutrients, TCE, either methane or propane as sources of C and energy, and a mixture of bacteria. These organisms degrade not only TCE but also vinyl chloride and PCE (Phelps et al., 1991).

DDT is cometabolized anaerobically, but the products of the cometabolic transformation are metabolized by aerobes (Pfaender and Alexander, 1972). This partnership is the basis for a means proposed for the degradation of DDT. The two-organism system is made up of *Enterobacter cloacae* that anaerobically dechlorinates DDT and forms 4,4'-dichlorodiphenyl-methane as it grows on and ferments lactose and a strain of *Alcaligenes* that grows aerobically on diphenylmethane but cometabolizes 4,4'-dichlorodiphenylmethane. The two bacteria, when immobilized in Ca alginate, are able to carry out the two steps in a single reactor (Beunink and Rehm, 1988).

WHITE-ROT FUNGI

Phanerochaete and related fungi that have the ability to attack wood possess a powerful extracellular enzyme that, differing from many enzymes, acts on a broad array of organic compounds. The enzyme is a peroxidase that, with H_2O_2 produced by the fungus, catalyzes a reaction that cleaves a surprising number of compounds. For example, cultures of *Phanerochaete chrysosporium,* the most widely studied of these fungi for its biodegradative capacity, can degrade a number of PAHs [including benz(*a*)pyrene, benz(*a*)anthracene, and pyrene], di- and trichlorobenzoic acids, several PCBs, 2,3,7,8-tetrachlorodibenzo-*p*-dioxin, DDT, lindane, and chlordane—compounds that few other microorganisms will decompose (Bumpus *et al.*, 1991). The explosives TNT and RDX are mineralized, albeit slowly (Fernando and Aust, 1991).

Bioremediation technologies using this fungus thus have considerable promise, especially for compounds not acted on readily, if at all, by bacteria. However, the transformations by the fungus are slow. Except for one evaluation of a *Phanerochaete* species for PCP biodegradation under field conditions (Lamar and Dietrich, 1990), the possibility of exploiting the catabolic activity of these fungi under realistic simulations of field conditions or industrial wastes has not been explored.

ANAEROBIC PROCESSES

Nearly all bioremediations in practical use are aerobic. This neglect of anaerobic bacteria is a gross error because it is now clear that they are able to catalyze many reactions and destroy many compounds that are resistant to aerobes. Particular attention in recent years has been given to chlorinated molecules, because of not only their toxicity but also their persistence in polluted environments. Anaerobes can reductively dehalogenate chlorinated molecules that persist and that are rarely attacked by aerobic bacteria (Mohn and Tiedje, 1992). Highly chlorinated PCBs, carbon tetrachloride, PCE, and many other chlorinated products, some of which are ubiquitous as well as persistent, are converted to less highly chlorinated compounds. Metabolism of the latter could probably be best accomplished by aerobes because aerobic conversions of the less chlorinated molecules are often faster. An example is a two-stage anaerobic–aerobic biofilm reactor proposed for treating groundwater and industrial effluents containing highly chlorinated molecules. In the anaerobic phase, the mixed culture reductively dechlorinates TCE, chloroform, and hexachlorobenzene in the presence of acetate to yield partially dehalogenated products. These less chlorinated products are then introduced into an aerobic reactor, the net result being that more than 93% of the three substrates are converted to nonvolatile products and CO_2 (Fathepure and Vogel, 1991). An anaerobic process has also been suggested for the on-site treatment of soil containing toxaphene, but the process, which involves addition of large amounts of organic matter to create anaerobic conditions, is slow and incomplete (Mirsatari *et al.*, 1987). One-carbon halogenated compounds are nearly completely transformed anaerobically following the onset of sulfate reduction in a biofilm column reactor inoculated with sewage microorganisms, suggesting a possible way of enhancing the transformation (Cobb and Bouwer, 1991).

Given the great biochemical potential of anaerobic processes, additional development is needed to convert the many microbiological studies of anaerobic bacteria into practical methods for bioremediation.

ENZYMATIC CONVERSIONS

Proposals have been made to use enzyme preparations to destroy individual pollutants or toxic chemicals. Some of the enzymes are quite stable during storage, and they thus could be used in emergency responses to spills since the catalysts would be immediately available in active form. It has also been suggested that they can be used to decontaminate soils having unwanted pesticides or other toxicants. They might even be useful for converting some persistent pollutants to products that are harmless. Both soluble and immobilized enzymes have been suggested for these purposes, but most research has focused on immobilized enzymes.

An enzyme immobilized on porous glass or porous silica beads was found to hydrolyze a number of organophosphate insecticides in solution (Munnecke, 1979, 1981), and a phosphotriesterase from *Escherichia coli* was immobilized on nylon membrane, powder, and tubing (Caldwell and Raushel, 1991). Both enzymes were suggested as means for detoxifying pesticides. Enzymes have also been added to soil, where they sorb to particulate matter, and the sorbed catalysts have been suggested as a means of hydrolyzing insecticides such as diazinon. Indeed, parathion hydrolase added to soil converts more than 90% of diazinon at 1.0 g/kg to nontoxic products in 4 h (Honeycutt et al., 1984). It has also been suggested that toxic aromatic compounds in water may be rendered innocuous by converting them to less soluble, high-molecular-weight products, such as by addition of peroxidase and H_2O_2 (Maloney et al., 1986) or by use of laccase to convert 2,4-dichlorophenol to water-insoluble oligomers and products that do not leach through soil (Ruggiero et al., 1989; Shannon and Bartha, 1988).

Practical technologies based on these innovations have yet to be applied.

REFERENCES

Aggarwal, P. K., and Hinchee, R. E., Environ. Sci. Technol. 25, 1178–1180 (1991).

Alvarez-Cohen, L., and McCarty, P. L., Environ. Sci. Technol. 25, 1387–1393 (1991).

Aronstein, B. N., and Alexander, M., Environ. Toxicol. Chem. 11, 1227–1233 (1992).

Aronstein, B. N., Calvillo, Y. M., and Alexander, M., Environ. Sci. Technol. 25, 1728–1731 (1991).

Beunink, J., and Rehm, H.-J., Appl. Microbiol. Biotechnol. 29, 72–80 (1988).

Block, R. N., Clark, T. P., and Bishop, M., in "Petroleum Contaminated Soils" (P. T. Kostecki and E. J. Calabrese, eds.), Vol. 3, pp. 167–175. Lewis Publishers, Chelsea, MI, 1990.

Bumpus, J. A., Milewski, G., Brock, B., Ashbaugh, W., and Aust, S. D., in "Innovative Hazardous Waste Treatment Technology Series" (H. M. Freeman and P. R. Sferra, eds.), Vol. 3, pp. 47–54. Technomic Publ. Co., Lancaster, PA, 1991.

Caldwell, S. R., and Raushel, F. M., Appl. Biochem. Biotechnol. 31, 59–73 (1991).

Canter, L. W., Streebin, L. E., Arquiaga, M. C., Carranza, F. E., Miller, D. E., and Wilson, B. H., "Innovative Processes for Reclamation of Contaminated Subsurface Environments." Publ. EPA/600/S2-90/017. Kerr Laboratory, U.S. Environmental Protection Agency, Ada, OK, 1989.

Cobb, G. D., and Bouwer, E. J., Environ. Sci. Technol. 25, 1068–1074 (1991).

Compeau, G. C., Mahaffey, W. D., and Patras, L., in "Environmental Biotechnology for Waste Treatment" (G. S. Sayler, R. Fox, and J. W. Blackburn, eds.), pp. 91–109. Plenum, New York, 1991.

Dasu, B. N., and Sublette, K. L., Biotechnol. Bioeng. 34, 405–409 (1989).

Devine, K., "Bioremediation Case Studies: An Analysis of Vendor Supplied Data." Publ. EPA/600/R-92/043. Office of Engineering and Technology Demonstration, U.S. Environmental Protection Agency, Washington, DC, 1992.

Dibble, J. T., and Bartha, R., Soil Sci. 128, 56–60 (1979).

Diks, R. M. M., and Ottengraf, S. P. P., Bioprocess Eng. 6, 93–99 (1991).

Ehrhardt, H. M., and Rehm, H.-J., Appl. Microbiol. Biotechnol. 30. 312–317 (1989).

Ermisch, O., and Rehm, H.-J., DECHEMA Biotechnol. Conf. 3, 905–908 (1989).

Fathepure, B. Z., and Vogel, T. M., Appl. Environ. Microbiol. 57, 3418–3422 (1991).

Fernando, T., and Aust, S. D., in "Gas, Oil, Coal, and Environmental Biotechnology III" (C. Akin and J. Smith, eds.), pp. 193–206. Institute of Gas Technology, Chicago, 1991.

Fogel, S., Findlay, M., and Moore, A., *in* "Petroleum Contaminated Soils" (E. J. Calabrese and P. T. Kostecki, eds.), Vol. 2, pp. 201–209. Lewis Publishers, Chelsea, MI, 1989.

Frye, R. J., Welsh, D., Berry, T. M., Stevenson, B. A., and McCallum, T., *Soil Biol. Biochem.* **24**, 607–612 (1992).

Fu, J. K., and O'Toole, R., *in* "Gas, Oil, Coal, and Environmental Biotechnology II" (C. Akin and J. Smith, eds.), pp. 145–169. Institute of Gas Technology, Chicago, 1990.

Hallas, L. E., Adams, W. J., and Heitkamp, M. A., *Appl. Environ. Microbiol.* **58**, 1215–1219 (1992).

Hartmans, S., deBont, J. A. M., Tramper, J., and Luyben, K. C. A. M., *Biotechnol. Lett.* **7**, 383–388 (1985).

Heitkamp, M. A., Carmel, V., Reuter, T. J., and Adams, W. J., *Appl. Environ. Microbiol.* **56**, 2967–2973 (1990).

Hildebrandt, W. W., and Wilson, S. B., *JPT, J. Pet. Technol.* **43**, 18–22 (1991).

Hinchee, R. E., Miller, R. N., and Dupont, R. R., *in* "Innovative Hazardous Waste Treatment Technologies" (H. M. Freeman and P. R. Sferra, eds.), Vol. 3, pp. 177–183. Technomic Publ. Co., Lancaster, PA, 1991.

Hirai, M., Ohtake, M., and Shoda, M., *J. Ferment. Bioeng.* **70**, 334–339 (1990).

Honeycutt, R., Ballantine, L., LeBaron, H., Paulson, D., Seim, V., Ganz, C., and Milad, G., *in* "Treatment and Disposal of Pesticide Wastes" (R. F. Krueger and J. N. Seiber, eds.), pp. 343–352. American Chemical Society, Washington, DC, 1984.

Hutchins, S. R., Downs, W. C., Wilson, J. T., Smith, G. B., Kovacs, D. A., Fine, D. D., Douglass, R. H., and Hendrix, D. J., *Ground Water* **29**, 571–580 (1991).

Hwang, J. S., and Chang, H. N., *Biotechnol. Bioeng.* **34**, 380–386 (1989).

Jerger, D. E., and Flatham, P. E., *in* "Gas, Oil, Coal, and Environmental Biotechnology II" (C. Akin and J. Smith, eds.), pp. 67–81. Institute of Gas Technology, Chicago, 1990.

Kaake, R. H., Roberts, D. J., Stevens, T. O., Crawford, R. L., and Crawford, D. L., *Appl. Environ. Microbiol.* **58**, 1683–1689 (1992).

Kampbell, D. H., and Wilson, J. T., *J. Hazard. Mater.* **28**, 75–80 (1991).

Kampbell, D. H., Wilson, J. T., Read, H. W., and Stocksdale, T. T., *J. Air Pollut. Control Assoc.* **37**, 1236–1240 (1987).

Lamar, R. T., and Dietrich, D. M., *Appl. Environ. Microbiol.* **56**, 3093–3100 (1990).

Lewandowski, G. A., Armenante, P. M., and Pak, D., *Water Res.* **24**, 75–82 (1990).

Lin, J.-E., and Wang, H. Y., *J. Ferment. Bioeng.* **72**, 311–314 (1991).

Litchfield, C. D., *in* "Environmental Biotechnology for Waste Treatment" (G. S. Sayler, R. Fox, and J. W. Blackburn, eds.), pp. 147–157. Plenum, New York, 1991.

Livingston, A. G., and Willacy, A., *Appl. Microbiol. Biotechnol.* **35**, 551–557 (1991).

Lynch, J., and Genes, B. R., *in* "Petroleum Contaminated Soils" (P. T. Kostecki and E. J. Calabrese, eds.), Vol. 1, pp. 163–174. Lewis Publishers, Chelsea, MI, 1989.

Madsen, E. L., *Environ. Sci. Technol.* **25**, 1662–1673 (1991).

Maloney, S. W., Manem, J., Mallevialle, J., and Fiessinger, F., *Environ. Sci. Technol.* **20**, 249–253 (1986).

Manilal, V. B., and Alexander, M., *Appl. Microbiol. Biotechnol.* **35**, 401–405 (1991).

McCarty, P. L., *in* "Environmental Biotechnology" (G. S. Omenn, ed.), pp. 143–162. Plenum, New York, 1988.

McCarty, P. L., *J. Hazard. Mater.* **28**, 1–11 (1991).

Miller, D. E., and Canter, L. W., *Environ. Prog.* **10**, 300–306 (1991).

Mirsatari, S. G., McChesney, M. M., Craigmill, A. C., Winterlin, W. L., and Seiber, J. N., *J. Environ. Sci. Health, Part B* **B22**, 663–690 (1987).

Mohn, W. W., and Tiedje, J. M., *Microbiol. Rev.* **56**, 482–507 (1992).

Morgan, P., and Watkinson, R. J., *FEMS Microbiol. Rev.* **63**, 277–299 (1989).

Morgan, P., and Watkinson, R. J., *Water Sci. Technol.* **22**(6), 63–68 (1990).

Mueller, J. G., Lantz, S. E., Blattmann, B. O., and Chapman, P. J., *Environ. Sci. Technol.* **25**, 1045–1055 (1991a).

Mueller, J. G., Lantz, S. E., Blattmann, B. O., and Chapman, P. J., *Environ. Sci. Technol.* **25**, 1055–1061 (1991b).

Munnecke, D. M., *Biotechnol. Bioeng.* **21**, 2247–2261 (1979).

Munnecke, D. M., *in* "Microbial Degradation of Xenobiotics and Recalcitrant Compounds" (T. Leisinger, A. M. Cook, R. Hütter and J. Nüesch, eds.), pp. 251–269. Academic Press, London, 1981.

Pfaender, F. K., and Alexander, M., *J. Agric. Food Chem.* **20**, 842–846 (1972).

Phelps, T. J., Niedzielski, J. J., Malachowsky, K. J., Schram, R. M., Herbes, S. E., and White, D. C., *Environ. Sci. Technol.* **25**, 1461–1465 (1991).

Prince, R. C., *in* "Microbial Control of Pollution" (J. C. Fry, G. M. Gadd, R. A. Herbert, C. W. Jones, and I. A. Watson-Craik, eds.), pp. 19–34. Cambridge Univ. Press, Cambridge, UK, 1992.

Pritchard, P. H., and Costa, C. F., *Environ. Sci. Technol.* **25**, 372–379 (1991).

Raymond, R. L., Hudson, J. O., and Jamison, V. W., *Appl. Environ. Microbiol.* **31**, 522–535 (1976).

Rittmann, B. E., and Johnson, N. M., *Water Sci. Technol.* **21**(4/5), 209–219 (1989).

Ruggiero, P., Sarkar, J. M., and Bollag, J.-M., *Soil. Sci.* **147**, 361–370 (1989).

Ryan, J. R., Loehr, R. C., and Rucker, E., *J. Hazard. Mater.* **28**, 159–169 (1991).

Shannon, M. J. R., and Bartha, R., *Appl. Environ. Microbiol.* **54**, 1719–1723 (1988).

Shoda, M., *in* "Biological Degradation of Wastes" (A. M. Martin, ed.), pp. 31–46. Elsevier Applied Science, London, 1991.

Song, H.-G., Wang, X., and Bartha, R., *Appl. Environ. Microbiol.* **56**, 652–656 (1990).

Stinson, M. K., Hahn, W., and Skovronek, H. S., *in* "Innovative Hazardous Waste Treatment Technology Series" (H. M. Freeman and P. R. Sferra, eds.), Vol. 3, pp. 163–167. Technomic Publ. Co., Lancaster, PA, 1991.

Tabak, H. H., Haines, J. R., Venosa, A. D., Glaser, J. A., Desai, S., and Nisamaneepong, W., *in* "Gas, Oil, Coal and Environmental Biotechnology III" (C. Akin and J. Smith, eds.), pp. 3–38. Institute of Gas Technology, Chicago, 1991.

Thomas, J. M., and Ward, C. H., *Environ. Sci. Technol.* **23**, 760–766 (1989).

Thomas, J. M., Gordy, V. R., Fiorenza, S., and Ward, C. H., *Water Sci. Technol.* **22**(6), 53–62 (1990).

Thomas, J. M., Marlow, H. J., Ward, C. H., and Raymond, R. L., *in* "Fate of Chemicals and Pesticides in the Environment" (J. L. Schnoor, ed.), pp. 211–227. Wiley (Interscience), New York, 1992.

Vail, R. L., *Oil Gas J.* **89**(45), 53–57 (1991).

Valo, R., and Salkinoja-Salonen, M., *Appl. Microbiol. Biotechnol.* **25**, 68–75 (1986).

Wang, X., and Bartha, R., *Soil Biol. Biochem.* **22**, 501–505 (1990).

Wang, X., Yu, X., and Bartha, R., *Environ. Sci. Technol.* **24**, 1086–1089 (1990).

Wilson, J. T., and Ward, C. H., *Dev. Ind. Microbiol.* **27**, 109–116 (1987).

Wilson, J. T., Fogel, S., and Roberts, P. V., *in* "Detection, Control, and Renovation of Contaminated Ground Water" (N. Dee, W. F. McTernan, and E. Kaplan, eds.), pp. 168–178. American Society of Civil Engineers, New York, 1987.

Woodward, R., and Ramsden, D., *in* "Gas, Oil, Coal and Biotechnology II" (C. Akin and J. Smith, eds.), pp. 59–66. Institute of Gas Technology, Chicago, 1990.

Ziegenfuss, P. S., Williams, R. T., and Myler, C. A., *J. Hazard. Mater.* **28**, 91–99 (1991).

16 RECALCITRANT MOLECULES

Many organic compounds, both low molecular weight and polymeric, persist for long periods in soils, subsoils, aquifers, surface waters, and aquatic sediments. Some of the sites containing these compounds or synthetic polymers are so rich in toxic substances that the persistence can be attributed to the inability of microorganisms to grow and bring about biodegradation in the presence of the toxins. However, many of the durable molecules are located in environments in which microorganisms are proliferating and actively metabolizing, so that the persistence cannot be ascribed to conditions inimical to microbial life. The low- and high-molecular-weight substances that thus resist biodegradation are known as *recalcitrant* molecules.

Recalcitrance is a property not commonly attributed to chemicals. The characteristic of stubbornness is more commonly associated with people, and the term usually denotes a resistance to guidance or authority. Nevertheless, even if the word stretches somewhat the limits of good English usage, the adjective "recalcitrant" aptly characterizes the resistance of the molecules to the expected omnivorous feeding habits of microbial communities in nature, and for better or worse, the term is now generally accepted.

Molecules that persist in nature are undesirable for many reasons. Some are intrinsically toxic and deleteriously affect humans, domesticated animals, agricultural crops, wildlife, fish and other aquatic animals, or microorganisms. The longer the molecule remains in nature, the greater is the exposure of susceptible individuals or populations and the greater is the risk of harmful effects. Some recalcitrant molecules are not toxic at the concentrations found in nature, but they reach hazardous levels because they are biomagnified in natural food chains. Persistent polymers are of no toxicological significance, but their obvious presence in forests or parks and on roadsides is aesthetically unpleasant. The enormous amount of nonbiodegradable packaging materials that end up in municipal wastes must ultimately be hauled away and buried, thereby not only contributing to the increasing costs of waste disposal but occupying much space in landfills. Moreover, a compound that is both mobile and persistent may be transported to previously uncontaminated areas, increasing thereby the potential exposure of susceptible individuals and populations and hence increasing the risk of harm; in contrast, a mobile compound that is readily degraded will tend to be destroyed and thus the extent of its transport is less.

Another reason for concern with persistent molecules is related to the imperfect state of knowledge of toxicology—as with all other sciences. The knowledge of toxicology is growing constantly, and what was once deemed to be safe is occasionally later found to be harmful, often as a result of assessments of previously unevaluated physiological processes. The broadening vista of toxicology has shown that compounds previously believed to be innocuous were in fact harmful to humans, animals, or plants. When this is found to be true, government regulatory agencies typically ban the chemicals. Such bans quickly result in a rapid reduction in exposure of susceptible organisms if the compounds are readily biodegraded. However, the cessation of production and use of the newly recognized toxicant does not lead to a quick diminution in exposure if the molecule is recalcitrant; it has already been released in nature, and a biological means for its rapid destruction does not exist. This problem is well illustrated by certain chlorinated hydrocarbon insecticides (such as DDT, dieldrin, heptachlor, and chlordane) that were widely used two, three, or four decades ago; although no longer manufactured or used, they remain in many soils and continue to pose problems of contamination.

A recalcitrant molecule may be wholly resistant to microbial modification. Some indeed are. Other recalcitrant molecules are destroyed, either by growth-linked processes or cometabolically, but the rates in nature are so slow that detectable levels are observed for years after the first introduction of the compound to the affected environment.

EXAMPLES OF RECALCITRANCE

PCBs are one of the more prominent groups of persistent compounds. At one time, they were widely used in transformers and capacitors, as hydraulic and heat-transfer fluids, and as solvents and plasticizers. The PCBs that are highly chlorinated are notably persistent, and sediments contaminated with these toxicants 20 or more years ago still show disturbingly high levels. Dibenzothiophene and alkylated dibenzothiophene from crude oil are known to persist in sea sediments for periods in excess of 10 years (Sinkkonen, 1989), a quaternary ammonium compound known as Amo-1618 persists in soil for more than 8 years (Marth and Mitchell, 1959), and tetralin and indane sulfonates, which are minor components of widely used linear ABS surfactants, are refractory to microbial attack (Field *et al.*, 1992). Also notably persistent in soil are such 5-ring PAHs as perylene and 1,2,5,6-dibenzanthracene (Bossert and Bartha, 1986) and the tetrachloro through the octachloro dibenzo-*p*-dioxins and dibenzofurans (Orazio *et al.*, 1992).

Data showing the persistence of a number of compounds in soil are presented in Table 16.1. Except for the polychlorinated dibenzofurans and dibenzo-*p*-dioxins, all the substances in the table are pesticides. The chemicals were still present at the sampling time shown, so that the actual longevities are longer than the times tabulated. Several of the insecticides, because of their resistance, contaminate the surfaces of vegetables grown

TABLE 16.1
Persistence of Several Compounds in Soil

Compound	Persistence (years)[a]	Reference
Azinphosmethyl	8	Staiff *et al.* (1975)
BHC	16	Nash and Harris (1973)
Chlordane	16	Nash and Harris (1973)
Chlorfenvinphos	4	Chisholm (1975)
DDT	21	Martin *et al.* (1993)
Dicamba	4	Burnside *et al.* (1971)
Dieldrin	21	Martin *et al.* (1993)
EDB	19	Steinberg *et al.* (1987)
Endrin	16	Nash and Harris (1973)
Heptachlor	16	Nash and Harris (1973)
Isodrin	16	Nash and Harris (1973)
Lindane	21	Martin *et al.* (1993)
Mirex	12	Carlson *et al.* (1976)
Monuron	3	Birk (1955)
Paraquat	6	Fryer *et al.* (1975)
Picloram	5	Burnside *et al.* (1971)
Polychlorinated dibenzofurans	8	Hagenmaier *et al.* (1992)
Polychlorinated dibenzo-*p*-dioxin	8	Hagenmaier *et al.* (1992)
Simazine	20	Scribner *et al.* (1992)
Tordon	5	Burnside *et al.* (1971)
Toxaphene	16	Nash and Harris (1973)
2,3,6-Trichlorobenzoic acid	4	Burnside *et al.* (1965)
Trifluralin	3	Golab *et al.* (1979)

[a]Compound still present in soil at the time indicated.

underground, enter eartheorms that are consumed by and hence affect birds, or are carried from treated fields into rivers and lakes long after their last application for insect control. The early dates of some of the references in the table attest to the public concern and clamor to have many of those insecticides removed from the market.

Among many of the less readily metabolized compounds that remain in soil years after their first introduction, a bi- or multiphasic rate of disappearance is evident. In the first few months, the degradation rate is rapid, but then it slows dramatically to a point that little is lost in succeeding years. This may reflect the sequestering of the compound in a manner that makes it less available to microorganisms, that is, the so-called "aging" of the molecule. For reasons that are unclear, but again possibly related to the aging effect, a readily metabolized substrate may be found in soil for unexpectedly long periods, as in the detection of approximately 0.1% of the parathion, an insecticide that is usually quickly degraded, applied to soil 16 to 20 years earlier (Stewart *et al.*, 1971).

The persistence of chemicals in waters is of special significance because of the use of surface and groundwaters for drinking and the frequently rapid and distant transport of pollutants in rivers and streams. A typical example is the finding of various chloroethenes, dichlorobenzenes, and alkylbenzenes as well as nonylphenol isomers in a groundwater contaminated originally with secondary sewage effluents, some of the contaminants undoubtedly persisting in the aquifer for more than 30 years (Barber *et al.*, 1988). The nonylphenols may have been derived from the nonylphenol polyethoxylate surfactants in some detergent preparations. Analysis of drinking water and samples from rivers also show the presence of a disturbing array of synthetic organic molecules (Meijers and van der Leer, 1976; Miller, 1973), some of which may have been recently introduced but many of which are known to be recalcitrant.

Molecular recalcitrance is not restricted to anthropogenic chemicals because natural organic materials often remain undecomposed not merely for years or decades but occasionally for millenia or even longer. The persistence of a few of these paleobiochemicals is depicted in Fig. 16.1. These ancient residues are found in fossils, sedimentary rocks, and coal—where their presence is not surprising—but also in lake and marine sediments, where their occurrence is anomalous. On the basis of radiocarbon dating techniques, even part of the humus fraction of soil appears to persist for hundreds to thousands of years, and the organic matter in some peat deposits has withstood microbial destruction for tens of thousands of years. Soils and peats support active microbial communities, so that the persistence of organic fractions or discrete organic molecules is noteworthy. Even resting structures of various fungi persist in viable form for many years—evidence of the resistance to biodegradation of the organic

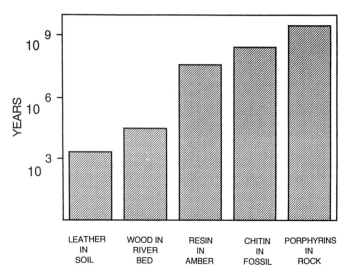

FIG. 16.1 Persistence of paleobiochemicals and other natural materials. (From Alexander, 1973.)

components of the outer surfaces of the sclerotia, chlamydospores, and other resting structures (Alexander, 1973).

Thus, despite the remarkable catabolic versatility of microorganisms, they are not omnivorous. Not all organic molecules are catabolized, at least not at reasonable rates. For one reason or another, microbial communities in nature are, unfortunately, not infallible.

PERSISTENT PRODUCTS

Products generated as bacteria and fungi transform organic molecules are sometimes persistent. The original substrate in some cases may itself be long-lived, but instances are known in which the original substrate is quickly destroyed yet yields a metabolite that is not. The products of the latter and sometimes of the former thus are found long after all traces of the parent compound have disappeared.

The longevity of products is well illustrated by some of the early chlorinated insecticides. DDT, for example, is itself not quickly degraded microbiologically, but it is slowly converted to a number of products, including DDE (Kuhr *et al.,* 1972), which remain in treated soil for years after the last pesticide application. The chlorinated insecticides aldrin and heptachlor are converted microbiologically in soil to their corresponding epoxides, which are known as dieldrin and heptachlor epoxide and are also insecticidal, and the products remain when little or none of the parents can be detected in treated fields (Wilkinson *et al.*, 1964). A location outside of Denver where aldrin was manufactured still contains dieldrin more than 20 years after industrial disposal of the chemical ceased.

Among the pesticides, 2,6-dichlorobenzamide formed from the herbicide dichlobenil (Verloop and Nimmo, 1970), phorate sulfone generated from the insecticide phorate (Lichtenstein *et al.*, 1973), and a keto derivative produced from the nematicide avermectin (Gullo *et al.*, 1983) are present in soil longer than their precursors. Many other examples exist among the pesticides deliberately applied to or inadvertently reaching soil. Analogous formation of long-lived products from less persistent or readily biodegradable parents occurs in surface, ground, and wastewaters. For example, nonylphenol polyethoxylate surfactants are metabolized during sewage treatment, but they are transformed to (nonylphenoxy)acetic acid and

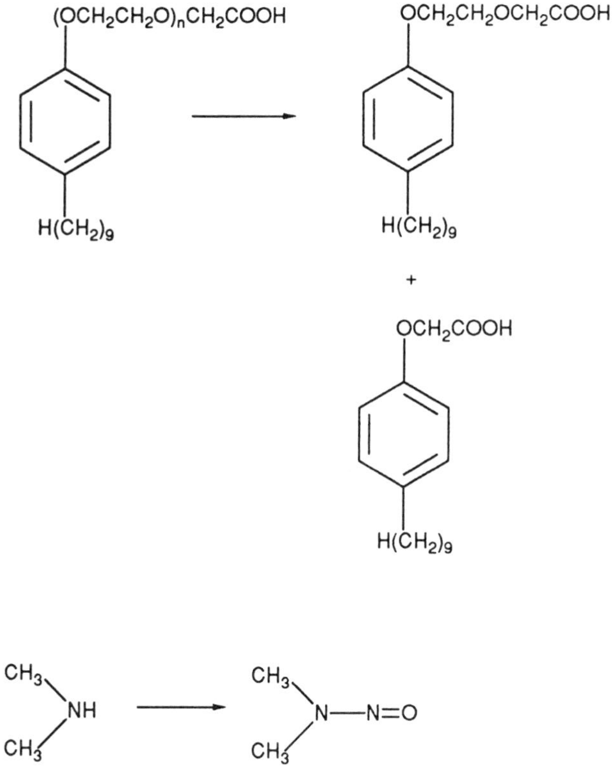

FIG. 16.2 Conversion of the readily degradable nonylphenoxycarboxylic acids and dimethylamine to persistent products, namely, (nonylphenoxy)acetic and [(nonylphenoxy)ethoxy]acetic acids (top) and *N*-nitrosodimethylamine (bottom).

[(nonylphenoxy)ethoxy]acetic acid, which persist in the river waters that receive effluent from the sewage-treatment plants (Ahel *et al.*, 1987). Similarly, at a waste-disposal site in Ontario into which dimethylamine was introduced, this readily degradable amine was converted to the carcinogen N-nitrosodimethylamine, which was still detected in adjacent groundwater more than 20 years after the last addition of dimethylamine to the waste-disposal pit (Fig. 16.2). Although this potent carcinogen can be metabolized by high cell densities of microorganisms under laboratory conditions, the same rapid conversion clearly does not occur in at least some natural environments.

SYNTHETIC POLYMERS

Industrialized countries use vast quantities of synthetic polymers. These polymers find uses as packaging materials, fabrics for clothing and carpets, insulation, and bedding, and large amounts are used in the construction of buildings and the manufacture of automobiles. An appreciation of the enormous quantities involved can be gained by an examination of the amounts produced in one year by the United States alone (Table 16.2).

In contrast to many other recalcitrant materials, the resistance of synthetic polymers is not a toxicological issue. Nevertheless, their durability in nature is the basis for concern. One reason is the large contribution

TABLE 16.2
Production of Synthetic Polymers in the United
States in 1992[a]

Polymer	kg $\times$ 10^9
Plastics	
Polyethylene	9.89
Poly(vinyl chloride)	4.53
Polypropylene	3.82
Polystyrene	2.29
Phenol and other tar resins	1.32
Urea resins	0.70
Unsaturated polyesters	0.54
Synthetic fibers	
Polyester	1.62
Nylon	1.16
Olefin	0.90
Acrylic	0.20

[a]From Reisch (1993).

they make to municipal solid wastes; in the United States, plastics account for 8% of the total weight and approximately 20% of the volume of municipal solid wastes (Palmisano and Pettigrew, 1992). This enormous tonnage represents much of the volume and occupies much of the space in landfills and, once buried in a landfill, such polymers do not undergo reduction in weight or volume because of biodegradation. Prominent in such wastes are polyethylene, poly(vinyl chloride), and polystyrene, which are major plastics used in packaging materials. A second basis for the public outcry is aesthetic: packaging materials that end up as litter in forests, parks, and roadsides detract from human enjoyment of the surroundings. Moreover, plastic particles composed of polyethylene, polystyrene, and polypropylene have been found in many locations in the ocean, sometimes even in remote areas (Colton et al., 1974; Morris, 1980), and these could cause blockages in the intestines of small fish. The plastic particles may come from solid wastes discarded by ships at sea, or they may originate from plastic-producing factories or processing facilities adjacent to coastal areas, rivers, or estuaries and be transported to remote locations because they are not destroyed microbiologically.

The synthetic polymers that represent the most commonly used plastics and fibers are wholly resistant to biodegradation, and they thus remain for as yet undetermined periods of time even in environments with highly diverse and physiologically active communities of aerobic or anaerobic microorganisms. Some of these refractory polymers are listed in Table 16.3. The recalcitrant polymers not only do not serve as C sources for

TABLE 16.3
Polymers Resistant to Microbial Degradation[a]

Acetate rayon (Estron)	Polyisobutylene (high molecular
Acrylonitrile-vinyl chloride	weight)
(Dynel)	Poly(methyl methacrylate)
Carboxymethyl cellulose (high	Polymonochlorotrifluorethylene
degree of substitution)	Polystyrene
Cellulose acetate (fully	Polytetrafluoroethylene (Teflon)
acetylated)	Polyurethane (polyether linked)
Cellulose acetate-butyrate	Poly(vinyl butyral)
Nylon	Poly(vinyl chloride)
Phenol-formaldehyde	Poly(vinyl chloride)-acetate
Polyacrylonitrile (Orlon)	Poly(vinylidene chloride)
Polydichlorostyrene	Resorcinol-formaldehyde
Polyethylene (high molecular	Silicone resins
weight)	Vinylidene chloride-vinyl
Poly(ethylene glycol)	chloride copolymer (Saran)
terephthalate (Dacron)	Zein formaldehyde (Vicara)

[a]From Alexander (1973).

any bacterium or fungus, but they are also not subject to cometabolism. The frequent reports that fungi or bacteria degrade these polymers have been discounted, and the spurious findings are attributed to microbial growth on substances added during processing, impurities, substances coating the surfaces of the polymers, or low-molecular-weight compounds added as plasticizers, lubricants, or stabilizers.

Nevertheless, a few synthetic polymers of commercial importance are biodegradable. These include high-molecular-weight polyethylene glycols (Obradors and Aguilar, 1991), poly(vinyl acetate) (Garcia Trejo, 1988), poly(vinyl alcohol) (Sakai et al., 1988), polyester polyurethane (Kay et al., 1991), polyester polycaprolactone (Cameron et al., 1988), and a number of other polylactones (Tanaka et al., 1976) and polyesters (Fields and Rodriguez, 1976). Also biodegradable are bacterial polymers that may have industrial importance, for example, poly(3-hydroxybutyrate) (Cain, 1992) and copolyesters of 3- and 4-hydroxybutyrates (Kunioka et al., 1989). Among a number of the nonbiodegradable high-molecular-weight polymers, the lower-molecular-weight counterparts are susceptible to microbial decomposition, but these smaller molecules are usually not of practical importance.

The desire of society for biodegradable plastics has prompted some companies to label their products as being biodegradable. In fact, however, the synthetic polymer in most of these products is not biodegradable. The recalcitrant polymer is formulated together with starch or gelatin. Once the starch or gelatin is destroyed, the product is converted to a fragmented material, but the original synthetic polymer is essentially unaltered. The physical structure of the product that is marketed is thereby altered, but the chemical integrity of the synthetic polymer is not modified.

MECHANISMS OF RECALCITRANCE

Because of the continued presence in nature of persistent compounds, both simple and polymeric, and the need for replacements having the same utility, it is important to establish why such substances are resistant to biodegradation, that is, to establish the mechanisms of recalcitrance. Modern society requires packaging materials, fabrics, pesticides, and other chemicals, but future materials should not create the environmental problems that the existing or earlier products have made.

The mechanisms of recalcitrance may be linked to structural features of the molecules of concern, physiological limitations of living organisms, or properties of the environment in which the compounds are found, and a consideration of the reasons for persistence thus must include assessments of the contributions of chemical, microbiological, and environmental factors.

The compounds that are long-lived in nature can be separated into several categories. (a) Molecules that appear to be wholly resistant to microbial attack and that are not metabolized under any conditions, at least based on present knowledge. (b) Compounds that are always slowly metabolized in nature. These may be rapidly destroyed by high cell densities of bacteria or large fungal biomasses in culture, but analogous rapid turnover is not characteristic of the natural habitats of the microorganisms. (c) Chemicals that are destroyed quickly in some environments or under certain circumstances but persist in other environments or circumstances. Many examples of each category are known.

A number of conditions must be satisfied for biodegradation to occur. (a) An enzyme must exist that can catalyze the transformation. Obviously coupled with this condition is the existence of an organism containing that enzyme. (b) The organism—presumably a bacterium or fungus—must be present in the same environment as the compound. (c) The molecule must be in a form that is available for microbial utilization. (d) If the enzyme is intracellular, as is the case for most low-molecular-weight substrates, the substrate must pass through the cell surface. (e) Should the enzyme be inducible, conditions must allow for induction to occur. (f) Environmental conditions must be suitable for microbial metabolism and, because the biomass in nature of organisms active in degrading many synthetic compounds is small, often for proliferation (Alexander, 1973). These conditions were first presented in Chapter 1. A consideration of these six requisites suggests several reasons for persistence. Some of these explanations pertain to the truly recalcitrant molecules, which are always long-lived. Others pertain only to substrates that are often persistent but that are also, in one environment or another, quickly transformed.

Nonexistence of an active organism.   Given the millions of compounds that have now been described, it is plausible that biochemical evolution has not resulted in the appearance of an enzyme capable of catalyzing a modification in many of these synthetic novelties. Enzymes are reasonably specific for the molecules on which they act, and despite the hundreds of millions of years of biochemical evolution, only a limited number of catabolic pathways have appeared. The enzymes important in catabolism are obviously critical for the provision of energy and building blocks for microrganisms, but these enzymes function on the substrates that microorganisms have encountered during their evolution and not necessarily on each and every novelty created in the laboratories of organic chemists. The nonabsolute specificity of enzymes is probably the basis for the transformations of many of the novel molecules created in recent times, but it is wishful thinking to expect that every new compound that has been synthesized has an enzyme able to catalyze its alteration. If no enzyme for a specific compound exists, an organism able to modify that molecule also will not exist.

Impermeability of the cell.   The enzymes responsible for catalyzing the biodegradation of many compounds are solely intracellular, and if a potential substrate does not cross the cell membrane and penetrate to the site in the organism where the enzyme is found, no reaction will occur. The initial stages in the metabolism of some molecules, especially those of high molecular weight, are catalyzed by extracellular enzymes, and these molecules can thus be transformed. However, if no extracellular enzyme exists to generate products to which the cell is permeable, the parent nonpenetrating molecule will be recalcitrant. It is thus possible that an intracellular enzyme of broad specificity never combines with a potential substrate simply because of the permeability barrier at the cell surface. Molecular weight is not the sole determinant of permeation through the cell membrane, and molecular shape and other properties of the chemical may prevent its transport into the organism. Impermeability of the cell may explain why high-molecular-weight polyethylenes are resistant to biodegradation whereas the low-molecular-weight polyethylenes are metabolized (Potts *et al.*, 1972): the enzymes initiating the catabolism of molecules composed of chains of $-CH_2-$ (which are called alkanes among the simpler ones and polyethylenes among the larger chains) are entirely intracellular.

Inaccessibility of site in molecule potentially acted on enzymatically.   A particular part of the enzyme (known as the active site) must combine with the substrate for a reaction to occur, and the specific site in the molecule where the reaction is to take place must be accessible. If that site is inaccessible, then no conversion will occur. Some compounds, such as alkanes or long-chain fatty acids, are acted on at the terminal ends of the molecules, but the terminal ends may be inaccessible; such inaccessibility may be a consequence of the folding or coiling of the ends of large molecules, a possible reason why some synthetic polymers may resist microbial degradation. Other compounds may be protected because they contain a substituent that sterically prevents the enzyme from combining with a molecule that otherwise might be a suitable substrate. Large molecules also have extensive cross-linkages, and these may mask the site on the potential substrate with which the enzyme must bind.

Lack of induction of requisite enzymes. Enzymes involved in a particular physiological process may be active in the organism regardless of the presence of the substrate (constitutive enzymes), or they may only be formed and active when the substrate or possibly a closely related molecule is present (inducible enzymes). The initial phase of many biodegradative reaction sequences entails the activity of inducible enzymes, but the conditions may not be suitable for induction. The common inducer is the substrate itself. However, the concentration of the substrate or another inducer in the aqueous phase may be too low to result in enzyme formation—which may occur because most of the compound is sorbed or present in a NAPL, leaving a concentration in the water that is too low to promote enzyme formation.

Cometabolism. Should the population or biomass of organisms active on a substrate be small and the substrate be one that is only acted on by cometabolism, that population or biomass will not increase in size and the initially slow conversion will be maintained. This contrasts with growth-linked biodegradation, in which the slow initial stage is replaced by a period of rapid conversion. Cometabolism may be rapid in the laboratory if the organisms are grown on another substrate, but such high cell densities do not characterize natural ecosystems but they may be found in bioreactors or result from particular remediation technologies in the field.

Environmental factors. Chemicals may persist not because they are intrinsically refractory but because of some factor in the environment that prevents rapid destruction. These are not properly refractory molecules. If growth is necessary for significant biodegradation, all nutrients must be present and available; these nutrients include inorganic ions, water, O_2 or another electron acceptor, and—for cometabolizing species—an organic compound to serve as a C and energy source. The environment must not contain toxins at levels sufficiently high to prevent growth or activity. The inhibitors may be organic, metallic cations, or salts at high concentration, but sometimes the lack of activity is the result of a pH that is too high or too low.

A well-known anomaly in biodegradation of natural products is the resistance of peats and mucks to biodegradation. These soil materials often contain 30 to 90% organic matter, and they may exist in nature for thousands of years provided that they are waterlogged and hence anaerobic. Once O_2 is introduced, which takes place when the excess water is removed, decomposition proceeds at reasonable rates. Experiments involving the addition of cellulose—which is a good substrate for both aerobes and anaerobic bacteria—show that this polysaccharide is not destroyed if placed below the surface of soil that has been flooded for several weeks (Fig. 16.3). This inhibition of cellulose degradation appears to be a result of two changes, the fall in pH and the appearance of acetic, butyric, and possibly other organic acids (Kilham and Alexander, 1984). Although moderately low pH per se often does not abolish the metabolism of a polysaccharide like cellulose and low levels of fatty acids at high pH are not inhibitory, the combination of low pH and fatty acids results in antimicrobial potency. The reason is probably that the protonated molecules (which appear in increasing concentration as the pH falls) are toxic but not the unprotonated anions.

The poor availability or the total lack of bioavailability of a compound may be a major determinant of persistence. The molecule may not be available because of its sorption to particulate matter in the environment, its presence in a NAPL, or its sequestration at microenvironmental sites

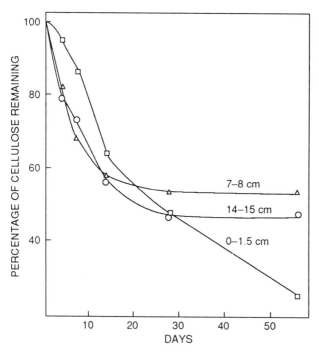

FIG. 16.3 Decomposition of cellulose added at three depths to a soil flooded at time of addition of the polysaccharide. [Reprinted with permission from Kilham, O. W., and Alexander, M. *Soil Sci.* **137**, 419–427. © Williams & Wilkins, 1984.]

not accessible to bacteria or fungi. The longevity of aged chemicals bears witness to the problems that microorganisms encounter with substrates that become less available.

A compound present at concentrations below the threshold for growth will also persist. Subthreshold concentrations appear to be characteristic of some pollutants that are found entirely in the aqueous phase of some environments. However, the equilibrium concentration in water of a chemical that is extensively sorbed or extensively partitioned into a NAPL may also be so low as to preclude growth of species able to metabolize that substrate.

REFERENCES

Ahel, M., Conrad, T., and Giger, W., *Environ. Sci. Technol.* **21,** 697–703 (1987).

Alexander, M., *Biotechnol. Bioeng.* **15,** 611–647 (1973).

Barber, L. B., II, Thurman, E. M., Schroeder, M. P., and LeBlanc, D. R., *Environ. Sci. Technol.* **22,** 205–211 (1988).

Birk, L. A., *Can. J. Agric. Sci.* **35,** 377–387 (1955).

Bossert, I. D., and Bartha, R., *Bull. Environ. Contam. Toxicol.* **37,** 490–495 (1986).

Burnside, O. C., Wicks, G. A., and Fenster, C. R., *Weeds* **13,** 277–278 (1965).

Burnside, O. C., Wicks, G. A., and Fenster, C. R., *Weed Sci.* **19,** 323–325 (1971).

Cain, R. B., *in* "Microbial Control of Pollution" (J. C. Fry, J. M. Gadd, R. A. Herbert, C. W. Jones, and I. A. Watson-Craik, eds.), pp. 293–338. Cambridge Univ. Press, Cambridge, UK, 1992.

Cameron, J. A., Bunch, C. L., and Huang, S. J., *in* "Biodeterioration" (D. R. Houghton, R. N. Smith, and H. O. Wiggins, eds.), pp. 553–561. Elsevier Applied Science, London, 1988.

Carlson, D. A., Konyha, K. D., Wheeler, W. B., Marshall, G. P., and Zaylskie, R. G., *Science* **194,** 939–941 (1976).

Chisholm, D., *Can. J. Soil Sci.* **55,** 177–180 (1975).

Colton, J. B., Jr., Knapp, F. D., and Burns, B. R., *Science* **185,** 491–497 (1974).

Field, J. A., Leenheer, J. A., Thorn, K. A., Barber, L. B., II, Rostad, C., Macalady, D. L., and Daniel, S. R., *J. Contam. Hydrol.* **9,** 55–72 (1992).

Fields, R. D., and Rodriguez, F., *in* "Proceedings of The Third International Biodegration Symposium" (J. M. Sharpley and A. M. Kaplan, eds.), pp. 775–784. Applied Science Publishers, London, 1976.

Fryer, J. D., Hance, R. J., and Ludwig, J. W., *Weed Res.* **15,** 189–194 (1975).

García Trejo, A., *Ecotoxicol. Environ. Saf.* **16,** 25–35 (1988).

Golab, T., Althaus, W. A., and Wooten, H. L., *J. Agric. Food Chem.* **27,** 163–179 (1979).

Gullo, V. P., Kempf, A. J., MacConnell, J. G., Mrozik, H., Arison, B., and Putter, I., *Pestic. Sci.* **14,** 153–157 (1983).

Hagenmaier, H., She, J., and Lindig, C., *Chemosphere* **25,** 1449–1456 (1992).

Kay, M. J., Morton, L. H. G., and Prince, E. L., *Int. Biodeterior. Bull.* **27,** 205–222 (1991).

Kilham, O. W., and Alexander, M., *Soil Sci.* **137,** 419–427 (1984).

Kuhr, R. J., Davis, A. C., and Taschenberg, E. F., *Bull. Environ. Contam. Toxicol.* **8,** 329–333 (1972).

Kunioka, M., Kawaguchi, Y., and Doi, Y., *Appl. Microbiol. Biotechnol.* **30**, 569–573 (1989).

Lichtenstein, E. P., Fuhremann, T. W., Schulz, K. R., Llang, T. T., *J. Econ. Entomol.* **66**, 863–866 (1973).

Marth, P. C., and Mitchell, J. W., *Plant Physiol.* **34**, Suppl., X (1959).

Martin, A., Bakker, H., and Schreuder, R. H., *Bull. Environ. Contam. Toxicol.* **51**, 178–184 (1993).

Meijers, A. P., and van der Leer, R. C., *Water Res.* **10**, 597–604 (1976).

Miller, S. S., *Environ. Sci. Technol.* **7**, 14–15 (1973).

Morris, R. J., *Mar. Pollut. Bull.* **11**, 164–166 (1980).

Nash, R. G., and Harris, W. G., *J. Environ. Qual.* **2**, 269–273 (1973).

Obradors, N., and Aguilar, J., *Appl. Environ. Microbiol.* **57**, 2383–2388 (1991).

Orazio, C. E., Kapila, S., Puri, R. K., and Yanders, Y. F., *Chemosphere* **25**, 1469–1474 (1992).

Palmisano, A. C., and Pettigrew, C. A., *Bioscience* **42**, 680–685 (1992).

Potts, J. E., Clendinning, R. A., Ackart, W. B., and Niegisch, W. D., *Polym. Prep.* **13**, 629–634 (1972).

Reisch, M. S., *Chem. Eng. News* **71**(15), 13–16 (1993).

Sakai, K., Hamada, N., and Watanabe, Y., *Kagaku to Kogyo (Osaka)* **62**(2), 39–47 (1988); *Chem. Abstr.* **109**, 3485 (1988).

Scribner, S. L., Benzing, T. R., Sun, S., and Boyd, S. A., *J. Environ. Qual.* **21**, 115–120 (1992).

Sinkkonen, S., *Chemosphere* **18**, 2093–2100 (1989).

Staiff, D. C., Comer, S. W., Armstrong, J. F., and Wolfe, H. R., *Bull. Environ. Contam. Toxicol.* **13**, 362–368 (1975).

Steinberg, S. M., Pignatello, J. J., and Sawhney, B. L., *Environ. Sci. Technol.* **21**, 1201–1208 (1987).

Stewart, D. K. R., Chisholm, D., and Ragab, M. T. H., *Nature (London)* **229**, 47 (1971).

Tanaka, H., Tonomura, K., and Kamibayashi, A., *Biseibutsu Kogyo Gijutsu Kenkyusho Kenkyu Hokoku* **48**, 75–79 (1976), *Chem. Abstr.* **87**, 65110 (1977).

Verloop, A., and Nimmo, W. B., *Weed Res.* **10**, 65–70 (1970).

Wilkinson, A. T. S., Finlayson, D. G., and Morley, H. V., *Science* **143**, 681–682 (1964).

APPENDIX
Abbreviations, Acronyms, and Structures

ABS	Alkylbenzene sulfonate
Alachlor	2-Chloro-2',6'-diethyl-*N*-(methoxymethyl)-acetanilide
Aldicarb	2-Methyl-2-(methylthio)propionaldehyde *O*-(methylcarbamoyl) oxime
Aldrin	1,2,3,4,10,10-Hexachloro-1,4,4a,5,8,8a-hexahydro-*exo*-1,4-*endo*-5,8-dimethanonaphthalene
Amitrole	3-Amino-1,2,4-triazole
Atrazine	2-Chloro-4-ethylamino-6-isopropylamino-1,3,5-triazine
Azinphosmethyl	*O,O*-Dimethyl *S*-[4-oxo-1,2,3-benzotriazin-3-(4H)yl]methyl phosphorodithioate
Benomyl	1-(Butylcarbamoyl)-2-benzimidazole carbamic acid, methyl ester
Benzoylprop-ethyl	Ethyl *N*-benzoyl-*N*-(3,4-dichlorophenyl)-DL-alaninate
BHC	Hexachlorocyclohexane
Bromoxynil	3,5-Dibromo-4-hydroxybenzonitrile
BTEX	Benzene, toluene, ethylbenzene, xylene
Butralin	4-(1,1-Dimethylethyl)-*N*-(1-methylpropyl)-2,6-dinitrobenzenamine
Butylate	*S*-Ethyl diisobutylthiocarbamate
Captan	*N*-Trichloromethylthio-3a,4,7,7a-tetrahydrophthalimide

Carbofuran	2,3-Dihydro-2,2-dimethylbenzofuran-7-yl-methyl carbamate
Chlordane	1,2,4,5,6,7,8,8-Octachloro-2,3,3a,4,7,7a-hexahydro-4,7-methanoindane
Chlorfenvinphos	2-Chloro-1-(2,4-dichlorophenyl)vinyl diethyl phosphate
Chlorobenzilate	Ethyl 4,4'-dichlorobenzilate
Chloroneb	1,4-Dichloro-2,5-dimethoxybenzene
Chlorothalonil	Tetrachloroisophthalonitrile
Chlorpropham	Isopropyl-N-3-chlorophenylcarbamate
Chlorpyrifos	O,O-Diethyl O-(3,5,6-trichloro-2-pyridyl) phosphorothioate
Chlorsulfuron	2-Chloro-N-[[(4-methoxy-6-methyl-1,3,5-triazin-2-yl)amino]carbonyl] benzenesulfonamide
CIPC	See Chlorpropham
CMC	Critical micelle concentration
2,4-D	2,4-Dichlorophenoxyacetic acid
Dalapon	2,2-Dichloropropionic acid
Dasanit	O,O-Diethyl O-[4-(methylsulfinyl)phenyl] phosphorothioate
4-(2,4-DB)	4-(2,4-Dichlorophenoxy)butyric acid
DBP	4,4'-Dichlorobenzophenone
DDD	1,1-Dichloro-2,2-bis(p-chlorophenyl)ethane
DDE	1,1-Dichloro-2,2-bis(p-chlorophenyl)ethylene
DDT	1,1,1-Trichloro-2,2-bis(p-chlorophenyl)ethane
DEHP	Di(2-ethylhexyl) phthalate
Diazinon	O,O-Diethyl O-(2-isopropyl-4-methyl-6-pyrimidinyl)phosphorothioate
Dicamba	3,6-Dichloro-2-methoxybenzoic acid
Dichlobenil	2,6-Dichlorobenzonitrile
Dichlorfop-methyl	Methyl 2-[4-(2,4-dichlorophenoxy)phenoxy]-propionate

Dieldrin	1,2,3,4,10,10-Hexachloro-6,7-epoxy-1,4,4a,5,6,7,8,8a-*endo, exo*-1,4:5,8-dimethanonaphthalene
Dietholate	*O,O*-Diethyl *O*-phenyl phosphorothioate
Dimethoate	*O,O*-Dimethyl *S*-(*N*-methylacetamide) phosphorodithioate
Dinoseb	2-*sec*-Butyl-4,6-dinitrophenol
Diphenamid	*N,N*-Dimethyl-2,2-diphenylacetamide
Diquat	1,1'-Ethylene-2,2'-dipyridylium
Disulfoton	*O,O*-Diethyl *S*-(2-ethylthioethyl)phosphorodithioate
Diuron	3-(3,4-Dichlorophenyl)-1,1-dimethylurea
DNAPL	Dense nonaqueous phase liquid
DNOC	4,6-Dinitro-*o*-cresol
DOC	Dissolved organic C
Dursban	See Chlorpyrifos
Dyfonate	*O*-Ethyl *S*-phenyl ethylphosphonodithioate
EDB	1,2-Dibromoethane (syn: ethylene dibromide)
Endothal	3,6-*endo*-Oxohexahydrophthalate
Endrin	1,2,3,4,10,10-Hexachloro-6,7-epoxy-1,4,4a,5,6,7,8,8a-octahydro-*endo, endo*-1,4:5,8-dimethanonaphthalene
Enthoprop	*O*-Ethyl *S,S*-dipropyl phosphorodithioate
EPTC	*S*-Ethyl dipropyldithiocarbamate
Fenitrothion	*O,O*-Dimethyl *O*-(4-nitro-*m*-tolyl)phosphorothioate
Fensulfothion	Diethyl 4-(methylsulfinyl)phenyl phosphorothionate
Flamprop-methyl	Methyl *N*-benzoyl-*N*-(3-chloro-4-fluorophenyl)-2-aminopropionate
Fluchloralin	*N*-(2-Chloroethyl)-α,α,α-trifluoro-2,6-dinitro-*N*-propyl-*p*-toluidine
Fonofos	*O*-Ethyl-*S*-phenyl ethylphosphonodithioate
Glyphosate	N-phosphonomethylglycine

Guthion	*See* Azinphosmethyl.
Heptachlor	1,4,5,6,7,8,8-Heptachloro-3a,4,7,7a-tetrahydro-4,7-methanoindene
Hexazinone	3-Cyclohexyl-6-(dimethylamino)-1-methyl-1,3,5-triazine-2,4(1H, 3H)-dione
HMX	Octahydro-1,3,5,7-tetranitro-1,3,5,7-tetraazocine
Ipazine	2-Chloro-4-diethylamino-6-isopropylamino-1,3,5-triazine
IPC	Isopropyl N-phenylcarbamate
Iprodione	3-(3,5-Dichlorophenyl)-N-1-methylethyl-2,4-dioxo-1-imidazolidinecarboxamide
Isodrin	An isomer of aldrin
Isofenphos	2-[[Ethyoxy[(1-methylethyl)amino] phosphinothioyl]oxy]benzoic acid 1-methyl-ethyl ester
Isoproturon	N-(4-Isopropylphenyl)-N', N'-dimethylurea
Kepone	Decachloro-octahydro-1,3,4-methano-2H-cyclobuta(cd)pentalene-2-one
Lindane	γ-1,2,3,4,5,6-Hexachlorocyclohexane
Linuron	3-(3,4-Dichlorophenyl)-1-methoxy-1-methylurea
Malathion	O,O-Dimethyl S-(1,2-bis-carbethoxy)ethyl phosphorodithioate
MBC	Methyl benzimidazol-2-yl carbamate
MCPA	2-Methyl-4-chlorophenoxyacetic acid
Mecoprop	2-(2-Methyl-4-chlorophenoxy)propionic acid
Metamitron	4-Amino-3-methyl-6-phenyl-1,2,4-triazin-5(4H)one
Methyl parathion	O,O-Dimethyl O-(p-nitrophenyl)phosphoro-thioate
Metolochlor	2-Chloro-N-(2-ethyl-6-methylphenyl)-N-(2-methoxy-1-methylethyl)acetamide

Metribuzin	4-Amino-6-(1,1-dimethylethyl)-3-(methylthio)-1,2,4-triazin-5(4H)one
Mirex	Dodecachlorooctahydro-1,3,4-metheno-2H-cyclobuta(cd)pentalene
Monocrotophos	Dimethyl-(E)-1-methyl-2-methylcarbamoyl-vinyl phosphate
Monolinuron	3-(4-Chlorophenyl)-1-methoxy-1-methylurea
Monuron	3-(4-Chlorophenyl)-1,1-dimethylurea
NAPL	Nonaqueous phase liquid
Nitrofen	2,4-Dichlorophenyl *p*-nitrophenyl ether
NTA	Nitrilotriacetic acid
Ordram	*S*-Ethyl hexahydro-1H-azepine-1-carbothioate
PAH	Polycyclic aromatic hydrocarbon
Paraoxon	Diethyl *p*-nitrophenyl phosphate
Paraquat	1,1'-Dimethyl-4,4'-bipyridinium
Parathion	*O*,*O*-Diethyl *O*-(*p*-nitrophenyl)-phosphorothioate
PCB	Polychlorinated biphenyl
PCE	Perchloroethylene (syn: tetrachloroethylene)
PCP	Pentachlorophenol
Phorate	*O*,*O*-Diethyl *S*-(ethylthio)methyl phosphorodithioate
Picloram	4-Amino-3,5,6-trichloropicolinic acid
Pirimiphos-methyl	*O*-(2-Dimethylamino-6-methylpyrimidin-4-yl) *O*,*O*-dimethyl phosphorothioate
Profluralin	*N*-(Cyclopropylmethyl)-α,α,α-trifluoro-2,6-dinitro-*N*-propyl-*p*-toluidine
Propachlor	2-Chloro-*N*-isopropylacetanilide
Propanil	*N*-(3,4-Dichlorophenyl)propionamide
Pyrazon	5-Amino-4-chloro-2-phenylpyridazine-3(2H)-one
RDX	Hexahydro-1,3,5-trinitro-1,3,5-triazine
SBR	Structure–biodegradability relationship

Sevin	1-Naphthyl N-methylcarbamate
Silvex	2-(2,4,5-Trichlorophenoxy)propionic acid
Simazine	2-Chloro-4,6-bis-ethylamino-1,3,5-triazine
Sucralose	4-Chloro-4-deoxy-α,D-galactopyranosyl-1,6-dichloro-1,6-dideoxy-β,D-fructofuranoside
2,4,5-T	2,4,5-Trichlorophenoxyacetic acid
TCA	Trichloroacetic acid
TCDD	2,3,7,8-Tetrachloro-dibenzo-p-dioxin
TCE	Trichloroethylene
TNT	2,4,6-Trinitrotoluene
Tordon	4-Amino-3,5,6-trichloropicolinic acid
Toxaphene	Chlorinated camphene
Triadimefon	1-(4-Chlorophenoxy)-3,3-dimethyl-1-(1,2,4-triazol-1-yl)butan-2-one
Trichloronat	O-Ethyl O-2,4,5-trichlorophenyl ethylphosphonothionate
Triclopyr	3,5,6-Trichloro-2-pyridinyloxyacetic acid
Trifluralin	α,α,α-Trifluoro-2,6-dinitro-N,N-dipropyl-p-toluidine
Vernolate	S-propyl dipropylthiocarbamate
Vinclozolin	3-(3,5-Dichlorophenyl)-5-ethenyl-5-methyl-2,4-oxazolidinedione
Zinophos	O,O-Diethyl 2-pyrazinyl phosphorothionate

INDEX